编委会

主　编
俞汉青

编　委
（以姓氏拼音排序）

陈洁洁	刘　畅	刘武军
刘贤伟	卢　姝	吕振婷
裴丹妮	盛国平	孙　敏
汪雯岚	王楚亚	王龙飞
王维康	王允坤	徐　娟
俞汉青	虞盛松	院士杰
翟林峰	张爱勇	张　锋

国家出版基金项目
NATIONAL PUBLICATION FOUNDATION

"十四五"国家重点出版物出版规划重大工程

污染物生物与化学转化中的界面电子转移机制

Interfacial Electron Transfer Mechanism
in Biological and Chemical Transformation of Pollutants

陈洁洁 　著
虞盛松

中国科学技术大学出版社

内 容 简 介

本书针对污染物生物转化和化学转化中界面电子转移的关键问题，采用理论计算与实验技术相结合的方法，阐明了微生物电子传递链终端氧化还原酶与电子受体以及化学催化剂与污染物分子的界面反应机制，实现了胞外电子传递途径的调控以及催化剂结构和表面特性的优化，提高了生物/化学转化系统的能量转换效率，促进了污染物的转化。本书以期为污染物化学处理系统的构建及优化提供依据，协助突破污染物资源化与能源化转化的新方法、新材料和新技术，为我国的环境保护和建立健全水污染控制系统贡献力量。

本书适合环境、化学、生物等专业及交叉学科的研究生和学者参考阅读。

图书在版编目(CIP)数据

污染物生物与化学转化中的界面电子转移机制/陈洁洁,虞盛松著. ——合肥:中国科学技术大学出版社,2022.3
(污染控制理论与应用前沿丛书/俞汉青主编)
国家出版基金项目
"十四五"国家重点出版物出版规划重大工程
ISBN 978-7-312-05396-2

Ⅰ.污…　Ⅱ.①陈…　②虞…　Ⅲ.水污染防治　Ⅳ.X52

中国版本图书馆 CIP 数据核字(2022)第 031951 号

污染物生物与化学转化中的界面电子转移机制
WURANWU SHENGWU YU HUAXUE ZHUANHUA ZHONG DE JIEMIAN DIANZI ZHUANYI JIZHI

出版	中国科学技术大学出版社 安徽省合肥市金寨路 96 号,230026 http://www.press.ustc.edu.cn https://zgkxjsdxcbs.tmall.com
印刷	安徽联众印刷有限公司
发行	中国科学技术大学出版社
开本	787 mm×1092 mm　1/16
印张	20.75
字数	395 千
版次	2022 年 3 月第 1 版
印次	2022 年 3 月第 1 次印刷
定价	128.00 元

总 序

建设生态文明是关系人民福祉、关乎民族未来的长远大计,在党的十八大以来被提升到突出的战略地位。2017年10月,党的十九大报告明确提出"污染防治"是生态文明建设的重要战略部署,是我国决胜全面建成小康社会的三大攻坚战之一。2018年,国务院政府工作报告进一步强调要打好"污染防治攻坚战",确保生态环境质量总体改善。这都显示出党和国家推动我国生态环境保护水平同全面建成小康社会目标相适应的决心。

当前,我国环境污染状况有所缓解,但总体形势仍然严峻,已严重制约了我国经济社会的持续健康发展。发展以资源回收利用为导向的污染控制新理论与新技术,是进一步推动污染物高效、低成本、稳定去除的发展方向,已成为国家重大战略需求和国际重要学术前沿。

为了配合国家对生态文明建设、"污染防治攻坚战"的一系列重大布局,抢占污染控制领域国际学术前沿制高点,加快传播与普及生态环境污染控制的前沿科学研究成果,促进相关领域人才培养,推动科技进步及成果转化,我们组织一批来自多个"双一流"大学、活跃在我国环境科学与工程前沿领域、有影响力的科学家共同撰写"污染控制理论与应用前沿丛书"。

本丛书是作者团队承担的国家重大重点科研项目(国家重大科技专项、国家863计划、国家自然科学基金)和获得的重大科技成果奖励(2014年国家自然科学奖二等奖、2020年国家科学技术进步奖二等奖)的系统总结,是作者团队攻读博士学位期间取得的重要的前沿学术成果(全国百篇优秀博士论文、中科院优秀博士论文等)的系统凝练,是一套系统反映污染控制基础科学理论与前沿高新技术研究成果的系列图书。本丛书围绕我国环境领域的污染物生化控制、转化机制、无害化处置、资源回收利用等亟须解决的一些重大科学问题与技术问题,将物理学、化学、生物学、材料学等学科的最新理论成果以及前沿高新技术应用到污染控制过程中,总结了我国目前在污染控制领域(特别是废水和固废领域)的重要研究进展,探索、建立并发展了常温空气阴极燃料电池、纳米材料、新兴生物电化学系统、新型膜生物反应器、水体污染物的化学及生物转化,以及固体废弃物污染控制与清洁转化等方面的前沿理论与技术,形成了具有广阔应用前景的新理论和新方法,为污染控制与治理提供了理论基础和科学依据。

"污染控制理论与应用前沿丛书"是服务国家重大战略需求、推动生态文明建设、打赢"污染防治攻坚战"的一套丛书。其出版将有利于促进最前沿的科研成果得到及时的传播和应用,有利于促进污染治理人才和高水平创新团队的培养,有利于推动我国环境污染控制和治理相关领域的发展和国际竞争力的提升;同时为环境污染控制与治理实践提供新思路、新技术、新材料,也可以为政府环境决策、强化环境管理、履行国际环境公约等提供科学依据和技术支撑,在保障生态环境安全、实施生态文明建设、打赢"污染防治攻坚战"中起到不可替代的作用。

编委会

2021 年 10 月

前　言

界面电子转移在生物体系、化学体系和环境修复中起着重要的作用，与能量的流动和转换密切相关。提高界面的电子传递效率，深入剖析电子转移机制，是促进界面反应的基础。本书针对污染物生物转化和化学转化中界面电子转移的关键问题，分析微生物电子传递链终端氧化还原酶与电子受体以及化学催化剂与能源小分子、污染物分子的界面反应，实现了胞外电子传递途径的调控以及催化剂结构和表面特性的优化，提高了系统的能量转换效率，促进了污染物的转化。

在环境工程领域中，生物转化在水中污染物的去除过程里扮演关键角色，且因其生物酶的作用而具有定向转化的优势。近年来，自然界中被发现存在一大类可以进行胞外呼吸的微生物，即电化学活性细菌（electrochemically active bacteria，EAB），它们可以通过细胞外膜上的关键蛋白进行跨膜电子转移，从而实现污染物的胞外转化。EAB 的胞外电子传递（extracellular electron transfer，EET），可以分为直接与间接两种形式，均有一定的优势。如果能够综合这两种电子传递的优势，将可以显著提高生物电化学系统的能量转换效率及污染物转化能力。但 EAB 缓慢的电子传递速率是限制其转化污染物效率的关键所在。针对该科学问题，笔者对 EAB 的跨膜电子传递过程的实时监测与机制解析开展了系统性的研究工作，揭示了 EET 在污染物生物转化中的作用机制，发展了强化 EET 降解多种污染物的新方法。在本书的第 2~7 章中，介绍了生物电子从 EAB 内部到胞外电子受体传递时各个阶段的分子机制及强化调控方法，从环境中 EAB 的筛选开始，到膜蛋白与矿物、电极材料的直接电子传递机制，提出了电极表面修饰强化 EET 的调控方案，再到基于间接电子传递的媒介分子的影响规律，最后通过媒介体的"分子桥梁"作用耦合直接与间接电子传递，提高了生物处理系统在水污染控制中的对污染物的处理能力。

另外,作为与污染物生物处理相辅相成的化学转化,污染物这种转化中的非均相催化技术可以解决均相催化需额外设置分离单元与回收工艺的难题,其核心过程包括表面化学吸附与活化,从而实现目标污染物分子的定向转化或矿化降解。因此,通过非均相催化剂表面结构的精准调控,将为研发先进的非均相体系、应对环境污染控制的挑战提供新的途径。但是,目前催化活性位点的性质以及它们在决定活性和选择性方面的作用仍未被清晰阐明。本书的第 8~12 章,具体展示了污染物化学转化方面的界面电子转移过程的机制解析与性能强化方法。以氧还原及析氧过程为模式反应,对化学催化界面的优化与调控进行研究,促进污染物的界面转化,是化学界面催化电子转移机制解析部分的基本思路。污染物化学转化系统中催化剂的界面或晶面调控技术包括助催化剂的负载工艺以及高活性晶面比例扩大的方案。污染物分子轨道与催化剂活性位点电子轨道的耦合是界面催化过程的第一步,也是催化的驱动力。基于界面优化思路的确定,选定了广泛应用的催化剂进行晶面晶相调控,将其在无机污染离子与芳香性有机污染物去除方面的应用过程进行优化与验证,为目标污染物的特点量身定制催化纳米材料提供了一个新的思路。

因此,本书从污染物多界面转化机制解析开始,逐步分析生物单元与化学单元两者界面上转化污染物的规律,揭示多界面电子转移机制与界面优化方法,从而构建高性能污染物转化系统,强化实际废水中污染物的处理能力。希望通过这些内容能够为污染物多界面转化机制解析提供参考,为微生物/载体材料界面耦合强化方法提供思路,为污染物化学处理系统的构建及优化提供依据,协助突破污染物资源化与能源化转化的新方法、新材料和新技术,为我国的环境保护和建立健全水污染控制系统贡献力量。

本书是研究团队基于课题研究成果撰写而成的,得到了国家自然科学基金项目(项目编号:51508545,51978637,52022093)的资助,得到了中国科学技术大学环境科学与工程系科研平台的支撑,也得到了中国科学技术大学出版社给予的大力支持。在此一并表示感谢!本书内容属于探索性研究,由于作者水平所限,错误与瑕疵难免,敬请广大专家学者指正。

著 者

2021 年 10 月

目 录

总序 —— i

前言 —— iii

第 1 章
污染物转化中的电子转移 —— 001

1.1 污染物转化中电子转移的重要性 —— 003

1.2 污染物生物转化中的电子传递 —— 004

1.3 污染物化学转化中的电子传递 —— 015

1.4 能量转换的电子传递机制解析 —— 020

1.5 电子转移的光谱电化学检测方法 —— 022

第 2 章
三氧化钨纳米团簇界面的微生物胞外电子传递机制 —— 039

2.1 基于光学的 EAB 高通量鉴定概述 —— 041

2.2 三氧化钨与微生物界面电子转移研究方法 —— 042

2.3 EAB 高通量鉴定机制 —— 046

第 3 章
EAB 外膜蛋白 OmcA 与赤铁矿界面电子传递机制 —— 065

3.1 EAB 对铁元素地球化学循环的影响 —— 067

3.2 外膜蛋白与赤铁矿界面电子传递的研究方法 —— 069

3.3 外膜蛋白在 EAB 与赤铁矿界面电子传递中的关键作用机制 —— 71

第 4 章
EAB 外膜细胞色素 c 与氧化石墨烯的界面电子传递机制 —— 081

4.1 胞外电子传递链与石墨烯材料相互作用的可能性
—— 083

4.2 细胞色素的提纯、氧化石墨烯制备与表征
—— 084

4.3 OmcA 蛋白与 GO 之间的电子转移过程解析
—— 087

第 5 章
生物电化学系统阳极碳材料表面修饰对微生物 EET 的调控 —— 099

5.1 生物电化学系统阳极材料的特征 —— 101

5.2 阳极材料表面 EET 过程研究方法 —— 102

5.3 阳极材料表面修饰对 EET 的调控机制 —— 105

第 6 章
EAB 外膜蛋白活性中心与媒介分子核黄素的电子传递机制 —— 121

6.1 外膜蛋白细胞色素 c 活性中心的特征 —— 123

6.2 氧化还原活性有机分子电子转移的研究方法
—— 124

6.3 核黄素与卟啉铁分子之间的电子转移 —— 129

第 7 章
氧化还原媒介分子结构对微生物 EET 路径的调控
—— 147

7.1 氧化还原媒介分子的质子耦合电子传递反应
—— 149

7.2 吩嗪分子衍生物氧化还原电位研究方法
—— 150

7.3 吩嗪分子取代基对电化学性质的影响 —— 153

第 8 章
内源性氧化还原媒介分子修饰的碳电极对 EAB 电子传递过程的调控 —— 167

8.1 内源性氧化还原媒介的优势 —— 169

8.2 内源性氧化还原媒介提升 EET 过程机制解析方法 —— 170

8.3 内源性氧化还原媒介界面电子转移 —— 175

第 9 章
金属硫化物 $FeNiS_2$ 电催化氧化过程的界面电子转移机制及电催化性能 —— 191

9.1 电催化转化模式反应与金属硫化物的电催化优势 —— 193

9.2 金属硫化物的制备、表征及界面电子转移性能测试方法 —— 194

9.3 金属硫化物电催化界面电子转移性能与机制 —— 196

第 10 章
金属硫化物/石墨烯纳米复合材料界面电子转移的自氧化调控 —— 209

10.1 金属硫化物与石墨烯的复合与调控作用 —— 211

10.2 金属硫化物与石墨烯复合材料的制备、表征与性能测试方法 —— 213

10.3 金属硫化物复合石墨烯对界面电子传递的调控作用 —— 215

第 11 章
多金属硫化物纳米复合材料界面电子传递过程中氧化/还原的双向调控 —— 235

11.1 金属硫化物中多金属组合的优势 —— 237

11.2 多金属硫化物纳米复合材料的制备与研究方法 —— 238

11.3 多金属硫化物界面电子传递催化氧化/还原反应的双向调控 —— 243

第 12 章
光催化硝化与反硝化过程中的界面电子转移 —— 271

12.1 光催化硝化现象与光电催化反硝化 —— 273

12.2 典型光催化剂界面硝化与反硝化机制解析方法 —— 274

12.3 典型光催化剂界面硝化与反硝化过程的电子转移 —— 275

第 13 章
Pt/TiO$_2$ 光催化降解硝基苯的界面电子转移 —— 293

13.1 贵金属与半导体复合催化剂的优势 —— 295

13.2 Pt/TiO$_2$ 催化剂制备与机制解析方法 —— 296

13.3 Pt/TiO$_2$ 催化剂对硝基苯的转化 —— 299

第 1 章

污染物转化中的电子转移

1.1 污染物转化中电子转移的重要性

界面电子转移在生物体系、化学体系和环境修复中起着重要的作用,提高界面电子传递效率,深入剖析电子转移机制,是推动能源转换、污染物转化和界面反应相关主题研究向前发展的基础。污染物的生物与化学转化涉及微生物酶/化学催化剂与污染物分子的界面电子转移,在生物、化学和环境等交叉学科日益呈现出关键性的作用。

对于污染物的生物转化,电子传递反应是微生物的电子传递链终端氧化还原酶与污染物分子相互作用的重要过程。例如,电化学活性菌能够在自身的呼吸代谢途径中将胞内电子传递到胞外的环境介质中[4-5],该过程是微生物体系与化学体系联系的纽带。对于污染物的化学转化,研究化学催化剂与污染物分子的界面电子传递有助于理解催化剂结构和表面特性对污染物化学转化的影响,从而强化对催化剂的设计与调控,进一步促进污染物的转化。电子转移不仅是生物体进行基础生物过程的重要步骤[5],也是无机纳米技术应用的关键,如构建功能器件特征表面、设计超分子或生物分子氧化还原传感器、优化电极表面结构以及操控仿生催化过程[8],都与电子转移过程密切相关。电子转移过程在物理体系、化学体系、生物体系以及复杂的复合体系中均存在,通过对电子转移机制的解析,能够获得有关热力学与动力学方面的重要参数。

1.2
污染物生物转化中的电子传递

1.2.1
污染物的生物转化

生物转化是指利用微生物将空气、水体及土壤中的污染物进行转化,从而生成容易被降解的化合物,在一个体系中可能会同时存在生物转化、生物降解和生物催化过程。目前,人类生产的新化学品与日俱增,这些新化学品不可避免会进入自然环境体系中,可能会对人类与环境都产生危害。早在40多年前,Alexander等[9]就提出人造的化学品或某些自然条件下产生的化学物质只有在适当的条件下才能自然降解,这个观点逐渐被广泛接受。许多化学品已经超出了环境系统自然净化的能力,从而在环境中累积,造成污染。而自然界中的微生物能够针对变化的目标污染物,演化出新的催化路径,以获得碳源、能源以及营养元素或只是为了解除它们的毒性。然而,揭示污染物的降解过程非常困难,因为污染物的数量众多,在环境中多数浓度较低,而且微生物在对其降解和转化中会产生许多未知的化学物质。从各种生物酶的作用出发,对生物转化与生物降解机制进行解析,不断揭示新的生物降解路径,分析物化条件与微生物自身的生理条件对生物转化降解能力的进化和影响,将促进生物修复在实际中的应用。

自然界中普遍存在的生物转化带动了元素的全球循环。图1.1给出了由微生物催化的碳、氢、氧、氮的生物循环,并给出了一些有代表性的微生物催化转化反应。较为简单的生物转化反应有加成反应,如氨基的乙酰化[11]或羟基基团的甲基化[12]以及各种取代和裂解型反应[13]。通常情况下,相对较弱的键,如酯类、有机磷或酰胺类[14-15],很容易进行这些生物转化。由于微生物的繁殖速度和多样性进化速度比其他的生物要快得多,细胞在不断分裂的过程中需要不断地从环境中摄取养分和能源进行代谢,这保证了微生物利用其生存环境中的多种污染物质的可能,其中明尼苏达大学的生物催化-生物降解数据库(UM-BBD)的微生物索引给出了每种微生物能够降解的污染物种类等相关信息。不同污染物包括重金属,多环芳烃(PAH),带有氯、氰基、硝基和磷酸基的脂肪族和芳香族化合物以及脂环族化合物。

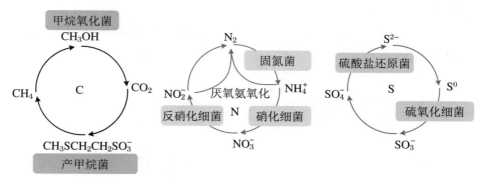

图 1.1　微生物转化促使元素的全球循环[10]

污染物的微生物转化一般都存在特定的酶的作用,或是涉及电子转移的氧化还原反应过程。Qin 等[16]报道了黄石嗜热真核藻类(表 1.1)可以将有毒的五价砷 As(Ⅴ)通过生物转化降低价态并进行甲基化,从而说明这些藻类能够容忍砷存在于它们的生存环境中,由此也可以通过 As(Ⅴ)的浓度变化进行藻类甲基化酶的表征。将环境中 As(Ⅴ)还原为 As(Ⅲ),涉及电子传递过程,是藻类导出电子的过程。由于 PAH 在环境中分布广泛,具有持久性、生物累积性和致癌性,故所有 PAH 衍生物都存在显著的环境问题[19,21]。具备 PAH 降解功能的细菌数量庞大,报道较为普遍的有 *Pseudomonas aeruginosa*,*Pseudomons fluoresens*,*Mycobacterium* spp.,*Haemophilus* spp.,*Rhodococcus* spp. 和 *Paenibacillus* spp.。在石油化工产生的污染物中,苯并芘(benzo(a)pyrene,BaP)被认为是致癌性和毒性最强的 PAH。Ye 等[20]发现在培养 *Sphingomonas paucimobilis* 菌株 168 h 后可以使 BaP 浓度降低 5%,而且通过检测 CO_2 浓度的变化发现存在羟基化以及苯并环的裂解。水杨酸(SA)是分解代谢 PAH 关键的中间体,而且 SA 及其衍生物,尤其是乙酰水杨酸,通常被作为有效的止痛剂,用于各种药剂中,最后随着生活废水和医药废水进入环境体系中[22]。*Pseudomonas* spp. 通过氧化脱羧过程来降解 SA,生成儿茶酚,起到关键作用的是水杨酸-1-羟化酶,这是一种黄素单加氧酶的模式酶[23]。Wahidullah 等[17]发现 SA 可以被海藻 *Bryopsis plumosa* 协同海洋细菌 *Moraxella* spp. MB1 在不发生环裂解的条件下转化为肉桂酸酯,其中苯基丙氨酸解氨酶在降解过程中起到关键作用(表 1.1)。除了以上提及的甲基转移酶、单加氧酶以及解氨酶外,还有一种能够进行污染物降解的重要蛋白,即微生物细胞色素 P450(microbial cytochrome P450,CYPs)[24],自从 20 世纪 70 年代被发现后引起了广泛的关注。微生物的 CYPs 不仅仅在合成天然产物中起到关键作用,部分真菌和细菌的 CYPs 在微生物系统中也起到生物修复的作用。在对 Rhodococcus 和 Gordonia 菌株降解汽油添加剂(如甲基叔丁基醚、乙基叔丁基醚和叔戊基叔丁基醚)的研究中发现有 CYP249 参与。除了可以对污染物进行生物转化,细胞色素 P450 还能生物催化正烷烃羟基化[25]。

表 1.1 微生物协助的污染物转化

污染物	微生物	机制
砷[16]	Yellowstone thermoacido-philic eukaryotic alga	将 As(V)还原为 As(III),并甲基化 As(III)以形成三甲基氧化砷和二甲基砷酸盐(甲基转移酶)
水杨酸[17]	Moraxella spp.	不开环的条件下转化为肉桂酸酯(苯胺裂解酶)
染料含氧物[18]	Rhodococcus,Gordonia	将乙基叔丁基醚转化为叔丁醇(微生物细胞色素 P450)
多环芳烃[19]	Pseudomonas spp.,Mycobacterium spp.	在细胞外通过氧化催化破坏化学键,形成自由基
苯并芘[20]	Sphingomonas paucimobilis	羟基化 苯环开环

Shewanella oneidensis MR-1 可以从甲基橙染料废水中获得能量并伴随着偶氮还原反应,如果阻断金属还原通道(metal-reducing,Mtr)呼吸途径,那么染料废水的脱色效率可降低 80%,因此,细胞色素 c(outer-membrane cytochrome c,OM c-Cyt)在其中起到了关键的作用。Mtr 途径是指电子传递链从内膜到外膜的过程,与金属氧化物还原反应以及 MFC 阳极电流的产生密切相关[27-28]。对 Shewanella oneidensis MR-1 在厌氧条件下降解硝基苯的机制的基因水平研究表明,omcA-mtrCAB 基因簇被认为是 Shewanella 细菌还原降解污染物最重要的相关基因,但并未参与到硝基苯的生物还原降解中[29]。而把 cymA 基因敲除后,硝基苯的降解效率与野生株相比降低了 60%,而 cymA 对应的蛋白是在周质空间中具有 4 个卟啉铁辅基的细胞色素 c。这说明 Shewanella oneidensis MR-1 对硝基苯的降解与其他污染物不同,可能是在胞内的某些蛋白共同参与的还原过程。

由 DNA 重组技术衍生的分子生物学技术对生物转化和降解的研究产生了深刻的影响,目前已经可以优化微生物降解各种污染物所需要的条件,构建新的代谢途径以消除产生非产物的副反应、减少有毒中间产物的产生以及提高酶的活性。大多数微生物降解机制都涉及电子传递,即氧化或还原过程。但微生物参与的降解过程通常会存在降解速率较低的问题,且对环境条件有较多的要求。

因此,需要将生物方法与物理化学方法各自在降解污染物过程中的优势有机地结合起来,提高去除环境污染物的效果。

1.2.2
生物电化学系统中的污染物转化

生物电化学系统(BES)是一种利用微生物在电极上进行氧化和/或还原的电化学系统,其中微生物燃料电池(MFC)能在微生物降解有机废物的同时产生电能[30],将废水处理和生物产电有机地结合起来。电化学活性细菌(EAB)可将电子转移到金属氧化物、电极等胞外的电子受体,在环境修复、生物地球大循环及 BES 中发挥了重要作用。目前研究最普遍的 EAB 是 *Shewanella* 和 *Geobacter*,此外也已经发现有多种细菌同样具有电化学活性[32]。EAB 几乎可以利用水中任何可以生物降解的有机物或无机物进行代谢,将细胞体内的电子通过外膜蛋白或者核黄素等氧化还原媒介转移到它所生存的环境中。在 MFC 中这些细菌将电子传至阳极,由外接的导线到达阴极,同时释放一个质子在阳极室的溶液中,通过质子交换膜到达阴极室,形成一个完整回路而产生电能。

通过 MFC 可以有效地去除污染湖水中的溶解性有机物(DOM)。在经过 MFC 处理后,被污染湖水的总有机碳降低了 50%。被污染湖水中类蛋白物质是 MFC 的主要底物,腐殖质类物质是不易被生物降解的。通过 *Salmonella typhimurium* Sal94 的毒性评估,发现基因毒性物质几乎被完全去除,MFC 的最高产电功率密度可达到 164 mW·m^{-3}。在磁场的作用下,*Shewanella* 接种的单室或双室 MFC 的最大电压可提高 20%~27%[35],这是由于在磁场的作用下通过氧化应激机制提高了微生物的生物电化学活性,促进了对底物的降解。因此,在 MFC 体系中引入磁场可以作为提高其能量转换效率的有效方法之一。而利用光刻胶技术制备平面型金阳极构建的微型 MFC[36]具有的最大电流密度为 2148 mA·m^{-2}。微型 MFC 在医学或通信领域具有潜在的应用价值。

将 MFC 技术与其他污染物降解方法进行耦合,提供了原位利用 MFC 输出电能的新思路,拓展了 MFC 的应用范围。其中,微生物电解电池(MEC)已经能够成功将纤维素、葡萄糖或乙酸盐转化为氢气[37],但 MEC 需要外加一个电势来克服热力学能垒,如果将 MFC 与 MEC 耦合,那么 MFC 正好可以作为外加电源提供能量促使 MEC 中的产氢反应发生[38]。电 Fenton 方法可以有效地对污染物进行降解,但是由于需要外加电压才能实现,电 Fenton 技术的推广受限于高昂的成本[39]，而 MFC 与 Fenton 技术的集成应用则可以有效地解决这个问题。将阳极 Fenton 反应器与双室空气阴极 MFC 组成的系统用于对污染物的降解，则 MFC 提供的能源可以用于驱动阳极 Fenton 反应的发生，是一种具有成本效益且能节约能源的水处理系统[40]。考虑膜生物反应器的优势，将之与 MFC 组成新型的生物电化学膜反应器（BEMR）[41]，在废水处理与能源回收方面具有广阔的应用前景。在 BEMR 中以不锈钢网附着生物膜后同时用作阴极以及过滤材料，附着的微生物将催化氧还原反应的发生。同时这个系统也表现出了高效的化学需氧量（COD）和 NH_4^+-N 去除效率，两者都超过了 90%。虽然目前构建的 BEMR 系统的产电能力和能源回收效率有限，但通过进一步优化电极材料与 MFC 构型以及解析微生物催化机制，将提升 BEMR 在废水处理与能源回收领域的实际应用价值。将序批式反应器（SBR）的优势与 MFC 相结合，构建 SBR-MFC 系统[42]，可以使 MFC 的最大输出功率密度达到 2.34 W·m^{-3}，可以为进一步改善现有活性污泥方法提供新的途径。结合光催化技术的优势，利用 MFC 阴极有效分离光电催化系统中 TiO_2 产生的电子，并使 MFC 阳极原位产生的电子到达光电催化系统的阴极构成 MFC 辅助光电催化系统（MPEC）[43]，利用 MFC 高效地减少光催化剂光生电子与空穴的复合，将有效地提高污染物降解效率。

但是，生物体系缓慢的电子传递速率仍然是 MFC 发展的瓶颈，而对于 MFC 中电子传递过程的机制解析，将有助于提高 MFC 输出功率密度以及污染物降解效率。

1.2.3 微生物胞外电子传递

研究者们提出了多种胞外电子传递（extracellular electron transfer，EET）机制[44-45]，包括外膜蛋白直接接触、可溶还原性媒介和菌毛传递。EET 又可以分为直接与间接两种情况：直接电子传递（direct electron transfer，DET）分为通过细胞膜直接与电极接触以及通过纳米导线与电极接触[44,46]，其中前者通过细胞外膜上的细胞色素 c 与固相受体的界面直接接触进行电子传递。间接电子传递（mediated electron transfer，MET）又称为媒介参与的电子传递[47]，也分为两种不同的形式，主要以媒介的不同加以区分。媒介可以是外界提供的氧化还原物质，也可以是微生物分泌的次级代谢物。DET 与 MET 的电子传递形式在能量效率上有各自的优缺点,如 DET 有较高的库仑效率,而 MET 则有较大的电流和功率密度。如果能够结合这两种电子传递的优势,将可以显著提高 MFC 的产电性能。EET 过程的解析在提高 MFC 产电能力以及污染物降解效果方面具有非常重要的作用。

1.2.3.1 直接电子传递

MFC 阳极作为产电微生物附着的载体,在进行直接电子传递时,其阳极材料与微生物直接接触,需要提高产电微生物附着量,并能够诱导电子从微生物向阳极的传递。因此,阳极材料的优化对提高 MFC 废水处理效率和产电性能有至关重要的作用。研究合适的阳极材料具备的结构特征,分析阳极材料和表面特性对微生物产电特性的影响,对提高 MFC 的产电能力有十分重要的意义。碳基材料由于其结构变化层出不穷,具备较高的比表面积、化学稳定性和生物相容性等优势,而成为普遍使用的 MFC 阳极材料(表 1.2)。

表 1.2 MFC 阳极碳基材料

组成	特点	电流密度/功率密度
分级微/纳米结构:CNT 改性毡(CNT-GF)[48]	使用 CNT 改性毡作为阳极,与裸石墨毡阳极相比,MFC 的功率输出增加了 7 倍	1680 mW·m^{-2}
单壁碳纳米角改性碳纤维(SWNH-CF)[49]	SWNH 的改性充当"电子线",加速电解质溶液和电极表面之间的电子转移	1400 mW·m^{-2}
石墨纤维刷[50-51]	具有高表面积和多孔结构的刷状阳极可产生高功率密度	2400 mW·m^{-2}
氧化石墨烯纳米带(GONR)[52]	通过在电极上仿生地制造 GONR 的网络结构,实现了 EET 过程的显著增强	34.2 mW·m^{-2}
3D 碳纳米管布(CNT-textile)[53-54]	这种新型阳极材料为生物膜的生长提供了开放的结构,使底物能够有效地运输并通过多种微生物群落进行内部定植	1098 mW·m^{-2}
石墨烯海绵[55]	低成本、节能的石墨烯涂层海绵阳极可以通过简单的制造工艺生产,并且这种阳极可以按比例放大	1570 mW·m^{-2}
多壁碳纳米管[56]	可持续设计的微型 MFC 的负极材料	19 mW·m^{-2}
碳纳米管网络[57]	增强细胞色素 c 和固体电子受体之间的相互作用	2.65 A·m^{-2}

续表

组成	特点	电流密度/功率密度
碳纤维无纺布（CFM）[58]	静电纺和溶液吹制三维碳纤维非织造布	30 A·m^{-2}
硝酸和乙二胺处理的碳纤维毡（ACF-AT，ACF-A，ACF-N）[59]	EDA 处理将 ACF-AT 的表面基团从吡咯变为内酰胺、酰亚胺和酰胺。这些官能团往往更利于细菌黏附，更容易进行电子转移	1304 mW·m^{-2} 1641 mW·m^{-2} 2066 mW·m^{-2}
丝瓜海绵碳[60]	大大增加了细菌的负载能力	（1090±72）mW·m^{-2}
氧化石墨烯-泡沫镍[61]	表面积大有利于细菌定植；均匀大孔支架有利于电子介质、培养基有效扩散	661 W·m^{-3}

分层的微/纳结构碳复合材料具有较大的优势，如氨气处理后碳纳米管修饰的多微孔石墨碳毡（CNT-GF）[48]材料制备的阳极将 MFC 的输出功率增加了 7 倍。单壁碳纳米角修饰的碳纤维（SWNH-CF）[49]起到了"电线"的作用，加速了电解液与电极表面的电子传递。以具有高比表面积和多孔结构的石墨刷为阳极，相比于其他材料 MFC 达到了很大的功率密度。微生物与电极的架桥作用是加快 EET 的重要途径之一，在电极表面修饰氧化石墨烯纳米带（GONR）网状结构能够有效地收集微生物的电子[52]。碳材料还能与其他种类的材料组成复合材料，从而可以进一步优化和发挥碳材料的优势。碳材料与聚合物的结合可以明显地提高 MFC 阳极材料的电化学活性，MFC 的功率密度比以普通的石墨电极为阳极的电池得到大幅度的提高。另外，石墨烯海绵制作工艺简单，可以降低 MFC 阳极材料的成本。微型 MFC 本质上是一个微型能量采集器，可以使用在微型电子芯片或传感器上。使用多壁碳纳米管（MWCNT）作为阳极，能够构建性能稳定、可持久运行的微型 MFC。进一步对 CNT 进行处理，形成 CNT 网络，则可以显著提高 MFC 的电流密度。

如果使用分子模拟技术首先对具有不同结构特征的碳材料进行电子传递的预测，就可以有效改进碳材料的结构，提高 MFC 能量转换效率。选择出具有高比表面积、低内阻以及低廉成本的阳极材料对 MFC 的实际应用具有促进作用。

1.2.3.2　媒介参与的电子传递

氧化还原媒介（redox mediator）或称为电子穿梭体（electron shuttle，ES）可以有效增强 EET 过程，它不断与微生物细胞及外界氧化物接触，并在此过程中传递电子。MFC 阳极室中的氧化还原媒介在传递电子的过程中，涉及的是质子耦合电子传递（PCET）反应[63]。PCET 反应分为电子传递和质子化两个连续或交替的基元反应过程。不同结构的氧化还原媒介分子可以调控电子传递的路径，并减少电子传递过程中的能量损失，有效提高电能输出[64]。蛋白酶的氧化还原活性中心与电极之间的距离导致了较高的过电势，从而会很大程度地降低电子传递速率[65]。而利用氧化还原媒介可以降低过电势，解决部分酶的活性中心无法与电极接触的问题，加速电子传递，并获得电极与酶活性中心之间的平衡。氧化还原媒介必须是水溶性的，且有较低的分子质量和较高的反应活性[66]。合适的氧化还原媒介可以显著提高电子传递过程中的能量转换效率。

Shewanella oneidensis MR-1 细胞能够分泌水溶性的核黄素与黄素单核苷酸（FMN），而这些加速了铁氧化物和电极的反应过程。目前黄素还原铁氧化物的确切机制仍不清楚，但是在 MtrAB 和 MtrC/OmcA 的突变株中，黄素介导还原铁氧化物的过程遭到了破坏，表明孔蛋白复合物和胞外十血红素（decaheme）辅基的细胞色素在与黄素的电子转移中扮演着重要角色。MtrC 和 MtrB 也需要用于还原人工的胞外 ES,蒽醌-2,6-二磺酸（AQDS）是一种在腐殖质中发现的具有氧化还原活性成分的类似物。胞外细胞色素具有较大的氧化还原电势范围，能够还原黄素和 AQDS 的中点电位的范围是 $-200\sim-100$ mV(pH 7 vs. SHE)。扩散性还原剂，如黄素，其优势是能够避开尺寸较小的氧化物颗粒的物理限制，因为黄素不会发生络合作用，是性能优异且能快速扩散的还原剂[70]。虽然加入可逆的氧化还原媒介可以加速胞外电子传递，但有可能会造成环境污染问题以及降低成本效益，而将氧化还原媒介固定在电极表面，则有可能克服这些问题。将中性红（NR）共价键合在 MFC 碳基阳极表面，显著地提高了 MFC 的功率输出以及对碳源的利用[71]。NR 被证明是一种有效的氧化还原媒介，可用于捕获微生物代谢产生的电子，这是由于其优异的电化学可逆性以及与主要的代谢电子载体（如 $NADH/NAD^+$）有兼容的氧化还原电位。在这个过程中，NR 的固定化过程是通过亚硫酰氯使羧基化表面酰氯化，接着与 NR 进行酰胺化。这种固定化过程能够显著增强 NR 分子在碳基表面的固定能力，而不改变其氧化还原性能。

Shewanella oneidensis MR-1 自身分泌的黄素作为外膜蛋白传导电子的氧化还原辅助因子，提高了 OM c-Cyt 转移电子的能力[72]，如图 1.2 所示。全细胞的差示脉冲伏安法（DPV）显示黄素的氧化还原电位可逆地正向移动了超过 100 mV，这与微生物产电增加的现象是相符的。更重要的是，试验结果表明黄素/OM c-Cyt 之间的相互作用加速了通过半醌发生的一电子（one-electron）氧化还

原反应,且反应速率比自由的黄素快了 $10^3 \sim 10^5$ 倍。这种机制虽然与之前描述的氧化还原媒介参与的电子传递机制不同,但是黄素/OM c-Cyt 的相互作用同时调节了胞外电子传递的程度以及胞内代谢活动。

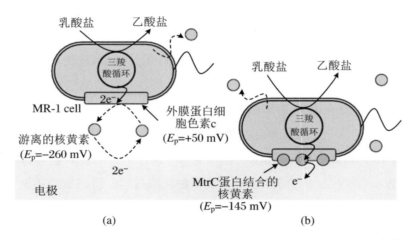

图 1.2　*Shewanella oneidensis* MR-1 自身分泌的黄素参与 EET 过程的解析[72]

除了 *Shewanella oneidensis* MR-1,其他细菌如 *Pseudomonas aeruginosa* 能产生蓝色的吩嗪色素——绿脓菌素作为 ES[73]。该化合物可以作为具有氧化还原活性的抗生素以及呼吸链辅助色素,而且更重要的是绿脓菌素不仅可以被 *Pseudomonas* 自身用于电子传递,也可以在 MFC 中被其他细菌如 *Lactobacillus amylovorus*,*Enterococcus faecium* 和 *Brevibacillus* spp. 利用。为了提高 MFC 中输出的电能,添加的氧化还原媒介包括自然条件下存在的化合物,如腐殖质(HS),或者人工合成的化合物,如 AQDS、NR、2-羟基-1,4-萘醌(HNQ)、硫堇、刃天青和亚甲基蓝等[75-77]。因此,需要深入了解这些有机小分子的物化性质,如其化学结构中点电位、水中的溶解度、稳定性和毒性等[78],才能更好地提高 MFC 的能量转换效率。在微生物能量代谢中,人工补充的 ES 能够产生新的电子流动路径,从而开辟出微生物技术应用的新的可能性。生物电化学反应器结合 ES 可能是一种调控微生物的生长和代谢的有效方法[78]。

Geobacter sulfurreducens 是一种革兰 δ-蛋白质菌,也能够通过呼吸作用将电子转移到包括阳极电极在内的固态胞外电子受体上。电子传递过程是通过表达在外膜上丰富的 OM c-Cyt 进行的。这类细菌可以不借助外加氧化还原媒介分子进行 EET 过程[79],也就不会因为电子从细菌到氧化还原媒介之间的传递损失生物所需的能量,因而逐渐引起了研究者们对其在化学能向电能转变领域潜在应用的探讨。

1.2.3.3 外膜蛋白细胞色素 c

无论是将电子直接传递至不同的固相受体或是传递至可溶性的氧化还原媒介，EAB 的 OM c-Cyt 都扮演着至关重要的角色[80]。不同的 OM c-Cyt 都呈现出类似的结构，由具有氧化还原活性的血红素分子(heme)以及氨基酸组成的肽链缠绕形成，其中血红素的中心位置 Fe 轴将配位了的双组氨酸残基接入肽链中。这些具有氧化还原活性中心的蛋白质位于细胞膜上，是电子从胞内到胞外的通道，这些蛋白质中暴露于溶剂中的 heme，在电子传递过程中可能会直接接触电子受体分子或界面，但其中的相互作用机制还有待研究。

Shewanella oneidensis 包含编码为 *mtrDEF-omcA-mtrCAB* 的基因簇。MtrA 与 MtrB 组成了跨外膜的电子传递复合体，包括 β-折叠的孔蛋白(MtrB)以及嵌在其中的十血红素辅基细胞色素(MtrA)[87-88]。MtrC 则是一个胞外十血红素辅基细胞色素，是这个复合体的终端。MtrF，MtrD 与 MtrE 分别是 MtrC，MtrA 与 MtrB 的同源蛋白。操纵子 MtrDEF 在生物膜的生长中呈现出最高表达，但是会在 MtrCAB 和 MtrFDE 组建之间形成杂化复合体[86,90]。OmcA 蛋白是 MtrC 和 MtrF 的同源蛋白，能够与 MtrC 或 MtrF 相互作用，从 MtrCAB 或 MtrFDE 复合体接受电子[91]，也可以在 MtrC 或 MtrF 突变株中替代这些蛋白[86]。

目前，已经对其中一种外膜电子传递通道中的十血红素辅基细胞色素 MtrF 进行了 X 射线晶体结构解析。根据这个结构模型，可以研究不同类型的 EET 或解析可能的 EET 发生机制。对 MtrF 晶体结构的解析第一次确定了 10 个血红素的空间排布构型，其中血红素以一种独特的交叉构型贯穿在 4 个结构域(Domains Ⅰ,Ⅱ,Ⅲ,Ⅳ)中(图 1.3)。这个结构可以提供分子水平研究的可能性，用于分析 EAB 如何还原不溶性底物，如矿物、可溶性底物(如黄素)以及与细胞表面不同氧化还原细胞色素终端之间形成的电子传递链。

根据 MtrF 的结构，可得外膜蛋白组成的电子传递通道复合体的分子结构示意图如图 1.4 所示，其中红色的是 MtrC，蓝色的是 OmcA，青色的 MtrA 是由两个五血红素辅基 NrfB 单体末端相连组成的，绿色的是孔蛋白 MtrB。这里 MtrC 与 MtrA 嵌入孔蛋白的深度是未知的，示意图只是展示了其中的一种可能情况。目前实验方法还无法研究蛋白内沿着血红素组成的通道进行的电子传递，而高性能计算则可以从分子水平解析 heme-to-heme 电子传递的热力学和动力学性质[93]。采用这些蛋白组成独特的分子机器进行长距离的电子传输，转移电子的距离可以远于100 Å,这对于生物纳米技术设备的设计具有显而易见的科学意义。

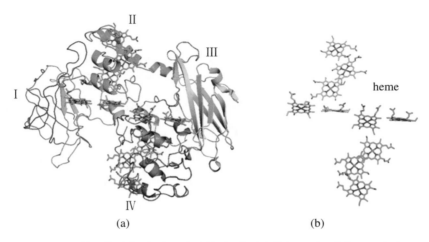

图1.3 MtrF的晶体结构(包含4个结构域,分别用蓝色、绿色、黄色和红色表示结构域Ⅰ,Ⅱ,Ⅲ和Ⅳ(a))以及MtrF中血红素(heme)的结构排布(b)

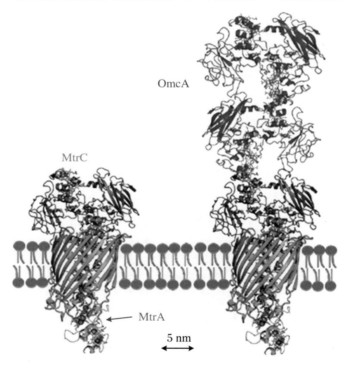

图1.4 外膜上MtrCAB孔蛋白-细胞色素c的结构示意图,以及胞外MtrC-OmcA细胞色素链。图中MtrC用MtrF的晶体结构来表示,MtrA是由两个五血红素辅基NrfB单体末端相连组成的细胞色素结构,OmcA用两个MtrF结构表示[94]

虽然目前对于 EET 过程已经有很深入的探索,但是对于该过程涉及的界面过程和控制因素仍然很不清楚,而计算化学可以为实验结果提供充分而有力的证明与解释,能够为实验研究提供有指导价值的成果。在电子传递过程中可以对电子给体分子与电子受体分子以及固相界面组成的体系进行量子化学计算[95],可以获得体系分子的电子结构与几何结构信息,从而分析分子轨道能级、原子的电子密度分布、电子传递反应过程的热力学和动力学性质[96]。自然体系中的电子传递过程一般发生在水溶液环境中,而分子动力学模拟可以构建较大的模型体系,以考虑真实环境体系中水溶液中近程水分子的影响,分析分子间作用力以及离子或分子在液体或固体中的扩散。通过有效的预测可以揭示其内在规律性。理论计算可以得到实验技术无法或者较难测定的物化性质。将计算模拟方法与精湛的实验技术有效结合,能够为研究提供更具优势的工具,加快研究成果的实际应用。理论计算方法可以为阳极材料的微观结构设计与电子传递系统的工作原理和影响因素提供理论基础,从而应用于 BES 的优化。

1.3
污染物化学转化中的电子传递

1.3.1
污染物的化学转化

化学体系中质子耦合电子转移(PCET)可以激活惰性小分子转变为更富含能量的物质[98],比如,CO_2 转化为甲酸是一个 2 电子/2 质子耦合过程[99],水的氧化需要累计损失 4 个电子和 4 个质子,而将 N_2 转变为 NH_3 是一个 6 电子/6 质子转移反应[101]。同样地,有机污染物的转化和降解过程中也会发生电子或质子的转移,电子的转移伴随着能量的流动。例如,光催化材料可以利用太阳能将水转化为氢能,降解环境中的有机污染物,是一种能够在光作用下诱发光氧化-还原反应的半导体材料,在解决当前能源与环境问题方面具有广阔的应用前景。

近年来,半导体光催化材料在清洁能源转换和环境污染控制等方面的应用已经成为研究的焦点。1976年,Garey报道了TiO_2光催化氧化法,成功地将水中多氯联苯脱氯[102]。此后,半导体非均相光催化在水处理领域引起了广泛的重视,被认为是极具发展潜力的新技术。

利用光催化剂实现的污染物的降解和转化,是从太阳能到化学键能的转变。金属氧化物半导体材料在能源转换和储存、环境修复、光电、记忆存储、光致发光等领域都有广泛的应用潜力。TiO_2是目前研究得非常多的金属氧化物光催化剂,具备特殊的稳定性和高效的光催化效率。这种材料廉价、丰富且无毒,能够有效地应对未来的环境和能源问题。分析TiO_2界面氧化还原过程,对污染物的光催化降解、H_2O的分解以及选择性的污染物转化都具有促进作用。最近的相关研究多集中在如何提高TiO_2的光催化降解效率方面,如用金属和非金属掺杂以及材料的复合等。贵金属等离子激元纳米结构与半导体的结合为未来的能源需求提供了一个可能的途径[103]。

1.3.2

贵金属沉积

TiO_2表面沉积的贵金属通过捕集电子提高界面电荷的转移能力,从而降低了电子-空穴的复合速率,提高了催化效率。其原因在于贵金属费米能级(E_F)较低,因而可以俘获电子,延长了电子-空穴对的寿命[104-105]。当金属纳米颗粒与TiO_2接触时,电子转移至金属粒子,使金属的E_F向导带移动,直至两者达到平衡。电子的积累增加了Au的负电势,使得Au的E_F趋向TiO_2的导带,这更利于TiO_2表面的电荷分离,从而提高整个催化剂的氧化还原能力。整个过程如图1.5所示。

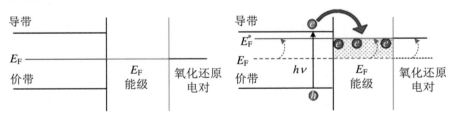

图1.5 紫外光照前、后半导体-金属纳米复合体系的能带结构[105]

研究者们通过实验证实了上述机制[104]。首先 TiO_2 胶体在紫外辐射下变成蓝色，这表明 TiO_2 表面积累了电子，这些电子导致 TiO_2 能够吸收红外光。接着向 TiO_2 胶体中加入定量的 Au 纳米颗粒，并记录吸收图谱，发现随着 Au 的加入红外区域的吸收峰一直在减小，这说明 TiO_2 在持续损失电子。在可见波段的吸收是由 Au 纳米颗粒的表面等离子体共振（SPR）造成的[104]。可以确定的是，在紫外辐射下激发态电子从 TiO_2 导带转移到 Au，降低了 TiO_2 本身的电子密度，因此红外区域吸收强度减弱。

光催化中 TiO_2 耦合贵金属的优势已有很多实验报道，例如，TiO_2 纳米片经 Au 颗粒修饰后对 X-3B 偶氮染料的光催化降解作用增强，利用 Ti 基底能导电的特点，将 Ti 片阳极氧化后再沉积 Au 纳米颗粒。光生电子激发后从 TiO_2 纳米管矩阵迁移到 Au 纳米颗粒，纳米管下的 Ti 基底因其良好的导电能力而进一步降低了电子-空穴复合速率[107]。Au/TiO_2 可用于苯酚的光催化降解，其中煅烧温度对合成 Au/TiO_2 的降解性能存在影响，经 400 ℃ 煅烧后其光催化效率得到提高，这可能是因为高温煅烧增强了金属和半导体纳米颗粒之间的接触，但是温度继续升高并不会进一步提高材料性能，相反会导致金属颗粒从 TiO_2 表面脱落从而降低了催化性能。

除了与 Au 和 Ag 耦合形成复合材料之外，Pt/TiO_2 材料在光催化降解或有机物转化中的性能较为优异[109-110]。将 Pt 纳米颗粒沉积在 TiO_2 表面，可以实现 4.4% 的苯转化为苯酚。对比 Pt/TiO_2 与 Ag/TiO_2 材料对甲苯和三氯乙烯的催化效果，结果发现 Pt/TiO_2 的催化性能优于 Ag/TiO_2，这是因为 Pt、Ag 相对于 TiO_2 逸出功不同（$W_{Pt} = 5.36 \sim 6.63$ eV，$W_{Ag} = 4.26 \sim 4.29$ eV，$W_{TiO_2} = 4.6 \sim 4.7$ eV）。Pt 与 TiO_2 的逸出功相差更多，因此能更好地捕获电子，促进电子-空穴的分离，而 Ag 与 TiO_2 的逸出功相差较小以至不能有效地捕获电子。另一个研究组同样对比了 Pt 和 Ag 在光催化降解中的影响，结果发现 Ag 不仅能产生更多超氧自由基，而且提高了空穴直接氧化底物的速率，因此，Ag/TiO_2 材料的催化效果更好[113]。尽管这两个研究组的结果相互冲突，但是他们提出的机制都有可能存在。Chen 等报道 TiO_2 与 Pt 之间电子达到平衡后 TiO_2 的 E_F 向上移动。在预先合成的 Pt 纳米管上利用化学气相沉积法沉积 TiO_2，即可制备 TiO_2-Pt 同轴纳米管材料。当 Pt 与 TiO_2 接触时，由于 Pt 的功函数更高，因此，Pt 的 E_F 会向上移动直到与 TiO_2 的 E_F 相等。当 TiO_2-Pt 受光激发时，由于光致电子的产生 TiO_2 的 E_F 会向上移动达到一个新的准费米能级，从而形成 Schottky 势垒，因此，电子从 TiO_2 迁移到 Pt，从而抑制了电子-空穴的复合[114]。Zhao 等将 Pt 纳米颗粒光催化沉积在 TiO_2 表面，研究了在可见光条件下 Pt-TiO_2 对酸性玫瑰红 B(SRB)的光催化降解。吸附于 TiO$_2$ 表面的 SRB 受可见光激发产生电子,然后电子通过 TiO$_2$ 转移到 Pt 上,导致 Pt 具有较高的电子密度,最终这些电子与被吸附的 O$_2$ 反应产生大量超氧自由基[115]。

半导体-等离子体金属纳米结构不仅能够提高材料在紫外光辐射下的光催化活性,而且可以通过表面等离子体共振效应(SPR)在可见光下起到光催化作用。Au-TiO$_2$ 或 Ag-TiO$_2$ 作为无机光敏剂,可增强对可见波段的吸收,通过改变等离子体纳米颗粒的大小、形貌和结构可以调整 SPR 的强度和波长。据报道,Au 纳米颗粒如果包裹在壳状 TiO$_2$ 内部,不仅能够防止对 Au 的腐蚀,反过来还能增强催化剂的整体稳定性[116]。

等离子体激元金属不仅可以通过捕获光生电子来提高电子-空穴分离效率,而且在可见光下可以诱导 SPR 效应,因此,可用它来研究整个太阳光波段下的光催化作用。紫外光谱可激发 TiO$_2$ 产生电子,该电子从 TiO$_2$ 的导带迁移到金属,从而实现光生电子和空穴有效分离。同时,可见光可以激发贵金属的表面等离子体共振,然后高能电子可转移至 TiO$_2$ 的导带(图 1.7)[115,117-118]。

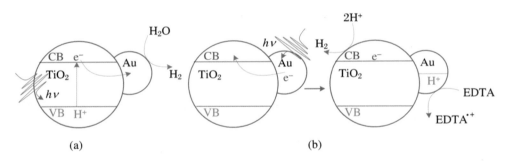

图 1.7 金/二氧化钛(Au/TiO$_2$)的光催化机制:紫外光激发(a);可见光条件下,乙二胺四乙酸(EDTA)是电子给体牺牲剂[118] (b)

Zhang 等通过 Au@TiO$_2$ 的核壳纳米结构研究了等离子体激元金属的双重效应。以覆有柠檬酸盐的 Au 纳米颗粒为核,采用 TiF$_4$ 为 TiO$_2$ 的前驱体在水热条件下将 TiO$_2$ 包裹在 Au 表面。以罗丹明 B(RhB)为研究对象,在紫外光和可见光条件下考察了该催化剂的催化性能,结果发现染料的降解机制与上述机制均不相同。在紫外光照射下有大量的·OH 产生,它有利于染料的降解;然而在可见光照射下没有检测到·OH。进一步的实验证明,可见光下空穴在 RhB 的降解中的作用大于·OH[115]。Wu 等以 Au 和 N-TiO$_2$ 复合材料为催化剂,分析了甲基橙的降解过程,最终证明了 Au 在其中所起的双重作用。除了电荷转移机制之外,Au 纳米颗粒的另一种促进作用,即表明等离子体激发产生的高能电子流向 TiO$_2$,促进电子-空穴的分离,然后这些被分离的电子-空穴对就可以进行光催化反应[119]。同太阳光照射相比,该材料的光催化降解能力在紫外光照射下更强。这是因为 TiO_2 的带隙宽度决定了只有能量更高的紫外光才能充分激发其价带的电子。

1.3.3
光催化过程中的电子传递

TiO_2 已被普遍视为金属氧化物半导体催化剂的基础材料,然而 TiO_2 相对较宽的能隙使它只能吸收紫外光,而且光生电子和空穴的复合速率也很快,这些都降低了它的催化效率。同纯 TiO_2 体系相比,TiO_2 异质结构不仅提高了电荷分离效率,而且可以吸收利用可见光。通过掺杂金属、非金属以及染料敏化作用可以提高 TiO_2 的光催化效率。在 TiO_2 上负载贵金属可从以下两方面提高其光催化性能:一是在紫外光照射下金属纳米粒子可作为电子捕集器捕获电子提高电子-空穴分离效率;二是可见光照射引发的表面等离子体共振效应(SPR)可增强光催化剂的催化作用。

表面等离子体共振诱导电子从金属转移至 TiO_2,该机制与染料敏化的机制相似,即固定在 TiO_2 表面的染料分子吸收可见光,并将激发电子转移至 TiO_2 的导带(CB)。表面等离子体是由金属表面的电荷密度波组成的,不能与半导体的价带(VB)和导带(VB)或染料的 HOMO-LUMO 能级相比。等离子体激元金属纳米颗粒吸收谐振光子,经 SPR 激发产生的高能电子转移至半导体[120-121]。SPR 的能带位置可通过改变纳米颗粒的大小和形状来进行调整,因此能够利用所有太阳光波段[103]。

不同于传统半导体的光电催化过程(电子由价带激发至导带),表面等离子体金属纳米颗粒在接收光照后激发产生的载流子(热电子与热空穴)为研究光催化反应提供了新思路。目前对可见光下表面等离子体参与的光催化反应的机制解释为:表面等离子体吸收可见光波段,受激发产生热电子;热电子流向 TiO_2 导带并还原 TiO_2 表面的溶解氧;氧化态的 Au 纳米颗粒从 H_2O 捕获电子以中和正电荷。热电子与热空穴在空间上得到了有效的分离,降低了复合率,从而增加了载流子寿命、大幅提升了反应性能[122]。

1.3.4
电催化过程中的电子传递

污染物转化的光催化方法与电子流动密切相关,电催化转化活性更易受到电子流动速率的直接影响。而且,电催化剂的活性可以通过尺寸限制、表面重建、界面调制和缺陷工程进行调控,电催化过程可以通过改变外加电位实时调节反应条件,同时可以通过电极表面双电层微区域的水合离子微调界面电势分布,增强电催化技术的可操作性[123];电催化技术也容易与膜技术、生物技术和Fenton技术等其他净化技术结合,提高实际应用的可能性[124]。电催化过程中催化剂基底与底物的吸附为初始步骤,其中吸附位点与底物原子的共价程度、界面上的电荷转移能力都是重要的描述符。因此,电催化剂的晶面暴露情况与界面上的原子组成将会直接影响电催化的活性,界面调制工程将在原子尺度上对底物分子在电极和电解液界面的吸附、活化和转化过程进行深入的阐释与有效的调控。

1.4
能量转换的电子传递机制解析

PCET 在普遍存在的能量转换过程中发挥着重要作用,包括光合作用和呼吸作用,也是多种类型的太阳能燃料电池和其他电化学装置的基础。PCET 在这些反应中如何发生是生物和化学体系中成功进行能量转换的关键问题[125]。电子和质子的耦合转移共同影响了反应的能量和机制,同时发生电子和质子的转移可以避免产生高能的中间产物。从对这些过程的研究中可以理解重要的生物化学反应,比如呼吸作用、固氮作用、光合作用以及人工的光合成反应或者燃料电池中的能量转换。

1.4.1
电子传递热力学与速率常数

反映热力学性质的参数包括氧化还原电势(E)、Gibbs 自由能变(ΔG)和平衡常数(K)。对于 PCET 反应,这些参数会受到 pH 的影响,可以通过 Nernst 方程进行预测。PCET 的电子转移部分是通过电子给体与受体之间的静电扰动而诱导的,这种扰动引起了给体(D)与受体(A)之间电子波函数的混合。在扩散的电子转移中,反应物 D 与 A 之间的预结合是通过电子转移和产物的分离,生成 D^+ 与 A^- 的。预结合和近程的接触促进了 D 与 A 之间电子波函数的混合,而且使溶剂媒介的能垒最小化。若 K_{AC} 是电子转移前反应物的结合常数,K'_{AC} 是电子转移发生后产物的常数,则电子转移步骤的自由能变可以表示为

$$\Delta G_{ET} = \Delta G^{\ominus} - RT\ln\left(\frac{K_{AC}}{K'_{AC}}\right) \tag{1.1}$$

在经典的 Marcus 电子传递理论中,电子传递速率是由电势差、重组能和电子供体与受体之间的距离决定的,电子传递速率常数(k_{ET})的计算基于自由能 $\Delta G^{\ominus}$,$\ln k_{ET}$ 随 $\lambda/4 + (\Delta G^{\ominus}/2)[1 + \Delta G^{\ominus}/(2\lambda)]$ 呈现抛物线形的变化,当 $-\Delta G^{\ominus} = \lambda$ 时,反应速率达到最大值,λ 是总的重组能,包括分子间和溶剂或媒介。根据 Marcus 两阶段模型,采用从头计算可以得到细胞色素与电极表面(上、下绝热面)的能量差(V_{AB})和重组能(λ),从而可以从下面的公式求得电子传递速率常数:

$$k_{ET} = \frac{2\pi}{h}|V_{AB}|^2\frac{1}{\sqrt{4\pi\lambda k_B T}}\exp\left[-\frac{(\Delta G^{\ominus} + \lambda)^2}{4\pi\lambda k_B T}\right] \tag{1.2}$$

在 PCET 反应中两个带电的离子在反应位点之间波动,它们溶剂化的微观细节可能存在区别,所以进一步发展为三维的抛物线方程[126]。

1.4.2
电极/溶液界面的 PCET

金属电极-溶液界面的 PCET 反应属于异相反应,其速率常数和电流密度可通过转移概率[127]进行计算。Hammes-Schiffer 研究组针对电化学系统,在考虑

了非绝热(弱电子耦合和快速的溶剂弛豫)以及溶剂控制的限制(强电子耦合和速率限制的溶剂弛豫)的条件下,对PCET过程的速率常数进行了推导,获得的表达式与Savéant研究组得出的结果相关,但是指前因子有很大的不同[128]。在Hammes-Schiffer的处理中,电流密度是通过对溶质-电极之间的距离积分得到的,适用于长程的电子传递。针对电化学体系的PCET,对Anderson-Newns-Schmickler Hamiltonian进行了扩展。新的Hamiltonian模型是基于电子-质子子系统密度矩阵的主方程,其优势是可以用于描述绝热和非绝热电化学PCET反应,甚至可以考虑化学键的断裂和形成的影响。

1.5 电子转移的光谱电化学检测方法

1.5.1 紫外可见光谱电化学

电化学是一个传统的学科,对于物理、化学和生物学/生理学等相关领域的研究而言都是基础手段,特别是对于其中一些相关的课题,如能源的转换和储存等至关重要。电化学过程在实验室和大型工业中的应用都十分广泛[129-130]。虽然传统的分析电化学提供了很好的方法来确定浓度(如传感器技术),以氧化还原电位的形式产生能量数据,并且通过动力学分析阐明反应的机制,但这些技术通常不能立即适合鉴定中间体形成的未知物质或氧化还原反应中的产物[131-133]。1964年,T. Kuwana制作了第一块光透电极[134],即在SnO_2玻璃板表面镀一层Sb,将光谱方法和电化学方法结合使用,由此,光谱电化学开始迅速发展[135-136]。

紫外-可见光谱(UV-Vis)法是分子光谱学中非常古老的方法。1852年,Bouguer-Lambert-Beer规则的最终制定为早期吸收测量的定量评估奠定了基础[137-138]。Bilal等[139]使用UV-Vis光谱电化学方法研究了邻苯二胺(OPD)、间甲苯胺（MT）的电聚合及两种物质的共聚反应，该方法使用的电极是掺杂了铟的氧化锡玻璃电极（ITO）。Shah 等[140]用同样的电极，使用 UV-Vis 光谱电化学方法研究邻氨基苯酚（OAP）与苯胺（ANI）共聚现象，结果揭示了通过 OAP 阳离子自由基和 ANI 阳离子自由基的交叉反应可在共聚初始阶段形成中间体，最终形成头-尾二聚体/低聚物。Petr 等[141]使用 UV-Vis 光谱和电化学方法结合的手段，测量了甲基取代的对苯二胺的有机氧化还原对。

1.5.2
荧光光谱电化学

荧光光谱学是分析科学许多领域的重要研究工具，由于其极高的灵敏度和选择性，它在化学、生物化学和医学研究中已成为必不可少的光谱手段[142-143]。Yilmaz 等[144]将荧光光谱和电化学方法结合，使用荧光和 UV-Vis 两种光谱电化学方法结合原位监测金属与酞菁的氧化还原过程。Dias 等[145]为了验证光谱电化学池的效率和多功能性，研究了可溶性苝衍生物的还原步骤，并且比较了 UV-Vis 光谱电化学和荧光光谱电化学两种方法的效率高低。Montilla 等[146]发现 poly-[2,7-(fluorene)-1,4-(phenylene)]的光致发光调控可以通过共轭链（p-或 n-掺杂）的电化学修饰来实现，采用可控电荷注入方式来控制共轭聚合物的荧光信号，荧光光谱电化学的手段被很好地应用于该研究。

1.5.3
衰减全反射傅里叶变换红外（ATR-FTIR）光谱电化学

ATR-FTIR 光谱是一种无标记的无损分析技术，可广泛用于研究不同条件下的各种分子[147]，并且是研究生物医学样本的强大工具[148]。2011 年，Kellenberger 等[149]使用 ATR-FTIR 光谱电化学方法研究了在聚苯胺（PANI）的 p-掺杂期间的电荷状态的结构和稳定性，实验研究了吩嗪类单体在苯胺和吩嗪衍生物（3,7-二氨基-5-苯基吩嗪氯化物，phenosafranine）的几种共聚物中的作用。次年，他们仍然使用光谱电化学手段研究具有不同分子量的翠绿亚胺盐和翠绿亚胺碱的电化学掺杂[150]，实验发现在 p-掺杂时，聚苯胺中载流子的形成和稳定依赖于物质中链的支化，在聚合物氧化时 IR 带的潜在依赖性清楚地证明了不同带电聚合物结构(π-二聚体、极化子和双极)的形成。Öhman 等[151]使用 ATR-FTIR 和电阻抗谱(EIS)的光谱电化学实验装置研究了铝/聚合物界面，将水和电解质通过聚合物膜转移到铝/聚合物界面并进行铝的氧化/腐蚀，然后用 ATR-FTIR 和 EIS 同时研究了该系统的阻抗特性。

1.5.4 二维相关光谱分析

二维(2D)相关光谱分析是一种基于光谱方法的分析技术，自从 1986 年首次引入基础概念，二维相关光谱已经成为一种流行和通用的分析工具。其中光谱强度被绘制为两个独立变量的函数，例如波长、频率或波数。将两个独立变量绘制于 2D 光谱平面的正交轴上，将光谱强度沿第三轴绘制[152]。Shen 等[153]使用 ATR-FTIR 光谱和二维相关分析手段研究水在间规聚丙烯中的扩散机制，发现聚集的水分子在扩散过程中首先扩散到聚合物中，具有强氢键的 H_2O 比具有中等强度氢键的 H_2O 分散得慢，在扩散过程的后期，聚丙烯基质中的一些聚集的 H_2O 被迫溶解成"游离水"形式。Wu 等[154]使用二维近红外光谱(2D FT-NIR)研究温度对无定型聚酰胺的影响，其中 6740 cm^{-1} 和 6780 cm^{-1} 相关峰的存在说明 N-H 基团的特征峰有两种不同的来源。

参考文献

[1] Cestellos-Blanco S, Zhang H, Kim J M, et al. Photosynthetic semiconductor biohybrids for solar-driven biocatalysis [J]. Nature Catalysis, 2020, 3(3): 245-255.

[2] Chen C L, Chen Y T, Demchenko A P, et al. Amino proton donors in excited-state intramolecular proton-transfer reactions [J]. Nature Reviews Chemistry, 2018, 2(7): 131-143.

[3] Scharl T, Ferrer-Ruiz A, Saura-Sanmartin A, et al. Charge transfer in graphene quantum dots coupled with tetrathiafulvalenes [J]. Chemical Communications, 2019, 55(22): 3223-3226.

[4] Liu Y, Ding M, Ling W, et al. A three-species microbial consortium for power generation [J]. Energy & Environmental Science, 2017, 10(7): 1600-1609.

[5] Bai Y, Mellage A, Cirpka O A, et al. AQDS and redox-active NOM enables microbial Fe(Ⅲ)-mineral reduction at cm-scales [J]. Environmental Science & Technology, 2020, 54(7): 4131-4139.

[6] Love J C, Estroff L A, Kriebel J K, et al. Self-assembled monolayers of thiolates on metals as a form of nanotechnology [J]. Chemical Review, 2005, 105(4): 1103-1169.

[7] Nakahata M, Takashima Y, Hashidzume A, et al. Redox-generated mechanical motion of a supramolecular polymeric actuator based on host-guest interactions [J]. Angewandte Chemie (International Edition), 2013, 52(22): 5731-5735.

[8] Fukushima M, Sawada A, Kawasaki M, et al. Influence of humic substances on the removal of pentachlorophenol by a biomimetic catalytic system with a water-soluble iron(Ⅲ)-Porphyrin complex [J]. Environmental Science & Technology, 2003, 37(5): 1031-1036.

[9] Alexander M. Environmental and microbiological problems arising from recalcitrant molecules [J]. Microbial Ecology, 1975, 2(1): 17-27.

[10] Wackett L P C, Hershberger D. Biocatalysis and biodegradation: Microbial transformation of organic compounds [M]. Washington D. C.: ASM Press, 2001.

[11] Chen H, Venkat S, Hudson D, et al. Site-specifically studying lysine acetylation of aminoacyl-tRNA synthetases [J]. ACS Chemical Biology, 2019, 14(2): 288-295.

[12] Moulay S. O-methylation of hydroxyl-containing organic substrates: A comprehensive overview [J]. Current Organic Chemistry, 2018, 22(20): 1986-2016.

[13] Posselt M, Mechelke J, Rutere C, et al. Bacterial diversity controls transformation of wastewater-derived organic contaminants in river-simulating flumes [J]. Environmental Science & Technology, 2020, 54(9): 5467-5479.

[14] Chen Z, Zhang J, Stamler J S. Identification of the enzymatic mechanism of nitroglycerin bioactivation [J]. Proceedings of the National Academy of Sciences of the United States of America, 2002, 99(12): 8306-8311.

[15] Lu P, Li Q, Liu H, et al. Biodegradation of chlorpyrifos and 3,5,6-trichloro-2-pyridinol by Cupriavidus sp. DT-1 [J]. Bioresource Technology, 2013, 127: 337-342.

[16] Qin J, Lehr C R, Yuan C, et al. Biotransformation of arsenic by a yellowstone thermoacidophilic eukaryotic alga [J]. Proceedings of the National Academy of Sciences of the United States of America, 2009, 106(13): 5213-5217.

[17] Wahidullah S, Naik D N, Devi P. Fermentation products of solvent tolerant marine bacterium *moraxella* spp. MB1 and its biotechnological applications in salicylic acid bioconversion [J]. PLoS One, 2013, 8(12):

[18] Malandain C, Fayolle-Guichard F, Vogel T M. Cytochromes P450-mediated degradation of fuel oxygenates by environmental isolates [J]. FEMS Microbiology Ecology, 2010, 72(2): 289-296.

[19] Haritash A K, Kaushik C P. Biodegradation aspects of polycyclic aromatic hydrocarbons (PAHs): A review [J]. Journal of Hazardous Materials, 2009, 169(1/2/3): 1-15.

[20] Ye D Y, Siddiqi M A, Maccubbin A E, et al. Degradation of polynuclear aromatic hydrocarbons by *Sphingomonas paucimobilis* [J]. Environmental Science & Technology, 1996, 30(1): 136-142.

[21] Tas D O, Pavlostathis S G. Occurrence, toxicity, and biotransformation of pentachloronitrobenzene and chloroanilines [J]. Critical Reviews in Environmental Science and Technology, 2014, 44(5): 473-518.

[22] Tao X Q, Lu G N, Dang Z, et al. A phenanthrene-degrading strain *Sphingomonas* spp. GY2B isolated from contaminated soils [J]. Process Biochemistry, 2007, 42(3): 401-408.

[23] Suzuki K, Katagiri M. Mechanism of salicylate hydroxylase-catalyzed decarboxylation [J]. Biochim Biophys Acta, 1981, 657(2): 530-534.

[24] Guengerich F P, Waterman M R, Egli M. Recent structural insights into Cytochrome P450 function [J]. Trends in Pharmacological Sciences, 2016, 37(8): 625-640.

[25] Jeffreys L N, Girvan H M, Mclean K J, et al. Characterization of cytochrome P450 enzymes and their applications in synthetic biology [C]// Scrutton N. Enzymes in Synthetic Biology, 2018: 189-261.

[26] Cai P J, Xiao X, He Y R, et al. Anaerobic biodecolorization mechanism of

methyl orange by *Shewanella oneidensis* MR-1 [J]. Applied Microbiology and Biotechnology, 2012, 93(4): 1769-1776.

[27] Li J, Tang Q, Li Y, et al. Rediverting electron flux with an engineered CRISPR-ddAsCpf1 system to enhance the pollutant degradation capacity of *Shewanella oneidensis* [J]. Environmental Science & Technology, 2020, 54(6): 3599-3608.

[28] Wang Y X, Li W Q, He C S, et al. Active N dopant states of electrodes regulate extracellular electron transfer of *Shewanella oneidensis* MR-1 for bioelectricity generation: Experimental and theoretical investigations [J]. Biosensors & Bioelectronics, 2020, 160:112231.

[29] Cai P J, Xiao X, He Y R, et al. Involvement of c-type cytochrome CymA in the electron transfer of anaerobic nitrobenzene reduction by *Shewanella oneidensis* MR-1 [J]. Biochemical Engineering Journal, 2012, 68: 227-230.

[30] Gul M M, Ahmad K S. Bioelectrochemical systems: Sustainable bio-energy powerhouses [J]. Biosensors & Bioelectronics, 2019, 142:111576.

[31] Logan B E, Rabaey K. Conversion of wastes into bioelectricity and chemicals by using microbial electrochemical technologies [J]. Science, 2012, 337(6095): 686-690.

[32] Logan B E, Rossi R, Ragab A, et al. Electroactive microorganisms in bioelectrochemical systems [J]. Nature Reviews Microbiology, 2019, 17(5): 307-319.

[33] Pant D, Van Bogaert G, Diels L, et al. A review of the substrates used in microbial fuel cells (MFCs) for sustainable energy production [J]. Bioresource Technology, 2010, 101(6): 1533-1543.

[34] He Y R, Xiao X, Li W W, et al. Electricity generation from dissolved organic matter in polluted lake water using a microbial fuel cell (MFC) [J]. Biochemical Engineering Journal, 2013, 71: 57-61.

[35] Li W W, Sheng G P, Liu X W, et al. Impact of a static magnetic field on the electricity production of Shewanella-inoculated microbial fuel cells [J]. Biosensors and Bioelectronics, 2011, 26(10): 3987-3992.

[36] Chen Y P, Zhao Y, Qiu K Q, et al. An innovative miniature microbial fuel cell fabricated using photolithography [J]. Biosensors and Bioelectronics, 2011, 26(6): 2841-2846.

[37] Kumar G, Bakonyi P, Zhen G Y, et al. Microbial electrochemical systems

for sustainable biohydrogen production: Surveying the experiences from a start-up viewpoint [J]. Renewable & Sustainable Energy Reviews, 2017, 70: 589-597.

[38] Sun M, Sheng G P, Zhang L, et al. An MEC-MFC-coupled system for biohydrogen production from acetate [J]. Environmental Science & Technology, 2008, 42(21): 8095-8100.

[39] Martinez-Huitle C A, Rodrigo M A, Sires I, et al. Single and coupled electrochemical processes and reactors for the abatement of organic water pollutants: A critical review [J]. Chemical Reviews, 2015, 115(24): 13362-13407.

[40] Liu X W, Sun X F, Li D B, et al. Anodic Fenton process assisted by a microbial fuel cell for enhanced degradation of organic pollutants [J]. Water Research, 2012, 46(14): 4371-4378.

[41] Wang Y K, Sheng G P, Li W W, et al. Development of a novel bioelectrochemical membrane reactor for wastewater treatment [J]. Environmental Science & Technology, 2011, 45(21): 9256-9261.

[42] Liu X W, Wang Y P, Huang Y X, et al. Integration of a microbial fuel cell with activated sludge process for energy-saving wastewater treatment: taking a sequencing batch reactor as an example [J]. Biotechnology and Bioengineering, 2011, 108(6): 1260-1267.

[43] Yuan S J, Sheng G P, Li W W, et al. Degradation of organic pollutants in a photoelectrocatalytic system enhanced by a microbial fuel cell [J]. Environmental Science & Technology, 2010, 44(14): 5575-5580.

[44] Shi L, Dong H L, Reguera G, et al. Extracellular electron transfer mechanisms between microorganisms and minerals [J]. Nature Reviews Microbiology, 2016, 14(10): 651-662.

[45] Wegener G, Krukenberg V, Riedel D, et al. Intercellular wiring enables electron transfer between methanotrophic archaea and bacteria [J]. Nature, 2015, 526(7574): 587-U315.

[46] Michelson K, Sanford R A, Valocchi A J, et al. Nanowires of *Geobacter sulfurreducens* require redox cofactors to reduce metals in pore spaces too small for cell passage [J]. Environmental Science & Technology, 2017, 51(20): 11660-11668.

[47] Oram J, Jeuken L J C. Tactic response of *Shewanella oneidensis* MR-1

toward insoluble electron acceptors [J]. Mbio, 2019, 10(1):e02490-18.

[48] Zhao Y, Watanabe K, Hashimoto K. Hierarchical micro/nano structures of carbon composites as anodes for microbial fuel cells [J]. Physical Chemistry Chemical Physics, 2011, 13(33): 15016-15021.

[49] Wen D, Xu X L, Dong S J. A single-walled carbon nanohorn-based miniature glucose/air biofuel cell for harvesting energy from soft drinks [J]. Energy & Environmental Science, 2011, 4(4): 1358-1363.

[50] Logan B, Cheng S, Watson V, et al. Graphite fiber brush anodes for increased power production in air-cathode microbial fuel cells [J]. Environmental Science & Technology, 2007, 41(9): 3341-3346.

[51] Lanas V, Logan B E. Evaluation of multi-brush anode systems in microbial fuel cells [J]. Bioresource Technology, 2013, 148: 379-385.

[52] Huang Y X, Liu X W, Xie J F, et al. Graphene oxide nanoribbons greatly enhance extracellular electron transfer in bio-electrochemical systems [J]. Chemical Communications, 2011, 47(20): 5795-5797.

[53] Xie X, Hu L B, Pasta M, et al. Three-dimensional carbon nanotube-textile anode for high-performance microbial fuel cells [J]. Nano Letters, 2011, 11(1): 291-296.

[54] Adachi M, Shimomura T, Komatsu M, et al. A novel mediator-polymer-modified anode for microbial fuel cells [J]. Chemical Communications, 2008(17): 2055-2057.

[55] Xie X, Yu G H, Liu N, et al. Graphene-sponges as high-performance low-cost anodes for microbial fuel cells [J]. Energy & Environmental Science, 2012, 5(5): 6862-6866.

[56] Mink J E, Hussain M M. Sustainable design of high-performance microsized microbial fuel cell with carbon nanotube anode and air cathode [J]. ACS Nano, 2013, 7(8): 6921-6927.

[57] Liu X W, Chen J J, Huang Y X, et al. Experimental and theoretical demonstrations for the mechanism behind enhanced microbial electron transfer by CNT network [J]. Scientific Reports, 2014, 4:3732.

[58] Chen S, Hou H, Harnisch F, et al. Electrospun and solution blown three-dimensional carbon fiber nonwovens for application as electrodes in microbial fuel cells [J]. Energy & Environmental Science, 2011, 4(4): 1417-1421.

[59] Zhu N, Chen X, Zhang T, et al. Improved performance of membrane free

single-chamber air-cathode microbial fuel cells with nitric acid and ethylene-diamine surface modified activated carbon fiber felt anodes [J]. Bioresource Technology, 2011, 102(1): 422-426.

[60] Yuan Y, Zhou S G, Liu Y, et al. Nanostructured macroporous bioanode based on polyaniline-modified natural loofah sponge for high-performance microbial fuel cells [J]. Environmental Science & Technology, 2013, 47(24): 14525-14532.

[61] Wang H Y, Wang G M, Ling Y C, et al. High power density microbial fuel cell with flexible 3D graphene-nickel foam as anode [J]. Nanoscale, 2013, 5(21): 10283-10290.

[62] Lan S, Wang X M, Yang P, et al. The catalytic effect of AQDS as an electron shuttle on Mn(II) oxidation to birnessite on ferrihydrite at circum-neutral pH [J]. Geochimica Et Cosmochimica Acta, 2019, 247: 175-190.

[63] Gentry E C, Knowles R R. Synthetic applications of proton-coupled electron transfer [J]. Accounts of Chemical Research, 2016, 49(8): 1546-1556.

[64] Chen J J, Chen W, He H, et al. Manipulation of microbial extracellular electron transfer by changing molecular structure of phenazine-type redox mediators [J]. Environmental Science & Technology, 2013, 47(2): 1033-1039.

[65] Aulenta F, Catervi A, Majone M, et al. Electron transfer from a solid-state electrode assisted by methyl viologen sustains efficient microbial reductive dechlorination of TCE [J]. Environmental Science & Technology, 2007, 41(7): 2554-2559.

[66] Torres C I, Marcus A K, Lee H S, et al. A kinetic perspective on extracellular electron transfer by anode-respiring bacteria [J]. FEMS Microbiology Reviews, 2010, 34(1): 3-17.

[67] Kees E D, Pendleton A R, Paquete C M, et al. Secreted flavin cofactors for anaerobic respiration of fumarate and urocanate by *Shewanella oneidensis*: Cost and role [J]. Applied and Environmental Microbiology, 2019, 85(16): e00852-19.

[68] Baron D, Labelle E, Coursolle D, et al. Electrochemical measurement of electron transfer kinetics by *Shewanella oneidensis* MR-1 [J]. Journal of Biological Chemistry, 2009, 284(42): 28865-28873.

[69] Lies D P, Hernandez M E, Kappler A, et al. *Shewanella oneidensis* MR-1 uses overlapping pathways for iron reduction at a distance and by direct

contact under conditions relevant for biofilms [J]. Applied and Environmental Microbiology, 2005, 71(8): 4414-4426.

[70] Light S H, Su L, Rivera-Lugo R, et al. A flavin-based extracellular electron transfer mechanism in diverse Gram-positive bacteria [J]. Nature, 2018, 562(7725): 140-144.

[71] Fathey R, Gomaa O M, Ali A E, et al. Neutral red as a mediator for the enhancement of electricity production using a domestic wastewater double chamber microbial fuel cell [J]. Annals of Microbiology, 2016, 66(2): 695-702.

[72] Okamoto A, Hashimoto K, Nealson K H, et al. Rate enhancement of bacterial extracellular electron transport involves bound flavin semiquinones [J]. Proceedings of the National Academy of Sciences of the United States of America, 2013, 110(19): 7856-7861.

[73] Huang Y, Zhou E Z, Jiang C Y, et al. Endogenous phenazine-1-carboxamide encoding gene PhzH regulated the extracellular electron transfer in biocorrosion of stainless steel by marine Pseudomonas aeruginosa [J]. Electrochemistry Communications, 2018, 94: 9-13.

[74] Pham T H, Boon N, Aelterman P, et al. Metabolites produced by *Pseudomonas* spp. enable a Gram-positive bacterium to achieve extracellular electron transfer [J]. Applied Microbiology and Biotechnology, 2008, 77(5): 1119-1129.

[75] Huang B, Fu G, He C W, et al. Ferroferric oxide loads humic acid doped anode accelerate electron transfer process in anodic chamber of bioelectrochemical system [J]. Journal of Electroanalytical Chemistry, 2019, 851: 113464.

[76] Martinez C M, Zhu X P, Logan B E. AQDS immobilized solid-phase redox mediators and their role during bioelectricity generation and RR2 decolorization in air-cathode single-chamber microbial fuel cells [J]. Bioelectrochemistry, 2017, 118: 123-130.

[77] Sund C J, Mcmasters S, Crittenden S, et al. Effect of electron mediators on current generation and fermentation in a microbial fuel cell [J]. Applied Microbiology & Biotechnology, 2007, 76(3): 561-568.

[78] Watanabe K, Manefield M, Lee M, et al. Electron shuttles in biotechnology [J]. Current Opinion in Biotechnology, 2009, 20(6): 633-641.

[79] Peng L, Zhang Y. Cytochrome OmcZ is essential for the current generation by *Geobacter sulfurreducens* under low electrode potential [J]. Electrochimica Acta, 2017, 228: 447-452.

[80] Kumar R, Singh L, Zularisam A W. Exoelectrogens: Recent advances in molecular drivers involved in extracellular electron transfer and strategies used to improve it for microbial fuel cell applications [J]. Renewable & Sustainable Energy Reviews, 2016, 56: 1322-1336.

[81] Leys D, Meyer T E, Tsapin A S, et al. Crystal structures at atomic resolution reveal the novel concept of "electron-harvesting" as a role for the small tetraheme cytochrome c [J]. The Journal of Biological Chemistry, 2002, 277(38): 35703-35711.

[82] Richardson D J, Butt J N, Fredrickson J K, et al. The 'porin-cytochrome' model for microbe-to-mineral electron transfer [J]. Molecular Microbiology, 2012, 85(2): 201-212.

[83] Clarke T A, Edwards M J, Gates A J, et al. Structure of a bacterial cell surface decaheme electron conduit [J]. Proceedings of the National Academy of Sciences of the United States of America, 2011, 108(23): 9384-9389.

[84] Shaik S. Biomimetic chemistry: Iron opens up to high activity [J]. Nature Chemistry, 2010, 2(5): 347-349.

[85] Shi L, Squier T C, Zachara J M, et al. Respiration of metal (hydr)oxides by *Shewanella* and *Geobacter*: a key role for multihaem c-type cytochromes [J]. Molecular Microbiology, 2007, 65(1): 12-20.

[86] Coursolle D, Gralnick J A. Modularity of the Mtr respiratory pathway of *Shewanella oneidensis* strain MR-1 [J]. Molecular Microbiology, 2010, 77(4): 995-1008.

[87] Ross D E, Ruebush S S, Brantley S L, et al. Characterization of protein-protein interactions involved in iron reduction by *Shewanella oneidensis* MR-1 [J]. Applied and Environmental Microbiology, 2007, 73(18): 5797-5808.

[88] Hartshorne R S, Reardon C L, Ross D, et al. Characterization of an electron conduit between bacteria and the extracellular environment [J]. Proceedings of the National Academy of Sciences of the United States of America, 2009, 106(52): 22169-22174.

[89] Mclean J S, Pinchuk G E, Geydebrekht O V, et al. Oxygen-dependent auto-aggregation in *Shewanella oneidensis* MR-1 [J]. Environmental Microbiology,

2008, 10(7): 1861-1876.

[90] Bucking C, Popp F, Kerzenmacher S, et al. Involvement and specificity of *Shewanella oneidensis* outer membrane cytochromes in the reduction of soluble and solid-phase terminal electron acceptors [J]. FEMS Microbiology Letters, 2010, 306(2): 144-151.

[91] Shi L, Chen B W, Wang Z M, et al. Isolation of a high-affinity functional protein complex between OmcA and MtrC: Two outer membrane decaheme c-type cytochromes of *Shewanella oneidensis* MR-1 [J]. Journal of Bacteriology, 2006, 188(13): 4705-4714.

[92] Clarke T A, Cole J A, Richardson D J, et al. The crystal structure of the pentahaem c-type cytochrome NrfB and characterization of its solution-state interaction with the pentahaem nitrite reductase NrfA [J]. Biochemical Journal, 2007, 406: 19-30.

[93] Chabert V, Babel L, Fueg M P, et al. Kinetics and mechanism of mineral respiration: how iron hemes synchronize electron transfer rates [J]. Angewandte Chemie-International Edition, 2020, 59: 1-7.

[94] Richardson D J, Butt J N, Fredrickson J K, et al. The 'porin-cytochrome' model for microbe-to-mineral electron transfer [J]. Molecular Microbiology, 2012, 85(2): 201-212.

[95] Vondrasek J, Bendova L, Klusak V, et al. Unexpectedly strong energy stabilization inside the hydrophobic core of small protein rubredoxin mediated by aromatic residues: correlated ab initio quantum chemical calculations [J]. Journal of the American Chemical Society, 2005, 127(8): 2615-2619.

[96] Zeng X C, Hu H, Hu X Q, et al. Calculating solution redox free energies with ab initio quantum mechanical/molecular mechanical minimum free energy path method [J]. The Journal of Chemical Physics, 2009, 130(16): 164111.

[97] Shaw D E, Maragakis P, Lindorff-Larsen K, et al. Atomic-level characterization of the structural dynamics of proteins [J]. Science, 2010, 330(6002): 341-346.

[98] Swords W B, Meyer G J, Hammarstrom L. Excited-state proton-coupled electron transfer within ion pairs [J]. Chemical Science, 2020, 11(13): 3460-3473.

[99] Concepcion J J, House R L, Papanikolas J M, et al. Chemical approaches to

artificial photosynthesis [J]. Proceedings of the National Academy of Sciences of the United States of America, 2012, 109(39): 15560-15564.

[100] Meyer T J, Huynh M H V, Thorp H H. The possible role of proton-coupled electron transfer (PCET) in water oxidation by photosystem II [J]. Angewandte Chemie International Edition, 2007, 46(28): 5284-5304.

[101] Howard J B, Rees D C. How many metals does it take to fix N_2? A mechanistic overview of biological nitrogen fixation [J]. Proc. Natl Acad Sci USA, 2006, 103(46): 17088-17093.

[102] Carey J, Lawrence J, Tosine H. Photodechlorination of PCB's in the presence of titanium dioxide in aqueous suspensions [J]. Bulletin of Environmental Contamination and Toxicology, 1976, 16(6): 697-701.

[103] Liu X Q, Iocozzia J, Wang Y, et al. Noble metal-metal oxide nanohybrids with tailored nanostructures for efficient solar energy conversion, photocatalysis and environmental remediation [J]. Energy & Environmental Science, 2017, 10(2): 402-434.

[104] Jakob M, Levanon H, Kamat P V. Charge distribution between UV-irradiated TiO_2 and gold nanoparticles: Determination of shift in the Fermi level [J]. Nano Letter, 2003, 3(3): 353-358.

[105] Subramanian V, Wolf E E, Kamat P V. Catalysis with TiO_2/gold nanocomposites. Effect of metal particle size on the Fermi level equilibration [J]. Journal of the American Chemical Society, 2004, 126 (15): 4943-4950.

[106] Liu Y, Chen L F, Hu J C, et al. TiO_2 nanoflakes modified with gold nanoparticles as photocatalysts with high activity and durability under near UV irradiation [J]. The Journal of Physical Chemistry C, 2010, 114(3): 1641-1645.

[107] Marelli M, Evangelisti C, Diamanti M V, et al. TiO_2 nanotubes arrays loaded with ligand-free Au nanoparticles: enhancement in photocatalytic activity [J]. ACS Applied Materials & Interfaces, 2016, 8 (45): 31051-31058.

[108] Su R, Tiruvalam R, He Q, et al. Promotion of phenol photodecomposition over TiO_2 using Au, Pd, and Au-Pd nanoparticles [J]. ACS Nano, 2012, 6(7): 6284-6292.

[109] Iervolino G, Vaiano V, Murcia J J, et al. Photocatalytic hydrogen produc-

tion from degradation of glucose over fluorinated and platinized TiO_2 catalysts [J]. Journal of Catalysis, 2016, 339: 47-56.

[110] Hejazi S, Mohajernia S, Osuagwu B, et al. On the controlled loading of single platinum atoms as a co-catalyst on TiO_2 anatase for optimized photocatalytic H_2 Generation [J]. Advanced Materials, 2020, 32(16):1908505.

[111] Park H W, Lee J S, Choi W Y. Study of special cases where the enhanced photocatalytic activities of Pt/TiO_2 vanish under low light intensity [J]. Catalysis Today, 2006, 111(3-4): 259-265.

[112] Young C, Lim T M, Chiang K, et al. Photocatalytic oxidation of toluene and trichloroethylene in the gas-phase by metallised (Pt, Ag) titanium dioxide [J]. Applied Catalysis B: Environmental, 2008, 78(1-2): 1-10.

[113] Moonsiri M, Rangsunvigit P, Chavadej S, et al. Effects of Pt and Ag on the photocatalytic degradation of 4-chlorophenol and its by-products [J]. Chemical Engineering Journal, 2004, 97(2-3): 241-248.

[114] Chen H, Chen S, Quan X, et al. Fabrication of TiO_2-Pt coaxial nanotube array Schottky structures for enhanced photocatalytic degradation of phenol in aqueous solution [J]. The Journal of Physical Chemistry C, 2008, 112(25): 9285-9290.

[115] Zhao W, Chen C, Li X, et al. Photodegradation of sulforhodamine-B dye in platinized titania dispersions under visible light irradiation: Influence of platinum as a functional co-catalyst [J]. The Journal of Physical Chemistry B, 2002, 106(19): 5022-5028.

[116] Bian Z, Zhu J, Cao F, et al. In situ encapsulation of Au nanoparticles in mesoporous core-shell TiO_2 microspheres with enhanced activity and durability [J]. Chemical Communications, 2009(25): 3789-3791.

[117] Zhang N, Liu S, Fu X, et al. Synthesis of $M@TiO_2$ (M = Au, Pd, Pt) core-shell nanocomposites with tunable photoreactivity [J]. The Journal of Physical Chemistry C, 2011, 115(18): 9136-9145.

[118] Gomes Silva C, Juárez R, Marino T, et al. Influence of excitation wavelength (UV or visible light) on the photocatalytic activity of titania containing gold nanoparticles for the generation of hydrogen or oxygen from water [J]. Journal of the American Chemical Society, 2010, 133(3): 595-602.

[119] Wu Y, Liu H, Zhang J, et al. Enhanced photocatalytic activity of nitrogen-doped titania by deposited with gold [J]. The Journal of Physical

Chemistry C, 2009, 113(33): 14689-14695.

[120] Ding Q Q, Li R, Chen M D, et al. Ag nanoparticles-TiO2 film hybrid for plasmon-exciton co-driven surface catalytic reactions [J]. Applied Materials Today, 2017, 9: 251-258.

[121] Dhandole L K, Mahadik M A, Kim S G, et al. Boosting photocatalytic performance of inactive rutile TiO_2 nanorods under solar light irradiation: synergistic effect of acid treatment and metal oxide co-catalysts [J]. ACS Applied Materials & Interfaces, 2017, 9(28): 23602-23613.

[122] Li H, Bian Z, Zhu J, et al. Mesoporous Au/TiO_2 nanocomposites with enhanced photocatalytic activity [J]. Journal of the American Chemical Society, 2007, 129(15): 4538-4539.

[123] Kong D Y, Liang B, Yun H, et al., Cathodic degradation of antibiotics: Characterization and pathway analysis [J]. Water Research, 2015, 72: 281-292.

[124] Seh Z W, Kibsgaard J, Dickens C F, et al., Combining theory and experiment in electrocatalysis: Insights into materials design [J]. Science, 2017, 355: 6321.

[125] Meyer T, Dongare P. Proton-coupled electron transfer. An overview [J]. Abstracts of Papers of the American Chemical Society, 2018, 255:1155.

[126] Zwickl J, Shenvi N, Schmidt J R, et al. Transition state barriers in multi-dimensional Marcus theory [J]. The Journal of Physical Chemistry A, 2008, 112(42): 10570-10579.

[127] Venkataraman C, Soudackov A V, Hammes-Schiffer S. Theoretical formulation of nonadiabatic electrochemical proton-coupled electron transfer at metal-solution interfaces [J]. The Journal of Physical Chemistry C, 2008, 112(32): 12386-12397.

[128] Costentin C, Robert M, Saveant J M. Adiabatic and non-adiabatic concerted proton-electron transfers. Temperature effects in the oxidation of intramolecularly hydrogen-bonded phenols [J]. Journal of the American Chemical Society, 2007, 129(32): 9953-9963.

[129] Miklos D B, Remy C, Jekel M, et al. Evaluation of advanced oxidation processes for water and wastewater treatment-A critical review [J]. Water Research, 2018, 139: 118-131.

[130] Wiebe A, Gieshoff T, Moehle S, et al. Electrifying organic synthesis [J].

Angewandte Chemie-International Edition, 2018, 57(20): 5594-5619.

[131] De La Escosura-Muniz A, Ambrosi A, Merkoci A. Electrochemical analysis with nanoparticle-based biosystems [J]. Trac-Trends in Analytical Chemistry, 2008, 27(7): 568-584.

[132] Chen K, Chou W, Liu L, et al. Electrochemical sensors fabricated by electrospinning technology: an overview [J]. Sensors, 2019, 19(17): 3676.

[133] Grattieri M, Minteer S D. Self-powered biosensors [J]. ACS Sensors, 2018, 3(1): 44-53.

[134] Kuwana T, Darlington R K, Leedy D W. Electrochemical studies using conducting glass indicator electrodes [J]. Analytical Chemistry, 1964, 36(10): 2023-2025.

[135] Lozeman J J A, Fuhrer P, Olthuis W, et al. Spectroelectrochemistry, the future of visualizing electrode processes by hyphenating electrochemistry with spectroscopic techniques [J]. Analyst, 2020, 145(7): 2482-2509.

[136] Zhai Y L, Zhu Z J, Zhou S S, et al. Recent advances in spectroelectrochemistry [J]. Nanoscale, 2018, 10(7): 3089-3111.

[137] Buschmann C. UV-Vis Spectroscopy and Its Applications [J]. Journal of Plant Physiology, 1995, 146(3): 380-381.

[138] Förster H. UV-Vis spectroscopy [M]. Berlin: Springer, 2004: 337-426.

[139] Bilal S, Holze R. In situ UV-Vis spectroelectrochemistry of poly(o-phenylenediamine-co-m-toluidine) [J]. Electrochimica Acta-ELECTROCHIM ACTA, 2007, 52: 5346-5356.

[140] Shah A U H A, Holze R J S M. In situ UV-Vis spectroelectrochemical studies of the copolymerization of o-aminophenol and aniline [J]. Synthetic Metals, 2006, 156(7-8): 566-575.

[141] Petr A, Dunsch L, Neudeck A J J O E C. In situ UV-Vis ESR spectroelectrochemistry [J]. Journal of Electroanalytical Chemistry, 1996, 412(1-2): 153-158.

[142] Liu J Y, Zeng L H, Ren Z H. Recent application of spectroscopy for the detection of microalgae life information: A review [J]. Applied Spectroscopy Reviews, 2020, 55(1): 26-59.

[143] Siraj N, El-Zahab B, Hamdan S, et al. Fluorescence, phosphorescence, and chemiluminescence [J]. Analytical Chemistry, 2016, 88(1): 170-202.

[144] Yilmaz I. In situ monitoring of metallation of metal-free phthalocyanine

via UV-Vis and steady-state fluorescence techniques. Thin-layer UV-Vis and fluorescence spectroelectrochemistry of a new non-aggregating and electrochromic manganese (3+) phthalocyanine [J]. New Journal of Chemistry, 2008, 32(1): 37-46.

[145] Dias M, Hudhomme P, Levillain E, et al. Electrochemistry coupled to fluorescence spectroscopy: a new versatile approach [J]. 2004, 6(3): 325-330.

[146] Montilla F, Pastor I, Mateo C R, et al. Charge transport in luminescent polymers studied by in situ fluorescence spectroscopy [J]. The Journal of Physical Chemistry B, 2006, 110(12): 5914-5919.

[147] Naseer K, Ali S, Qazi J. ATR-FTIR spectroscopy as the future of diagnostics: a systematic review of the approach using bio-fluids [J]. Applied Spectroscopy Reviews, 2021, 56(2): 85-97.

[148] Chan K L A, Kazarian S G. Attenuated total reflection Fourier-transform infrared (ATR-FTIR) imaging of tissues and live cells [J]. Chemical Society Reviews, 2016, 45(7): 1850-1864.

[149] Kellenberger A, Dmitrieva E, Dunsch L. The stabilization of charged states at phenazine-like units in polyaniline under p-doping: an in situ ATR-FTIR spectroelectrochemical study [J]. Physical Chemistry Chemical Physics, 2011, 13(8): 3411-3420.

[150] Kellenberger A, Dmitrieva E, Dunsch L. Structure dependence of charged states in "Linear" polyaniline as studied by in situ ATR-FTIR spectroelectrochemistry [J]. Journal of Physical Chemistry B, 2012, 116(14): 4377-4385.

[151] Öhman M, Persson D, Leygraf C. A Spectroelectrochemical study of metal/polymer interfaces by simultaneous In situ ATR-FTIR and EIS [J]. Electrochemical and Solid-state Letters, 2007, 10: C27-C30.

[152] Huang H. "Sequential order" rules in generalized two-dimensional correlation spectroscopy [J]. Analytical Chemistry, 2007, 79(21): 8281-8292.

[153] Shen Y, Wu P. Two-dimensional ATR-FTIR spectroscopic investigation on water diffusion in polypropylene film: Water bending vibration [J]. The Journal of Physical Chemistry B, 2003, 107(18): 4224-4226.

[154] Wu P, Yang Y, Siesler H W. Two-dimensional near-infrared correlation temperature studies of an amorphous polyamide [J]. Polymer, 2001, 42(26): 10181-10186.

> # 第 2 章
>
> ## 三氧化钨纳米团簇界面的微生物胞外电子传递机制

2.1 基于光学的 EAB 高通量鉴定概述

纳米级钨氧化物由于其良好的稳定性、光敏性和电子传递特性,被广泛应用在电致变色材料[1-2]、气体传感[3]、多相催化[4]、光电极[5]、氢气传输和储存[6]以及一些生物领域[7]中。在钨氧化物中,六方晶系三氧化钨(h-WO_3)具有独特的管道结构[8],其每层 WO_6 八面体共顶点的骨架结构存在六方和三方通道,提供了捕获电子和离子的空间[9]。实验室曾利用 h-WO_3 纳米晶体探针具有电致变色的特点,构建了一种基于光学的 EAB 高通量鉴定方法[10]。在该方法中,h-WO_3 纳米晶体能够在捕获微生物胞外电子的同时,自身的颜色也会从白色变成深蓝色,因而可以通过颜色变化来快速分离和筛选 EAB[11]。

之前已经提到 EAB 能够将电子传递给胞外的不溶性电子受体[12-13],这个特点使得 EAB 在产电、环境修复和生物地球化学过程中具有重要作用[14-16]。因此,EAB 在环境、能源、微生物学和生物地球化学等领域受到广泛的关注。然而,EAB 的分离往往需要通过很长时间 MFC 的运行才能实现[17],此外还需要特定的仪器装置[18-19]。目前缺乏能够快速有效鉴定 EAB 和胞外电子传递(EET)能力的方法。之前已经证实了 h-WO_3 纳米晶体能够用于快速有效鉴定 EAB 和评估其电子传递能力。此外,复合了聚苯胺的介孔 WO_3 材料也被证实可作为一种有效的 MFC 非铂阳极电催化剂[20]。

EAB,如 *Shewanella oneidensis* MR-1,有多种 EET 途径,包括通过氧化还原介体简介传递[21],通过外膜蛋白(含多个卟啉铁的细胞色素)直接传递[22]和通过导电菌毛或"纳米导线"[23]进行传递。海底沉积物中的 EAB 可能通过"电源线"连接着相距较远的化学反应,将电子传递跨过厘米宽的区域编织成电子传递网络,形成一个巨大的天然电池,从而产生电流[24-25]。通过实验证实,*Shewanella oneidensis* MR-1 与 h-WO_3 纳米晶体之间的电子传递是通过外膜的细胞色素 c(OM c-Cyt)直接进行的[11]。但是,h-WO_3 纳米晶体团簇如何从 EAB 的外膜蛋白捕获电子的分子机制仍然不清楚。而分子模拟可以为这个过程提供原子层面的研究手段,并建立结构与功能的关系以解释实验结果[26]。

在本章的研究中,采用分子模拟来阐释纳米团簇和微生物之间的胞外电子传递过程,并通过实验对胞外蛋白在金属氧化物界面上的电化学响应进行测定,以验证计算结果。假设电子是从 OM c-Cyt 中还原态的卟啉铁 Fe(Ⅱ)传递到

h-WO$_3$ 纳米晶体(001)晶面的 W(Ⅵ),并用分子动力学(MD)模拟研究了卟啉铁 heme 周围多肽链对 OM c-Cyt-WO$_3$ 系统构型的影响。还使用量化计算方法计算 heme 与 h-WO$_3$(001)晶面之间的相互作用能,并证实细胞色素和纳米团簇之间进行直接电子传递的可行性,也通过过渡态搜索对电子传递的能垒进行了计算。此外,还将理论计算与实验数据进行比较以证实提出的电子传递分子机制,从而提供了产电微生物 OM c-Cyt 与 h-WO$_3$ 纳米晶体界面电子传递的相关信息,有利于更好地理解蛋白-纳米材料之间的相互作用,为设计高效的功能纳米材料提供依据。

2.2 三氧化钨与微生物界面电子转移研究方法

2.2.1 微生物的培养

从 Luria-Bertani(LB)固体培养基上挑选单克隆 *Shewanella oneidensis* MR-1,接种在灭菌的 1 mL LB 培养基中,在 30 ℃恒温摇床中摇 12 h,再以 1∶1000 比例稀释转入 100 mL 灭菌 LB 中,培养 16 h 至稳定期,然后把菌液在 5000g 转速下离心 5 min,菌体用乳酸钠无机盐培养基重悬后再次离心,重复 3 次离心后重悬待用。

2.2.2
电化学实验

玻碳电极使用前要进行预处理,以获得洁净光滑的表面。预处理方法如下:用 $0.05~\mu m$ 的氧化铝抛光粉打磨 15 min,然后在超纯水中超声 3 次,每次 1 min。清洗之后的电极在 $0.5~mol \cdot L^{-1}$ 的硫酸溶液中用循环伏安法活化,参比电极为 Ag/AgCl,扫速为 $100~mV \cdot s^{-1}$,扫描范围为 $-1.0 \sim 1.0~V$,共扫描 20 圈。活化之后玻碳电极再在铁氰化钾溶液($5~mmol \cdot L^{-1}$ 铁氰化钾、$5~mmol \cdot L^{-1}$ 亚铁氰化钾、$0.2~mol \cdot L^{-1}$ 硝酸钾)中检测,扫描范围为 $-0.2 \sim 0.6~V$,扫速为 $50~mV \cdot s^{-1}$,参比电极为 Ag/AgCl。

WO_3 在乙醇中超声 10 min 后,取 $5~\mu L$ $0.3~mg \cdot mL^{-1}$ 的 WO_3 溶液滴加在预处理后的玻碳电极上,避光室温干燥,干燥后再滴加 $5~\mu L$ 0.05% 的 Nafion 溶液,室温干燥,记为 $GC-WO_3$。在 $GC-WO_3$ 电极上继续滴加卟啉铁、细胞色素可得到修饰后的电极,分别记为 $GC-WO_3$-hemin 和 $GC-WO_3$-Cyt。

玻碳电极修饰完成之后,在 $100~mmol \cdot L^{-1}$ 的 PBS(pH 7.0)中进行循环伏安实验,Ag/AgCl 作参比电极,Pt 丝作对电极,扫描范围为 $-0.6 \sim 0.4~V$,扫速分别为 $10~mV \cdot s^{-1}$、$20~mV \cdot s^{-1}$、$50~mV \cdot s^{-1}$、$100~mV \cdot s^{-1}$、$150~mV \cdot s^{-1}$、$200~mV \cdot s^{-1}$ 和 $250~mV \cdot s^{-1}$。

2.2.3
外膜和周质细胞色素 c 的破胞提取

外膜蛋白的提取采用的是文献中报道的方法[27],并进行了一些改进。首先,分别将 Shewanella oneidensis MR-1 和 $\Delta mtrC/\Delta omcA$ 突变株细胞 $5000g$ 离心 5 min 并用 PBS 溶液清洗 $1 \sim 2$ 遍(片剂配方:$2.68~mmol \cdot L^{-1}$ KCl、$1.47~mmol \cdot L^{-1}$ KH_2PO_4、$136.89~mmol \cdot L^{-1}$ NaCl、$8.1~mmol \cdot L^{-1}$ Na_2HPO_4);接着,每 10 mL 菌液收集的细胞用 1 mL SL 溶液[配方:$20\%(w/V)$ 蔗糖、$1~mg/mL$ 溶菌酶、$0.01~mol \cdot L^{-1}$ Tris pH 8.4]重悬,25 ℃ 恒温摇床轻轻晃动 1 h;然后,菌液 $5000g$ 离心 5 min,收集细胞沉淀,每 10 mL 原菌液收集细

胞沉淀重悬于 200 μL Tris-Mg^{2+} 溶液（0.01 mol·L^{-1} $MgCl_2$、0.01 mol·L^{-1} Tris pH 8.4）；25 ℃恒温摇床轻轻晃动 20 min；最后，5000g 离心 5～10 min，收集上清液，获得外膜和周质混合蛋白。

2.2.4
量子力学/分子力学（QM/MM）计算

最小尺寸的 QM 区域包括卟啉铁 heme、Na^+ 以及接受电子的 h-WO_3 纳米晶面中心。密度泛函理论（DFT）通过 $DMol^3$ 模块[28]，采用 Perdew-Wang 91（PW91）泛函结合包含有 p 轨道极化函数的双精度数值基组（DNP）[29]进行计算。对于 MM 区域，采用 Universal 力场通过 GULP[30]进行计算。QM/MM 相互作用则使用色散边界电荷电子嵌入处理。体系的能量可以通过以下方程计算：

$$E_{\text{Total}} = E_{\text{QM}} + E_{\text{MM}} + E_{\text{QM/MM}} \quad (2.1)$$

其中，E_{QM} 和 E_{MM} 分别表示 QM 和 MM 子系统的能量，$E_{\text{QM/MM}}$ 表示上述两个区域之间的相互作用能，并使用线性同步度越/二次同步度越方法[31]计算能垒。

2.2.5
细胞色素 c 模型

两种典型的 EAB——*Shewanella* 和 *Geobacter* 的外膜和菌毛上存在含多个血红素的 c-Cyts[32-34]。这些 c-Cyts 在引导电子从胞内代谢途径流动到电极上起到了关键作用，并且也可以在胞外电子受体暂时不存在时，起到电容器的作用以储存电子来维持呼吸作用[35-36]。含多个血红素辅基的 c-Cyts 可以作为跨越很长距离的固定的电子流通渠道，如对于 *Shewanella* 存在 CymA-MtrA-MtrC 系统[37-38]，而 OmcZ 分布在整个细胞周围的胞外基质中。另外，虽然含 4 个血红素辅基的小分子细胞色素（STC）不属于外膜上的细胞色素，但也是一种 c-Cyts，结构中血红素轴向配位双组氨酸[39-41]，而且 STC 也参与了铁氧化物的呼吸作用[38]。这些含 10 个血红素辅基的 c-Cyts 与 STC 的氧化还原活性中心是相同的，也就是血红素 heme，含铁的卟啉结构（P-Fe）[42]。因此，可以利用 STC

作为 MtrC 和 OmcZ 等 OM c-Cyts 的代表模型,来加快计算速度[43]。*Shewanella oneidensis* MR-1 在蛋白数据库中的 STC 识别码是 1M1R。

STC[图 2.1(c)]含有 91 个氨基酸,4 个血红素辅基在其中排列,hemes Ⅰ、Ⅱ、Ⅲ 和 Ⅳ 在取向上相互垂直,而 hemes Ⅱ 和 Ⅲ 相互平行。晶体结构中存在两个硫酸盐阴离子,在体系中引入 15 Na^+ 以中和蛋白所带的电荷。与铁原子轴向配位的都是双组氨酸残基[图 2.1(d)],垂直的配位键与组氨酸(His)上的 N 原子连接。模型包含了水溶液(3500 H_2O 分子)中的 STC 和 15 个 Na^+,位于 $h-WO_3$ 纳米晶体的(001)晶面上。在模型中未考虑多肽链,以便于分析氨基酸残基对 heme 与 WO_3 纳米簇之间电子传递的影响。

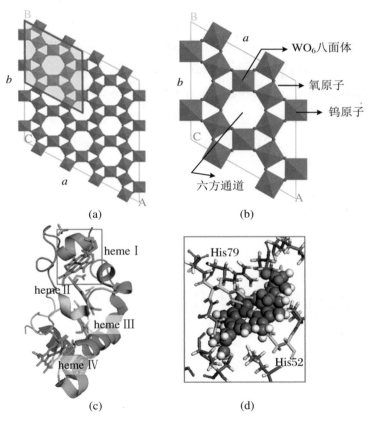

图 2.1 六方晶系 WO_3 的晶胞结构和 STC 构型:$4 \times 4 \times 1$ $h-WO_3$ 超晶胞:$a = b = 29.1928$ Å,$c = 3.8992$ Å(a);$2 \times 2 \times 1$ 超晶胞:$a = b = 14.5964$ Å,$c = 3.8992$ Å,$\alpha = \beta = 90°$,$\gamma = 120°$(b);从蛋白数据库(PDB)获得的 *Shewanella oneidensis* MR-1 的 STC 的分子构型,编码为 1M1R(c);STC 的局部方法,双组氨酸轴向配位卟啉铁平面(高亮标注,His52 与 His79)(d)

2.2.6
分子动力学(MD)模拟

水溶液中 STC-WO$_3$ 和游离态 heme-WO$_3$ 的 MD 盒子先进行能量最小化步骤,然后在 298 K 的 NPT 系综中弛豫 50 ps,接着在 298 K 下的 NVT 系综中弛豫 50 ps,最后在 298 K 下的 NVE 系综中弛豫 1500 ps。能量最小化步骤是使用智能最小化方法进行计算的,这个方法结合了最速下降算法、共轭梯度算法以及 Newton 算法,收敛标准分别设定为 0.01 kcal·(mol·Å)$^{-1}$、0.01 和 0.1 kcal·(mol·Å)$^{-1}$。非键相互作用(即静电作用和范德华力)通过 Ewald 加和方法[44]进行计算,精度是 0.01 kcal·mol^{-1}。MD 模拟的时间步长是 1.0 fs,以此来平衡电子转移模型,并获得最终结构。STC-WO$_3$ 和游离态 heme-WO$_3$ 电子传递系统采用了 cvff 力场[45]。Andersen 算法[46]的碰撞率设定为 1.0,Berendsen 算法[47]的衰减常数设定为 0.1 ps,以便于控制每个模型的温度和压强。

2.3
EAB 高通量鉴定机制

2.3.1
WO$_3$ 纳米晶体结构捕获电子的过程

根据已被普遍接受的电致变色机制、Faughnan 着色模型[48],h-WO$_3$ 的变色归因于该材料的隧道结构中同时捕获电子和阳离子,因此,EAB 与 h-WO$_3$ 之间的电子传递过程可以描述如下:

$$P\text{-}Fe^{2+} \longrightarrow P\text{-}Fe^{3+} + e^{-} \tag{2.2}$$

$$M^+ + e^- + WO_3 \longrightarrow MWO_3 \quad (2.3)$$

式中，P-Fe^{2+}和P-Fe^{3+}分别表示还原态、氧化态的含铁卟啉分子，M$^+$表示水溶液中的阳离子。在上述的电子传递过程中，OM c-Cyts 的活性中心（P-Fe）能够通过Fe(Ⅲ)到Fe(Ⅱ)的还原反应转移一个电子；之后，顶点共享的 WO$_6$ 八面体[图2.1(a)和(b)]骨架结构中的六方通道可以同时捕获电子和水溶液中的阳离子而形成钨青铜（tungsten bronzes）结构 MWO$_3$。这里水溶液中的阳离子是实验中大量存在的 Na$^{+[11]}$，而电子则是通过 OM c-Cyts 的活性中心氧化所产生的。对于 h-WO$_3$ 元胞[图2.1(b)]，总的电致变色反应可以用如下反应式描述：

$$P\text{-}Fe^{2+} + Na^+ + WO_3 \longrightarrow P\text{-}Fe^{3+} + NaWO_3 \quad (2.4)$$

2.3.2
界面电子传递过程中 heme 的结构特征

hemes 是细胞色素 c 的活性中心，也是一种大型的杂环化合物，其中心由铁（Fe）原子和附近的4个吡咯配体组成。十血红素辅基 c-Cyts 位于外膜的外表面[37]，这个特点也使得其能够与外界不溶性底物相互作用。图2.1(c)中显示了 *Shewanella oneidensis* MR-1 的 STC（PDB ID：1M1R）结构，图2.1(d)图示说明了 STC 的化学结构，并显示了多肽链中的化学键的类型和氢原子，图2.1(d)是图2.1(c)的局部放大图，显示了两个组氨酸残基轴向配位于活性中心卟啉平面的铁原子上的结构。

P-Fe 的几何结构变化会对 h-WO$_3$ 捕获细胞表面电子过程产生明显变化。表2.1比较了原卟啉平面和 porphyrin-WO$_3$ 复合体系中的 P-Fe 的几何结构参数（图2.2）。结果显示在电子传递过程中，当 P-Fe 接近 h-WO$_3$ 纳米晶体团簇的(001)晶面时，其结构会发生改变。复合体系 porphyrin-WO$_3$ 结构和 porphyrin-biHis-WO$_3$ 结构中的 Fe—N 键长（表2.1）要长于最初的 P-Fe 中的键长，并且在氧化态的（P-Fe^{3+}）和还原态（P-Fe^{2+}）结构中的 Fe—N 键长也不相同。不仅 Fe—N 键长（l）发生了改变，而且 N—Fe—N 键角（θ）也发生了变化。这些结果表明，在电子供体接近纳米晶体团簇后，高度共轭的卟啉环结构被破坏，导致了 Fe—N 键长的增加，这有利于电子传递过程的发生。

表 2.1　电子捕获过程中卟啉铁几何结构参数的变化，初始 P-Fe，porphyrin-WO_3 和 porphyrin-biHis-WO_3 中卟啉铁几何结构的比较

系统	$l(Fe—N_1)$ (Å)	$l(Fe—N_2)$ (Å)	$l(Fe—N_3)$ (Å)	$l(Fe—N_4)$ (Å)	$\theta(N_1—Fe—N_3)(°)$	$\theta(N_2—Fe—N_4)(°)$
初始 P-Fe	1.986	1.983	1.979	1.979	176.290	176.850
P-Fe^{2+} (porphyrin-WO_3)	2.155	2.034	2.154	2.074	167.604	172.032
P-Fe^{3+} (porphyrin-WO_3)	2.032	2.034	2.027	2.029	174.604	170.672
P-Fe^{2+} (Porphyrin-biHis-WO_3)	2.142	2.034	2.015	2.122	178.068	177.762
P-Fe^{3+} (Porphyrin-biHis-WO_3)	2.161	2.018	1.996	2.114	176.872	176.757

考虑了电子捕获过程中卟啉环（porphyrin）和 h-WO_3 表面可能的方向，图 2.2 展示了对于 porphyrin-WO_3 和 porphyrin-biHis-WO_3 电子传递过程的最低能量构象。最初，卟啉环平行于 h-WO_3 晶面[图 2.2(a)]，而在发生电子传递之后，卟啉环与该晶面成 45°[图 2.2(b)]。在 porphyrin-biHis-WO_3 系统中，轴向配位的两个组氨酸残基存在空间位阻效应，所以不论是处于氧化态还是还原态，复合物体系中卟啉环与晶面都成 45°角，如图 2.2(c) 和 (d) 所示。

对于 EAB 与胞外电子受体之间的电子传递反应速率主要受到传递距离的影响，而上述结构可以减少两个氧化还原中心的距离，因此也使得电子传递速率增加。在 porphyrin-WO_3 系统中的 Fe-W 距离为 7.093 Å（氧化态），比 porphyrin-biHis-WO_3 系统中的 Fe-W 距离（氧化态时是 9.618 Å）仅仅小 2.5 Å。结果说明组氨酸残基对电子传递过程可能会产生影响，但效果并不显著；轴向配体可能对于在自然环境中保持卟啉环中心金属离子的稳定性有重要作用，而中心金属离子的稳定性有助于维持和增强 Fe(Ⅱ)/Fe(Ⅲ) 氧化还原循环[49]。

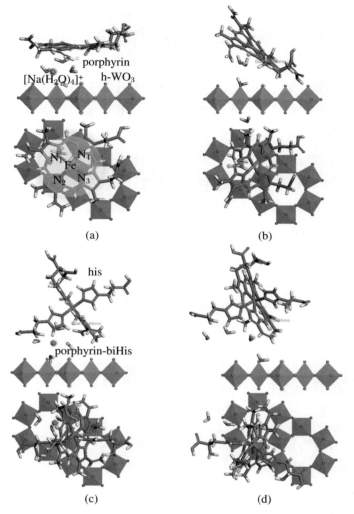

图 2.2 电子捕获系统优化的几何结构：porphyrin-WO_3 还原系统包括 WO_3，$[Na(H_2O)_4]^+$ 和 P-Fe^{2+} 经过 QM/MM 计算后的前视图和俯视图(a)；porphyrin-WO_3 氧化系统包含 P-Fe^{3+} (b)；porphyrin-biHis-WO_3 还原系统(c)；porphyrin-biHis-WO_3 氧化系统，包含 WO_3，$[Na(H_2O)_4]^+$ 和双组氨酸轴向配位的 P-Fe(d)

2.3.3
纳米晶体团簇在捕获电子过程中的电子结构

为考察 h-WO_3 的 DOS 和 Na^+ 插入结构中的影响,采用了第一性原理密度泛函理论(DFT)分别对纳米晶体团簇 h-WO_3 与钠离子钨青铜($NaWO_3$)的态密度(DOS)进行了计算。能量轴的零点位置对应于系统的费米能级,图 2.3 中显示了相关原子的局部态密度(PDOS),并且可以用来进一步说明价带和导带的主要贡献。W 5d 和 O 2p 的 PDOS 相互覆盖,形成成键的状态,属于价带,而反键态属于导带。在低能量区,DOS 主要由 Na 2p 轨道贡献,未呈现在图中。总态密度(TDOS)和 PDOS 在 Na^+ 嵌入后会逐渐移向低能量区,并且相对应的费米能级会偏移至导带。费米能级附近 h-WO_3 纳米晶体团簇的电子结构主要受到 Na^+ 和 EAB 产生的电子嵌入的影响。

2.3.4
轴向双组氨酸对 h-WO_3 纳米晶体团簇电子捕获行为的影响

为了研究从还原态卟啉铁(P-Fe^{2+})到 h-WO_3 纳米结构的 EET 过程中的能量变化,采取 QM/MM 方法进行计算。在卟啉环平面上的中心铁原子上引入轴向配位的两个组氨酸残基以解释氨基酸残基对电子传递过程的影响。图 2.2 中展示了 porphyrin-WO_3 和 porphyrin-biHis-WO_3 复合物结合一个水和钠离子的优化结构。水合钠离子[$Na(H_2O)_4$]$^+$ 第一水化层中被认为水分子数目接近 4,因为当 $n=4$ 时,会呈现最稳定的四面体结构[50]。在 QM/MM 算法中,总能量等于 QM 区域的 QM 能量和 MM 区域的 MM 能量以及 QM/MM 耦合所具备的能量之和。另外,QM/MM 的静电势能包括在 QM 能量之内。

电子传递过程中总能量变化为负($\Delta E_{Total} = -8.872$ eV)(表 2.2),说明了 P-Fe^{3+}-WO_3 氧化态体系[图 2.2(b)]比 P-Fe^{2+}-WO_3 还原态体系[图 2.2(a)]更加稳定。因此,当还原态 heme 暴露在溶液中时,就更容易将电子传递给 h-WO_3 纳米晶体团簇。而在 porphyrin-biHis-WO_3 体系中的能量变化 ΔE_{Total} 为正,则

说明两个轴向配位的组氨酸残基对 P-Fe 与 h-WO$_3$ 间电子传递过程产生了影响,当有组氨酸配位时,还原态体系更加稳定。总能量变化的贡献主要来自于 QM 能量,计算结果证实活性中心 Fe 和 heme 在细胞色素 c 和 h-WO$_3$ 间的电子传递中起到了重要作用。

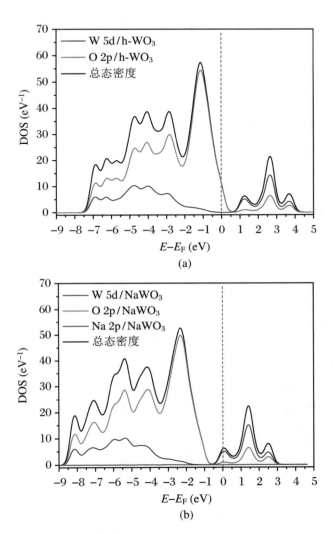

图 2.3　纳米晶体团簇 h-WO$_3$ 随着 Na$^+$ 和电子的嵌入而发生的电子结构的变化:h-WO$_3$ 中 W 和 O 原子的局部态密度(PDOS)和总态密度(TDOS)(a);NaWO$_3$ 中 W,O 和 Na 原子的 PDOS 和 TDOS 图 (b)

根据 DFT 计算的结果分析了标准条件下卟啉环氧化的 Gibbs 自由能变化（$\Delta G_{ox}^{\ominus}$, 表 2.2）。对于 porphyrin-WO$_3$ 系统，$\Delta G_{ox}^{\ominus}$ 值为负说明：游离态 heme 能够自发将电子传递给纳米晶体团簇，但 porphyrin-biHis-WO$_3$ 系统计算出的 $\Delta G_{ox}^{\ominus}$ 值为正。另外，实验测得 *Shewanella* 的 OM c-Cyt 与不溶性 Fe(Ⅲ) 化合物相互作用的还原电势相对于标准氢电势（NHE）为 $-0.32 \sim -0.10$ V，这说明了 OM c-Cyt 的氧化反应在热力学上并非是自发进行的。这可以通过其半反应的 $E_{red}^{\ominus}$ vs. NHE 来解释：

$$E_{red}^{\ominus} = \frac{-\Delta G_{red}^{\ominus}}{nF} - E_H^{\ominus} \tag{2.5}$$

其中，n 表示反应中的电子转移数，F 是法拉第常数 [23.06 kcal·(mol·V)$^{-1}$]，$E_H^{\ominus}$ 是 NHE 的标准还原电势（4.28 V）。因此，当 $E_{red}^{\ominus}$ 大于 -4.28 V 时，$\Delta G_{red}^{\ominus}$ 就为负值。

表 2.2　从细胞色素氧化还原活性中心捕获电子的能量变化和系统电子传递过程的能量变化，包括热力学性质（298.15 K, 1 atm）

	ΔE_{Total} (eV)	ΔE_{QM} (eV)	ΔE_{MM} (eV)	$\Delta G^{\ominus}$ (eV)	$\Delta H^{\ominus}$ (eV)	$\Delta S^{\ominus}$ (10^{-3} eV·K^{-1})
porphyrin-WO$_3$	-8.872	-0.169	-8.703	-0.326	-0.258	0.229
porphyrin-biHis-WO$_3$	2.733	2.599	0.134	2.661	2.783	0.410

实验报道的 OM c-Cyt 的 $E_{red}^{\ominus}$ 值表明 $\Delta G_{red}^{\ominus}$ 为负值，因此，其逆反应的 $\Delta G_{ox}^{\ominus}$ 为正值，porphyrin-biHis-WO$_3$ 系统的计算结果与实验数据非常符合，并证实了有组氨酸轴向配位时这个模型能够用来描述 OM c-Cyt 活性中心与 h-WO$_3$ 的界面电子传递过程。

上述计算结果与电化学实验结果相符，同空白玻碳（GC）电极和 WO$_3$ 修饰的 GC 相比，只有 WO$_3$-hemin 修饰的 GC 有一对明显的氧化还原峰 [图 2.4(a)]，说明在 h-WO$_3$ 界面 heme 能表现出良好的氧化还原响应，且峰电位在 -0.4 V（vs. Ag/AgCl）左右，即 -0.2 V（vs. NHE）。该值与文献中报道的 OM c-Cyt 的还原电位（$-0.32 \sim -0.1$ V）相符合[51-52]，说明细胞色素的还原电势主要受到了活性中心的影响。图 2.4(b) 研究了该电极在不同扫描速度下（$10 \sim 250$ mV·s^{-1}）的循环伏安（CV）响应，发现氧化峰和还原峰的峰电位基本保持不变，但氧化峰和还原峰的峰电流均随扫速的减小而减小，进一步分析可以发现 heme 的氧化峰和还原峰的峰电流与扫描速率均呈良好的线性关系 [图 2.4(c)]，说明该体系下的电化学反应是受表面反应控制的氧化还原反应。图 2.4(d) 表示的是在厌氧条件下马心 Cyt c 的 CV 图，其氧化峰在 0.10 V（vs. Ag/AgCl），还原峰在-0.05 V（vs. Ag/AgCl）。该值与文献中报道的 OM c-Cyt 的还原电位[51-52]有差异，是由于实验中使用的是马心 Cyt，其分子量比 OM c-Cyt 小，且卟啉环周围的肽链结构较简单，轴向配位的并不是组氨酸，因此，卟啉铁配位的氨基酸结构会影响整个 Cyt 的氧化还原过程。对比 GC-WO$_3$-heme 和 GC-WO$_3$-Cyt c 的还原电位，发现马心 Cyt c 的还原电位明显高于 heme 的还原电位，即存在肽链时，还原电位发生明显的正移，说明肽链可以维持卟啉铁结构在水溶液体系中的稳定性，影响了活性中心与 WO$_3$ 之间的电子传递。对于 OM c-Cyt 中的 MtrC，其还原电位为-0.1 V（vs. NHE）[52-53]，具有相同的趋势，可能是由于氧化活性中心被埋藏在复杂的肽链内部的原因。但对于另一种分子量比 MtrC 更大的外膜蛋白 OmcA，其还原电位为-0.32～-0.24 V（vs. NHE）[51]。由于其结构复杂，所以其晶体结构还未得到解析，其可能在蛋白的某些位点有利于活性位点的暴露，从而促进了胞外电子传递，这需要进一步的探索。

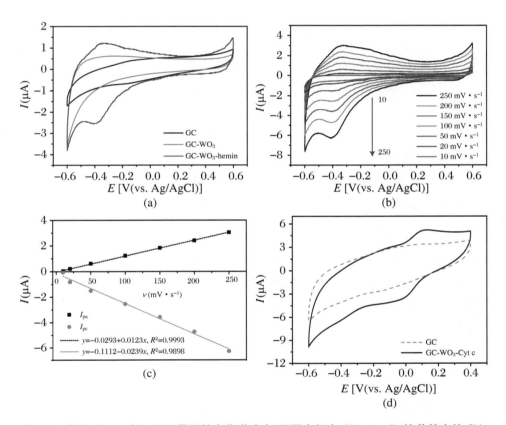

图 2.4　与马心 Cyt 在 h-WO$_3$ 界面的电化学响应：不同电极在 Shewanella 培养基中的 CV(a)；不同扫描速度下的 CV(b)；CV 的峰电流-扫描速度(c)；两种电极在 0.1 mol·L^{-1} PBS(pH 7.0) 的 CV(d)

电子传递速率也对 h-WO$_3$ 的接受电子能力具有重大影响。动力学分析表明电子从游离态 heme 转移到 h-WO_3 的能量势垒为负值(E_a)。轴向配位的两个组氨酸残基对电子传递过程具有一定的空间位阻效应,但是 E_a 值仍然小于零(表2.3)。负的 E_a 值说明了发生的反应不存在能量势垒。基元反应呈现负的 E_a 值是典型的 Barrierless 反应[54],这种反应的发生表明受体的表面形成了势阱,有利于对粒子进行捕获[55]。电子传递的速率常数(k_{ET})可以通过如下方程进行计算:

$$k_{ET} = A\exp\left(-\frac{E_a}{RT}\right) = \frac{k_B T}{h}E(c^{\ominus})^{1-n}\exp\left(\frac{\Delta^{\neq}S^{\ominus}}{R}\right)\exp\left(-\frac{E_a}{RT}\right) \quad (2.6)$$

其中,k_{ET} 是电子传递的速率系数[$mol^{1-n} \cdot (L^{1-n} \cdot s)^{-1}$,$n$ 为反应级数],A 是指前因子[$mol^{1-n} \cdot (L^{1-n} \cdot s)^{-1}$],$k_B$ 是玻尔兹曼常数[$1.381 \times 10^{-23}\ m^2 \cdot kg \cdot (s^2 \cdot K)^{-1}$],$h$ 是普朗克常数($6.626 \times 10^{-34}\ m^2 \cdot kg \cdot s^{-1}$),$c^{\ominus}$ 是标准摩尔浓度($1\ mol \cdot L^{-1}$),n 是反应的级数,$\Delta^{\neq}S^{\ominus}$ 是活化熵[$J \cdot (mol \cdot K)^{-1}$]。

表 2.3　纳米团簇从卟啉铁捕获电子的动力学性质,包括能垒(E_a),过渡态热力学数据($\Delta^{\neq}G^{\ominus}$,$\Delta^{\neq}H^{\ominus}$,$\Delta^{\neq}S^{\ominus}$)和电子传递速率常数(k_{ET})

System	E_a (eV)	$\Delta^{\neq}G^{\ominus}$ (eV)	$\Delta^{\neq}H^{\ominus}$ (eV)	$\Delta^{\neq}S^{\ominus}$ ($10^{-3}\ eV \cdot K^{-1}$)	$k_{ET}[mol^{1-n} \cdot (L^{1-n} \cdot s)^{-1}]$
porphyrin-WO_3	-6.640	1.963	2.205	0.812	3.70×10^{129}
porphyrin-biHis-WO_3	-1.959	4.176	4.304	0.430	3.25×10^{48}

卟啉环相对于 h-WO_3 表面的构型对于降低 E_a 至关重要,并可能使 E_a 由正值变成负值[56]。结合热力学和动力学分析结果发现,电子传递能在 EAB 底物代谢产生很低的热力学驱动力下发生[57,58],并且 porphrin-biHis-WO_3 体系能够迅速达到反应平衡[$k_{ET} = 3.25 \times 10^{48}\ mol^{1-n} \cdot (L^{1-n} \cdot s)^{-1}$]。计算得到的电子传递速率常数与实际体系中卟啉铁的浓度有关,而氧化还原活性中心周围存在更多的氨基酸残基。肽链的位阻效应隔开的电子传递距离以及细菌自身的保护作用会使电子传递速率常数降低。

2.3.5
外膜与周质混合 c-Cyt 的电化学响应

对 *Shewanella oneidensis* MR-1 和 $\Delta mtrC/\Delta omcA$ 突变株细胞分别进行破胞处理,提取了外膜与周质中混合 c-Cyt 蛋白,用处理马心 Cyt 相同的方法,将其固定在负载了 WO_3 的 GC 电极上进行 CV 测试,结果如图 2.5 所示。可以发

现从 $\Delta mtrC/\Delta omcA$ 基因的突变株中提取的混合蛋白,相比于野生株的蛋白提取物,在 $-0.2\sim-0.1$ V(vs. Ag/AgCl)之间缺少了电化学响应。由于这是混合蛋白的电化学响应,且是在 h-WO$_3$ 界面上的电子传递过程,所以会与文献[51-53]中提纯的 OM c-Cyt 在 Fe(Ⅲ)还原过程中得到的电位存在差异。图 2.5 中呈现的是较宽的峰,可能是由于多种 c-Cyt 蛋白的电化学信号相互影响造成的。

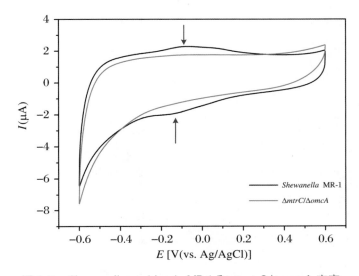

图 2.5 *Shewanella oneidensis* MR-1 和 $\Delta mtrC/\Delta omcA$ 突变株的外膜与周质中混合 c-Cyt 蛋白在 h-WO$_3$ 工作电极上的 CV 图

由于 CymA 等周质空间中的 c-Cyt 也具有电化学响应[51],但提取的混合蛋白中周质空间的 c-Cyt 蛋白浓度较低,可能用 CV 测试检测不到这些蛋白的氧化还原反应,也有可能是在胞外电子受体 h-WO$_3$ 表面,周质空间的 c-Cyt 并不表现出氧化还原活性,而野生株 *Shewanella oneidensis* MR-1 中的氧化还原峰可能是多种 OM c-Cyt 在 h-WO$_3$ 界面上同时进行电子传递过程的反映,说明了外膜蛋白在胞外电子传递中的重要性。

此外,从微生物中直接提取的 c-Cyt 蛋白的还原峰的位置也比游离的 heme 的还原峰更正,证明肽链的存在会影响活性中心与胞外电子受体的电子传递。但由于实际体系中其他因素也会对电子传递过程造成影响,肽链的影响程度无法判断。可以通过分子动力学模拟的方法,去除其他的影响因素,仅仅考察肽链对胞外电子传递的影响。

2.3.6
水溶液中细胞色素 c 在 WO_3 表面的电子传递

MD 模拟在等温(298 K)等压(0.1 MPa)条件下,进行 STC-WO_3 和游离态 heme-WO_3 体系结构的弛豫。通过一系列 MD 模拟计算可以找出 STC 在 h-WO_3 纳米晶体团簇表面的最优构型。在 MD 盒子中(图 2.6),最开始设置 STC 的质量中心(COM)与 h-WO_3 表面的距离为 30 Å。

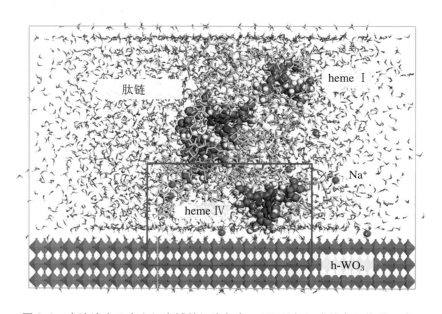

图 2.6 水溶液中 4 个血红素辅基细胞色素 c(STC)与组成的电子传递系统

heme Ⅳ 暴露在溶液中,链接着 His19,65,位于蛋白与晶体表面的界面上,对电子传递速率存在重要影响。STC 接近 h-WO_3 晶面使得 heme Ⅳ 成为了最有效的电子传递位点,这与实验结果一致[59]。STC-WO_3 体系的平衡态表明,在大多数情况下细胞色素与纳米晶体团簇表面的接触面积非常小,这是因为能够与晶面基础的残基数量是有限的。对于 STC-WO_3 体系,heme Ⅳ 在平衡的构型中与 h-WO_3 晶面呈 45°角[图 2.7(a)],这与氧化态和还原态的 porphyrin-bi-His-WO_3 体系[图 2.2(c)和(d)]中卟啉环方向一致。为了反映肽链的影响,也利用 MD 模拟计算了游离态 heme-WO_3 系统[图 2.7(b)]。

活性位点之间的径向分布函数(RDFs)反映了接近 h-WO_3 纳米晶体团簇晶面的 STC 微观构型。图 2.7(c)展示了 STC-WO_3 系统与不含肽链的游离态

heme-WO$_3$ 系统对位点(site-site) RDF[$g(r)$]的区别。RDF 中的最高峰显示了出现概率最大的 heme 中 Fe 原子与 h-WO$_3$ 中晶面顶层的 W 原子之间的距离。对于游离态 heme-WO$_3$ 系统,最高峰(r_2)为 7.3 Å,小于 STC-WO$_3$ 系统的 r_1(10.4 Å)。这说明暴露在溶剂中的 heme 与晶面之间的出现概率最大的距离可能会被肽链稍稍扩大。并且游离态 heme-WO$_3$ 的第一个峰($r_4 = 6.4$ Å)是指最靠近晶面的 heme 与晶面的最短距离,而 STC-WO$_3$ 系统(r_3)为 9.6 Å。这个值与 porphyrin-biHis-WO$_3$ 系统[图 2-2(d),在氧化态下为 9.618 Å]的 QM/MM 计算结果非常一致。同时,RDFs 通过 $g(r)$ 曲线提供了关于晶面上电子接收位点数量的相关信息。以上结果说明细胞色素动力学、构型变化和卟啉环扭曲均在界面电子传递中起到了重要作用。

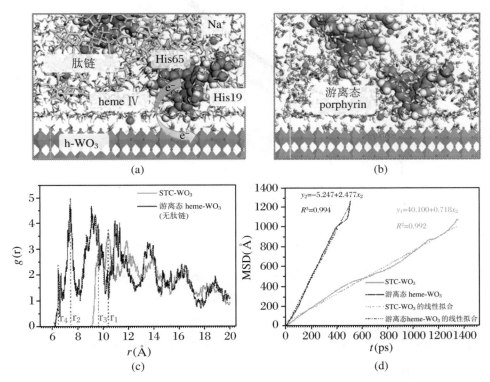

图 2.7 STC-WO$_3$ 和游离态 heme-WO$_3$ 系统平衡状态下的微观结构信息:溶剂化的 STC-WO$_3$ 系统中 STC 与 h-WO$_3$ 界面的放大图,高亮标出的是肽链(a);溶剂化的游离态 heme-WO$_3$ 系统(b);对于离晶面最近的 heme 中的 Fe 原子与 h-WO$_3$ 片最顶层的 W 原子之间的径向分布函数(RDFs)(c);STC-WO$_3$ 和游离态 heme-WO$_3$ 系统中 Na$^+$ 扩散的均方位移(MSD, Å^2)与时间 t(ps)的关系(d)

采用 Na$^+$ 的均方位移(MSDs)对时间的函数来表征 STC-WO$_3$ 系统的动力学(dynamic)特征。STC-WO$_3$ 系统中 Na$^+$ 的自扩散系数(D)可以用如下的

Einstein 关系式进行计算[60,61]：

$$D = \frac{1}{6N_\alpha} \lim_{t \to \infty} \frac{d}{dt} \sum_{i=1}^{N_\alpha} \left\langle [r_i(t) - r_i(0)]^2 \right\rangle \quad (2.7)$$

其中，N 表示体系中的扩散离子数量，$r_i(t) - r_i(0)$ 表示 i 类原子经过时间 t 后的矢量距离，矢量大小的平方是在滚动的时间间隔内的平均值，括号中的量表示 MSD。

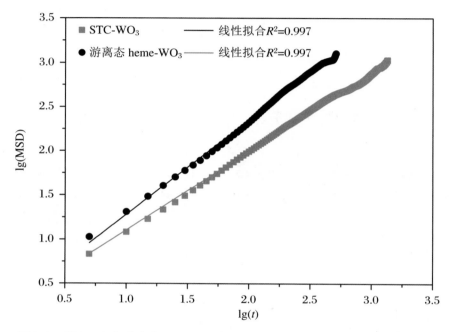

图 2.8　STC-WO$_3$ 与游离态 heme-WO$_3$ 与电子传递系统 lg(MSD) 与 lg(t) 的关系

图 2.7(d) 中展示了 STC-WO$_3$ 系统与游离态 heme-WO$_3$ 系统中 Na$^+$ 的总 MSD 随时间的变化情况。总 MSD 值为单独的 x,y,z 轴上的 MSD 之和，其中 lg(MSD) 与 lg(t) 呈线性关系（图 2.8），这说明计算结果是可靠的。由于这些体系的 MSDs 与时间呈线性关系，钠离子的扩散系数能够通过线性区域的斜率获得。Na$^+$ 在游离态 heme-WO$_3$ 体系中的扩散系数 (4.045×10^{-5} cm$^2 \cdot$ s^{-1}) 高于 STC-WO$_3$ 体系的 (1.197×10^{-5} cm$^2 \cdot$ s^{-1})，这说明了 Na$^+$ 在前者中呈现出更大的扩散运动，但仍在同一个数量级，所以，肽链对 Na$^+$ 扩散和电子传递速率的影响程度是有限的[62]。

参考文献

[1] Huang Z F, Song J J, Pan L, et al. Tungsten oxides for photocatalysis, elec-

trochemistry, and phototherapy [J]. Advanced Materials, 2015, 27(36): 5309-5327.

[2] Koo B R, Ahn H J. Fast-switching electrochromic properties of mesoporous WO_3 films with oxygen vacancy defects [J]. Nanoscale, 2017, 9(45): 17788-17793.

[3] Wang Y, Liu J, Cui X, et al. NH_3 gas sensing performance enhanced by Pt-loaded on mesoporous WO_3 [J]. Sensors and Actuators B-Chemical, 2017, 238: 473-481.

[4] Liu C, Zhang C, Sun S, et al. Effect of WO_x on Bifunctional Pd-WO_x/Al_2O_3 catalysts for the selective hydrogenolysis of glucose to 1,2-propanediol [J]. ACS Catalysis, 2015, 5(8): 4612-4623.

[5] Dias P, Lopes T, Meda L, et al. Photoelectrochemical water splitting using WO_3 photoanodes: the substrate and temperature roles [J]. Physical Chemistry Chemical Physics, 2016, 18(7): 5232-5243.

[6] Rather S U. Preparation, characterization and hydrogen storage studies of carbon nanotubes and their composites: A review [J]. International Journal of Hydrogen Energy, 2020, 45(7): 4653-4672.

[7] Deng Z, Gong Y, Luo Y, et al. WO_3 nanostructures facilitate electron transfer of enzyme: Application to detection of H_2O_2 with high selectivity [J]. Biosensors & Bioelectronics, 2009, 24(8): 2465-2469.

[8] Wang F G, Di Valentin C, Pacchioni G. Electronic and structural properties of WO_3: A systematic hybrid DFT study [J]. Journal of Physical Chemistry C, 2011, 115(16): 8345-8353.

[9] Szilagyi I M, Madarasz J, Pokol G, et al. Stability and controlled composition of hexagonal WO_3 [J]. Chemistry of Materials, 2008, 20(12): 4116-4125.

[10] Yuan S J, Li W W, Cheng Y Y, et al. A plate-based electrochromic approach for the high-throughput detection of electrochemically active bacteria [J]. Nature Protocols, 2014, 9(1): 112-119.

[11] Yuan S J, He H, Sheng G P, et al. A photometric high-throughput method for identification of electrochemically active bacteria using a WO_3 nanocluster probe [J]. Scientific Reports, 2013, 3:1315.

[12] Xie X, Criddle C, Cui Y. Design and fabrication of bioelectrodes for microbial bioelectrochemical systems [J]. Energy & Environmental Science, 2015,

8(12): 3418-3441.

[13] Logan B E, Rossi R, Ragab A A, et al. Electroactive microorganisms in bioelectrochemical systems [J]. Nature Reviews Microbiology, 2019, 17(5): 307-319.

[14] Zhang Y, Liu M, Zhou M, et al. Microbial fuel cell hybrid systems for wastewater treatment and bioenergy production: Synergistic effects, mechanisms and challenges [J]. Renewable & Sustainable Energy Reviews, 2019, 103: 13-29.

[15] Wang H, Lu L, Chen H, et al. Molecular transformation of crude oil contaminated soil after bioelectrochemical degradation revealed by FT-ICR mass spectrometry [J]. Environmental Science & Technology, 2020, 54(4): 2500-2509.

[16] Liu J, Pearce C I, Shi L, et al. Particle size effect and the mechanism of hematite reduction by the outer membrane cytochrome OmcA of *Shewanella oneidensis* MR-1 [J]. Geochimica et Cosmochimica Acta, 2016, 193: 160-175.

[17] Rosenbaum M, Zhao F, Schroder U, et al. Interfacing electrocatalysis and biocatalysis with tungsten carbide: a high-performance, noble-metal-free microbial fuel cell [J]. Angewandte Chemie (International ed in English), 2006, 45(40): 6658-6661.

[18] Biffinger J, Ribbens M, Ringeisen B, et al. Characterization of electrochemically active bacteria utilizing a high-throughput voltage-based screening assay [J]. Biotechnology and Bioengineering, 2009, 102(2): 436-444.

[19] Hou H J, Li L, Cho Y, et al. Microfabricated microbial fuel cell arrays reveal electrochemically active microbes [J]. PLoS One, 2009, 4(8): e6570.

[20] Wang Y Q, Li B, Zeng L Z, et al. Polyaniline/mesoporous tungsten trioxide composite as anode electrocatalyst for high-performance microbial fuel cells [J]. Biosensors & Bioelectronics, 2013, 41: 582-588.

[21] Ye J, Hu A, Cheng X, et al. Response of enhanced sludge methanogenesis by red mud to temperature: Spectroscopic and electrochemical elucidation of endogenous redox mediators [J]. Water Research, 2018, 143: 240-249.

[22] Yu S S, Chen J J, Liu X Y, et al. Interfacial electron transfer from the outer membrane cytochrome OmcA to graphene oxide in a microbial fuel cell: Spectral and electrochemical insights [J]. ACS Energy Letters, 2018,

3(10): 2449-2456.

[23] Edwards M J, White G F, Butt J N, et al. The crystal structure of a Biological insulated Transmembrane Molecular Wire [J]. Cell, 2020, 181(3): 665-673.

[24] Pfeffer C, Larsen S, Song J, et al. Filamentous bacteria transport electrons over centimetre distances [J]. Nature, 2012, 491(7423): 218-221.

[25] Nielsen L P, Risgaard-Petersen N, Fossing H, et al. Electric currents couple spatially separated biogeochemical processes in marine sediment [J]. Nature, 2010, 463(7284): 1071-1074.

[26] Tu Y S, Lv M, Xiu P, et al. Destructive extraction of phospholipids from Escherichia coli membranes by graphene nanosheets [J]. Nature Nanotechnology, 2013, 8(8): 594-601.

[27] Martinez-Gil M, Yousef-Coronado F, Espinosa-Urgel M. LapF, the second largest pseudomonas putida protein, contributes to plant root colonization and determines biofilm architecture [J]. Molecular Microbiology, 2010, 77(3): 549-561.

[28] Delley B. From molecules to solids with the DMol^3 approach [J]. The Journal of Chemical Physics, 2000, 8(6): 361-364.

[29] Perdew J P, Chevary J A, Vosko S H, et al. Atoms, molecules, solids, and surfaces: Applications of the generalized gradient approximation for exchange and correlation [J]. Physical Review B, Condensed Matter, 1992, 46(11): 6671-6687.

[30] Gale J D, Rohl A L. The general utility lattice program (GULP) [J]. Molecular Simulation, 2003, 29(5): 291-341.

[31] Halgren T A, Lipscomb W N. The synchronous-transit method for determining reaction pathways and locating molecular transition states [J]. Chemical Physics Letters, 1977, 49(2): 225-232.

[32] Lies D P, Hernandez M E, Kappler A, et al. *Shewanella oneidensis* MR-1 uses overlapping pathways for iron reduction at a distance and by direct contact under conditions relevant for Biofilms [J]. Applied and Environmental Microbiology, 2005, 71(8): 4414-4426.

[33] Bond D R, Lovley D R. Electricity production by *Geobacter sulfurreducens* attached to electrodes [J]. Applied and Environmental Microbiology, 2003, 69(3): 1548-1555.

[34] Inoue K, Qian X L, Morgado L, et al. Purification and characterization of OmcZ, an outer-surface, octaheme c-Type cytochrome essential for optimal current production by *Geobacter sulfurreducens* [J]. Applied and Environmental Microbiology, 2010, 76(12): 3999-4007.

[35] Bonanni P S, Schrott G D, Robuschi L, et al. Charge accumulation and electron transfer kinetics in *Geobacter sulfurreducens* biofilms [J]. Energy & Environmental Science, 2012, 5(3): 6188-6195.

[36] Xiong Y, Shi L, Chen B, et al. High-affinity binding and direct electron transfer to solid metals by the *Shewanella oneidensis* MR-1 outer membrane c-type cytochrome OmcA [J]. Journal of the American Chemical Society, 2006, 128(43): 13978-13979.

[37] Hartshorne R S, Reardon C L, Ross D, et al. Characterization of an electron conduit between bacteria and the extracellular environment [J]. Proceedings of the National Academy of Sciences of the United States of America, 2009, 106(52): 22169-22174.

[38] Shi L, Rosso K M, Clarke T A, et al. Molecular underpinnings of Fe(Ⅲ) oxide reduction by *Shewanella oneidensis* MR-1 [J]. Frontiers in Microbiology, 2012, 3: 50.

[39] Leys D, Meyer T E, Tsapin A S, et al. Crystal structures at atomic resolution reveal the novel concept of 'electron-harvesting' as a role for the small tetraheme cytochrome c [J]. The Journal of Biological Chemistry, 2002, 277(38): 35703-35711.

[40] Carlson H K, Iavarone A T, Gorur A, et al. Surface multiheme c-type cytochromes from thermincola potens and implications for respiratory metal reduction by gram- positive bacteria [J]. Proceedings of the National Academy of Sciences of the United States of America, 2012, 109(5): 1702-1707.

[41] Smith D M A, Dupuis M, Vorpagel E R, et al. Characterization of electronic structure and properties of a Bis(histidine) heme model complex [J]. Journal of the American Chemical Society, 2003, 125(9): 2711-2717.

[42] Shaik S. Biomimetic Chemistry iron opens up to high activity [J]. Nature Chemistry, 2010, 2(5): 347-349.

[43] Kerisit S, Rosso K M, Dupuis M, et al. Molecular computational investigation of electron-transfer kinetics across cytochrome-iron oxide interfaces [J]. The Journal of Physical Chemistry C, 2007, 111(30): 11363-11375.

[44] Essmann U, Perera L, Berkowitz M, et al. A smooth particle mesh Ewald method [J]. The Journal of Chemical Physics, 1995, 103: 8577-8593.

[45] Dauber-Osguthorpe P, Roberts V A, Osguthorpe D J, et al. Structure and energetics of ligand binding to proteins: Escherichia coli dihydrofolate reductase-trimethoprim, a drug-receptor system [J]. Proteins, 1988, 4(1): 31-47.

[46] Andersen H. Molecular dynamics simulation at constant pressure and/or temperature [J]. The Journal of Chemical Physics, 1980, 72: 2384-2393.

[47] Berendsen H, Postma J P M, Van Gunsteren W, et al. Molecular-dynamics with coupling to an external bath [J]. The Journal of Chemical Physics, 1984, 81: 3684-3690.

[48] Faughnan B, Crandall R, Heyman P. Electrochromism in WO_3, amorphous films [J]. RCA Review, 1975, 36: 177-197.

[49] Kopf S H, Henny C, Newman D K. Ligand-enhanced abiotic iron oxidation and the effects of chemical versus biological iron cycling in anoxic environments [J]. Environmental Science & Technology, 2013, 47(6): 2602-2611.

[50] Hashimoto K, Morokuma K. Ab initio molecular orbital study of $Na(H_2O)_n$ ($n = 1 \sim 6$) clusters and their ions. comparison of electronic structure of the 'surface' and 'interior' Complexes [J]. Journal of the American Chemical Society, 1994, 116(25): 11436-11443.

[51] Field S J, Dobbin P S, Cheesman M R, et al. Purification and magneto-optical spectroscopic characterization of cytoplasmic membrane and outer membrane multiheme c-type cytochromes from *Shewanella frigidimarina* NCIMB400 [J]. The Journal of Biological Chemistry, 2000, 275 (12): 8515-8522.

[52] Marsili E, Baron D B, Shikhare I D, et al. Shewanella secretes flavins that mediate extracellular electron transfer [J]. Proceedings of the National Academy of Sciences of the United States of America, 2008, 105(10): 3968-3973.

[53] Hartshorne R S, Jepson B N, Clarke T A, et al. Characterization of *Shewanella oneidensis* MtrC: a cell-surface decaheme cytochrome involved in respiratory electron transport to extracellular electron acceptors [J]. Journal of Biological Inorganic Chemistry, 2007, 12(7): 1083-1094.

[54] Alvarez-Idaboy J R, Mora-Diez N, Vivier-Bunge A. A quantum chemical

and classical transition state theory explanation of negative activation energies in OH addition to substituted ethenes [J]. Journal of the American Chemical Society, 2000, 122(15): 3715-3720.

[55] Mozurkewich M, Benson S W. Negative activation energies and curved Arrhenius plots. 1. theory of reactions over potential wells [J]. The Journal of Physical Chemistry, 1984, 88(25): 6429-6435.

[56] Silverstein T P. Falling enzyme activity as temperature rises: Negative activation energy or denaturation? [J]. Journal of Chemical Education, 2012, 89(9): 1097-1099.

[57] Summers Z M, Gralnick J A, Bond D R. Cultivation of an obligate Fe(II)-oxidizing *Lithoautotrophic* bacterium using electrodes [J]. Mbio, 2013, 4(1): e00420-12

[58] Beard D A, Qian H. Relationship between thermodynamic driving force and one-way fluxes in reversible processes [J]. PloS One, 2007, 2(1): e144.

[59] Harada E, Kumagai J, Ozawa K, et al. A directional electron transfer regulator based on heme-chain architecture in the small tetraheme cytochrome c from Shewanella oneidensis [J]. FEBS Letters, 2002, 532(3): 333-337.

[60] Einstein A. Über die von der molekularkinetischen Theorie der Wärme geforderte Bewegung von in ruhenden Flüssigkeiten suspendierten Teilchen [J]. Annalen der Physik 1905, 322(8): 549-560.

[61] Meunier M. Diffusion coefficients of small gas molecules in amorphous cis-1, 4-polybutadiene estimated by molecular dynamics simulations [J]. The Journal of Chemical Physics, 2005, 123(13): 134906.

[62] Cowley A B, Kennedy M L, Silchenko S, et al. Insight into heme protein redox potential control and functional aspects of six-coordinate ligand-sensing heme proteins from studies of synthetic heme peptides [J]. Inorganic Chemistry, 2006, 45(25): 9985-10001.

第 3 章

EAB 外膜蛋白 OmcA 与赤铁矿界面电子传递机制

3.1
EAB 对铁元素地球化学循环的影响

  Shewanella 和 *Geobacter* 是研究最为普遍的两大类模式 EABs，其能够将生物电子传递给不溶性电子受体，在环境、能源、微生物学和生物地球化学领域发挥着重要作用[1-2]。第 2 章已经证明了基于胞外电子传递机制，金属氧化物 WO_3 可用于筛选 EAB。EABs 又被称为异化金属还原菌（dissimilatory metal-reducing bacteria，DMRB），可以与其他金属氧化物，如铁、锰的氧化物进行电子转移，这个过程作为地球上非常古老的微生物代谢形式，在金属元素的生物地球化学循环中起了非常关键的作用[3]。这种金属元素循环的方式，特别是 Fe（Ⅲ）氧化物分解还原为 Fe 的过程，被认为有可能也存在于其他富含铁矿物的行星，是一种可能的地球外代谢过程[4]。另外，EAB 也可以与广泛使用的光催化剂二氧化钛存在协同作用[5]，EAB 在其中扮演着"能隙工程师"的角色，对光催化剂的能带结构进行有效的调控。这些过程都与 EAB/金属氧化物的界面协同作用有关，对生物体系与化学体系界面相互作用机制的解析，将有助于发挥电活性微生物-无机金属氧化物协同体系在生物环境修复和生物能源转换中的作用[6]。

  EAB 与金属氧化物存在多种界面电子转移方式，包括通过分布在外膜上的蛋白（含多个卟啉铁的细胞色素 c，OM c-Cyts）[7]和通过导电菌毛或"纳米导线"[8]进行直接接触的传递，也包括通过氧化还原介体的间接传递[9]，氧化还原媒介分子的扩散连接着 EAB 与不溶性电子受体。多种电子转移方式将 EABs 编织成电子传递网络，在海底沉积物中的 EABs 可能通过"电源线"连接着相距为厘米尺度区域的氧化还原反应，形成一个巨大的天然电池，从而产生电流[10-11]。其中，*Shewanella oneidensis* MR-1 的纳米导线是外膜的延伸[图 3.1（a）]，也分布着大量的外膜蛋白 OM c-Cyts[12]，向外突起的氧化还原功能外膜不仅适用于远距离的电子传输，也有利于与金属氧化物的充分接触。外膜蛋白组成的电子传递通道负责生物电子跨越细胞膜，连接着胞内与胞外，是不同电子转移方式的关键步骤。不同种类的 OM c-Cyts，如 MtrC，MtrF，OmcA 等，都呈现出类似的结构，由具有氧化还原活性的血红素分子以及氨基酸组成的肽链缠绕形成，其中血红素的中心位置 Fe，以轴向配位双组氨酸残基接入肽链[12-15]。而另一种模式 EAB 为 *Geobacter sulfurreducens*，其纳米导线是由多个 OmcS 蛋白按一定方式聚集组装形成的[图 3.1（b）]，但内部卟啉铁的配位方式仍然与

Shewanella oneidensis MR-1 的 OM c-Cyts 蛋白相同[16]，暴露于膜外环境且接近蛋白表面的血红素以及其周围的氨基酸，在电子传递过程中可能会直接接触金属氧化物界面。因此，OM c-Cyts 蛋白与金属氧化物之间的直接电子转移，无论是通过外膜蛋白还是纳米导线，在微界面处可能都是 OM c-Cyts 蛋白环境端与金属氧化物之间的界面电子转移。

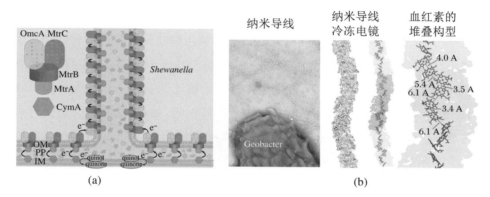

图 3.1　EAB 的纳米导线结构：Shewanella oneidensis MR-1 的纳米导线是外膜的延伸，导线上与外膜的环境相同，也分布着外膜蛋白[12] (a)；Geobacter sulfurreducens 的纳米导线冷冻电镜图片，OmcS 蛋白子单元的组装及 3 个子单元中血红素的堆叠构型[16] (b)

赤铁矿（$\alpha\text{-}Fe_2O_3$）是地球上土壤和沉积物中可被 EAB 用于呼吸的最丰富的电子受体[17]，且在火星上也分布广泛，由于其可以通过含水矿物沉淀形成，可能记录了火星生命的微观证据[18]。赤铁矿具有刚玉结构，属于 D_{3d}^6 空间群，铁原子与 6 个氧原子配位形成八面体结构，是一种 n 型半导体，具有光催化活性，由于易合成且对环境无害[19]，可以广泛使用于水环境中对污染物进行降解与转化；另外，赤铁矿也是一种反铁磁体，具有一定的弱亚铁磁性，与自然环境的磁学研究密切相关[20]。据报道，赤铁矿纳米颗粒在外膜蛋白存在的环境中会发生不同程度的聚集，这取决于不同尺寸的纳米颗粒、溶剂环境以及蛋白的性质等[21]。金属氧化物 $\alpha\text{-}Fe_2O_3$ 与 EAB 之间可能同时存在范德华力、静电作用力、空间排斥力等作用力，但是对于该过程涉及的界面过程和控制因素仍然很不清楚。

因此，本章针对微生物/金属氧化物协同体系机制不明确的问题，通过光谱技术与分子模拟相结合的方法，从分子水平上获得 $\alpha\text{-}Fe_2O_3$ 表面的外膜蛋白取向和两者界面处形成氢键作用的结构信息以及外膜蛋白活性中心至金属氧化物界面电子转移的距离分布。这些结果能够为研究外膜蛋白结合金属氧化物的过程对活性中心微观环境的改变提供有用的信息，为进一步阐明 EABs 的胞外电子流动对元素地球化学循环和环境效应的影响提供理论上的依据，促进 EAB/金属氧化物在环境修复和能源转换领域中的应用。

3.2 外膜蛋白与赤铁矿界面电子传递的研究方法

3.2.1 OmcA 的表达和提纯

将 Shewanella MR-1 3332 菌种转接到无菌 LB 培养基中，在 30 ℃下培养至 OD600 为 0.6 左右，加入终浓度为 1 mmol·L^{-1} 的阿拉伯糖进行诱导表达。14 h 后 8000g 离心 30 min 分离细菌和上清。Ni 柱（Ni Sepharose GE Healthcare Life Sciences, USA）用至少 3 倍柱体积的 buffer B（20 mmol·L^{-1} Hepes，pH 7.8，10%甘油，10 mmol·L^{-1}咪唑）预平衡。将离心后的上清液和预平衡的 Ni 柱混合，4 ℃下缓速搅拌 2 h。重新填装 Ni 柱并用 5 倍柱体积的洗脱液 buffer C（20 mmol·L^{-1} Hepes，pH 7.8，10%甘油，250 mmol·L^{-1}咪唑）洗脱蛋白。目标蛋白组分通过超滤管浓缩并去除咪唑，−80 ℃冷冻保存备用。

3.2.2 OmcA/α-Fe$_2$O$_3$ 相互作用过程的红外监测及二维相关分析

配置浓度为 1 mg·mL^{-1} 的 α-Fe$_2$O$_3$（尺寸为 30 nm，购买于中国阿拉丁公司）水溶液，超声分散半小时以上。部分 α-Fe$_2$O$_3$ 溶液冷冻干燥后进行 FTIR 分析，表征其表面官能团。取 200 μL 溶液滴加在 ZnSe 晶体上，自然干燥。在 α-Fe_2O_3 膜上添加空白缓冲溶液，采集 FTIR 背景信号。将缓冲溶液更换为 OmcA 溶液（11 μmol·L^{-1}），每 2 min 采集一次信号。ATR-FTIR 用带有碲化汞镉检测器的 Vertex 70 光谱仪（德国布鲁克公司）记录，检测器使用前需用液氮冷却至 77 K，红外晶体为 ZnSe。将 800 μL GO 溶液（2 g·L^{-1}）滴涂在 ZnSe 晶体表面，自然干燥。GO 膜在缓冲盐中平衡 2 h，采集光谱作为背景，理论上包括 ZnSe、GO 膜、缓冲盐和 H_2O 等 4 种成分的混合信号。将缓冲盐替换为 15 μmol·L^{-1} 的 OmcA 溶液（0.6 mL），每 3 min 采集红外光谱，扫描范围为 1800～1000 cm^{-1}，扫速为 4 cm^{-1}。所获得的红外光谱经基线校正、平滑、去背景后用于 2DCOS 分析（时间为外部变量），分析软件为 2D Shige（Kwansei-Gakuin 大学，日本）。根据同步相关峰 $\Phi(\nu_1,\nu_2)$ 和异步相关峰 $\Psi(\nu_1,\nu_2)$ 的正负来判定键 ν_1 和 ν_2 的变化顺序：当 $\Phi(\nu_1,\nu_2)$ 和 $\Psi(\nu_1,\nu_2)$ 同时为正或同时为负时，键 ν_1 先于 ν_2 发生变化；当 $\Phi(\nu_1,\nu_2)$ 和 $\Psi(\nu_1,\nu_2)$ 正负号相反时，键 ν_2 先于 ν_1 发生变化；当 $\Psi(\nu_1,\nu_2)$ 为 0 时，键 ν_1 和 ν_2 同时发生变化。

3.2.3
OmcA/α-Fe_2O_3 界面电子转移计算方法

由于已有晶体结构的 *Shewanella oneidensis* 外膜蛋白中，仅有 MtrF（PDB：3PMQ）是直接从 *Shewanella oneidensis* 提纯并解析的，而 MtrC 和 OmcA 则是在大肠杆菌中表达后再进行解析的，结构上也许存在一定的差异，且目前报道的理论计算工作大都采用 MtrF 作为典型的多血红素细胞色素的蛋白模型[22-24]。MtrF 是 MtrC 的同源蛋白，功能与结构都非常类似，都含有结构域Ⅰ，Ⅱ，Ⅲ 和 Ⅳ 以及 10 个活性中心血红素结构，为了方便与文献中的计算结果进行对比，本章选取 MtrF 作为外膜蛋白的模型。同时，蛋白膜伏安法证实 MtrF 有与固态电子受体表面进行快速电子交换的能力，可以作为外膜蛋白的模型蛋白进行理论分析与实验测试[22,25]。将 MtrF 结构域Ⅱ与 α-Fe_2O_3 的（001）晶面组成界面电子转移体系进行分子动力学模拟，获得结构域Ⅱ表面与固态电子受体之间相互作用的结构特征及有效的电子传递距离。

在分子动力学模拟中，整个体系包含 MtrF 的结构域Ⅱ及 α-Fe_2O_3（001）晶面，并加入水分子反映溶液环境。选取（001）晶面的依据是其在赤铁矿中暴露比例比较高，且可以采用氧封端保持较好的稳定性[26]，而铁原子封端更易接受生物电子,在实际环境中可吸附水分子降低表面能而稳定存在,所以电化学活性细菌在自然条件下进行直接电子转移时有较大的概率接触 α-Fe$_2$O$_3$(001)晶面。首先进行能量最小化步骤,然后在 298.15 K 的 NVT 系综中弛豫 5 ns,最后在 298.15 K 下 NVE 系综中弛豫 5 ns。体系中的静电作用通过 Ewald 加和方法[27]进行计算,精度是 $1.0×10^{-5}$ kcal·mol^{-1};而范德华作用力则通过基于原子的加和方法进行。外膜蛋白/α-Fe$_2$O$_3$ 电子传递系统的分子动力学模拟采用 Universal 力场,时间步长设为 1.0 fs,以此来平衡电子转移模型,并获得最终结构。Andersen algorithm[28]的碰撞率设定为 1.0,以便于控制每个模型的温度。

3.3 外膜蛋白在 EAB 与赤铁矿界面电子传递中的关键作用机制

3.3.1

OmcA 在 α-Fe$_2$O$_3$ 膜表面的红外性质

OmcA 在裸 ZnSe 晶体和 α-Fe$_2$O$_3$ 膜表面均表现出明显的 Amide Ⅰ峰和 Amide Ⅱ峰[图 3.2(a)]。在同样实验条件下,OmcA 在 α-Fe$_2$O$_3$ 膜表面的信号强度远高于蛋白在裸晶体表面的峰强,意味着在相同时间内 α-Fe$_2$O$_3$ 膜吸附了更多的 OmcA,表明 α-Fe$_2$O$_3$ 对 OmcA 具有良好的亲和力[21,29-31]。在 Amide Ⅰ 峰范围内(1700~1600 cm^{-1})可观察到,峰 1655 cm^{-1} 和 1644 cm^{-1} 分别对应着蛋白的 α 螺旋结构以及 β 折叠结构[32,33]。经 α-Fe$_2$O$_3$ 修饰后,蛋白的 Amide Ⅰ 主峰从 1655 cm^{-1} 迁移至 1644 cm^{-1},预示着 OmcA 吸附在 α-Fe$_2$O$_3$ 表面后部分 α 螺旋进一步转变为 β 折叠,与圆二色光谱结果一致[21]。共振拉曼光谱表征发现蛋白 α 螺旋到 β 折叠的转变会导致血红素更加暴露,进而使细胞色素 c 表现出更高的活性[34]。对于 OmcA 而言,它可以通过蛋白表面位于血红素 10 附近的三肽(苏氨酸 725-脯氨酸 726-丝氨酸 727)与 α-Fe$_2$O$_3$ 表面形成氢键[35-36],该氢键引起了蛋白部分二级结构的转变。晶体结构分析发现血红素轴向配位的氨基酸 724 与苏安酸 725 直接相连[图 3.2(b)],进一步佐证了氢键引起的结构变化可缩短血红素活性中心铁原子与 α-Fe$_2$O$_3$ 之间的距离。除此之外,该结构变化可能有利于蛋白以最大化接触面积的构象吸附在 α-Fe$_2$O$_3$ 表面[29]。

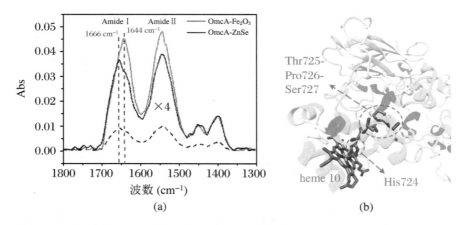

图 3.2 OmcA 在 α-Fe$_2$O$_3$ 膜(红色)和 ZnSe(黑色)表面的红外信号:黑色实线为黑色虚线放大 4 倍的结果(a);OmcA 中结合 α-Fe$_2$O$_3$ 的三肽苏氨酸 725-脯氨酸 726-丝氨酸 727。组氨酸 724 是血红素 10 的轴向配位氨基酸,组氨酸 724 与三肽苏氨酸 725-脯氨酸 726-丝氨酸 727 直接相连。苏氨酸 725 和丝氨酸 727 的自由羟基(红色棍状)朝向远离蛋白的一侧,利于结合 α-Fe$_2$O$_3$ (b)

3.3.2
OmcA 在 α-Fe$_2$O$_3$ 膜表面的红外光谱演变以及二维相关分析

随着时间的延长,OmcA 在 α-Fe$_2$O$_3$ 膜上的信号逐渐增强直至稳定[图 3.3(a)],说明 OmcA 在较短时间内(15 min)达到吸附平衡。对图 3.3(a)进行 2D COS 分析,得到同步相关谱[图 3.3(b)]和异步谱[图 3.3(c)],进而根据同步相关峰和异步相关峰的正负性(表 3.1),得出键变化先后顺序为 1400 cm^{-1}→1542 cm^{-1}→1642 cm^{-1}(峰的归属见表 3.2),暗示着 OmcA 依次通过羧基基团、酰胺Ⅱ和酰胺Ⅰ与 α-Fe$_2$O$_3$ 发生相互作用。由于 OmcA 含有大量天冬氨酸和谷氨酸,它们可以作为游离羧基的供体与 α-Fe$_2$O$_3$ 发生第一步的相互作用。红外

图谱发现 α-Fe_2O_3 表面含有大量羟基[37],进一步推测 OmcA 支链游离的羧基与 α-Fe_2O_3 表面羟基形成氢键,从而引发进一步的相互作用。

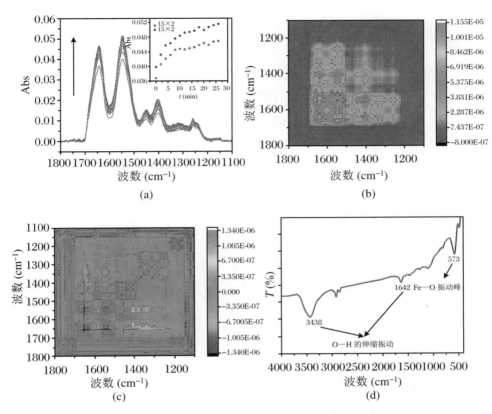

图 3.3 OmcA 在 α-Fe_2O_3 膜上红外信号随时间的演变:1642 cm^{-1} 和 1542 cm^{-1} 处峰强随时间的变化(a);基于图(a)的以时间为外部变量的同步 2D 谱(b);异步 2D 谱(c); α-Fe_2O_3 的 FTIR 信号(d)

表 3.1 同步 2D 谱和异步 2D 谱的相关峰的正负性

	~1642 cm^{-1}	~1542 cm^{-1}	~1400 cm^{-1}
~1642 cm^{-1}	+	+(−)	+(−)
~1542 cm^{-1}		+	+(−)
~1400 cm^{-1}			+

3.3.3
水溶液中外膜蛋白/$\alpha\text{-}Fe_2O_3$界面电子转移分子机制

纳米晶体结构 $\alpha\text{-}Fe_2O_3$ 从外膜蛋白活性中心卟啉铁捕获生物电子的过程可以用如下方程描述：

$$P\text{-}Fe^{2+} \longrightarrow P\text{-}Fe^{3+} + e^{-} \quad (3.1)$$

$$\alpha\text{-}Fe_2O_3 + xe^{-} \longrightarrow \alpha\text{-}Fe_{2-x}O_3 + xFe^{2+} \quad (3.2)$$

在方程式(3.1)和(3.2)描述的电子传递过程中，OM c-Cyts 的活性中心（P-Fe）能够通过 Fe(Ⅱ)到 Fe(Ⅲ)的氧化反应转移一个电子；而 $\alpha\text{-}Fe_2O_3$(001)晶面的 Fe(Ⅲ)获得电子进行还原反应。值得注意的是，$\alpha\text{-}Fe_2O_3$(001)晶面进行 Fe 的还原时，基元反应过程包括还原反应、离子溶出和金属单质形成的过程，且还原后的 Fe^{2+} 或继续还原为 Fe^0 可能会短时间地停留在(001)晶面上。由元胞结构可知，$\alpha\text{-}Fe_2O_3$ 中含有 6 个 O 配位的 Fe(Ⅲ)形成八面体结构[图3.4(a)]，而铁封端的(001)晶面 Fe(Ⅲ)都呈现不饱和配位的特征[图3.4(b)]，可结合水溶液中

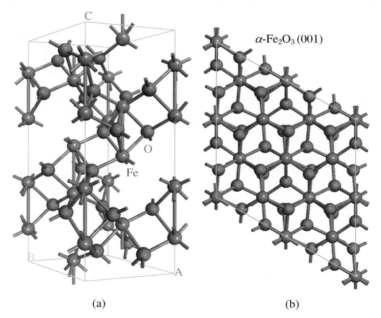

图3.4　$\alpha\text{-}Fe_2O_3$的晶体结构与表面原子排布：$\alpha\text{-}Fe_2O_3$的元胞结构，其中 Fe 与 6 个 O 配位形成八面体结构(a)；$\alpha\text{-}Fe_2O_3$的(001)晶面，Fe 原子呈现不饱和配位的状态(b)

的氢氧根或水分子降低表面能,由于可能存在不同的覆盖度,且在实际体系中无法预估,因此本章计算中暂未考虑这种情况,直接采用铁封端的(001)晶面作为接受电子的表面进行分子动力学模拟。

对构建的初始结构进行能量最小化及分子动力学模拟,发现外膜蛋白结构域Ⅱ在水溶液中运动并靠近(001)晶面[图3.5(a)],其中蛋白表面距离(001)晶面最近(2.4~3.3 Å)的氨基酸残基的侧链官能团为羟基与氨基[图3.5(b)和(c)],而这些氨基酸都是以终端活性中心血红素5的质心为圆心,半径为5 Å以

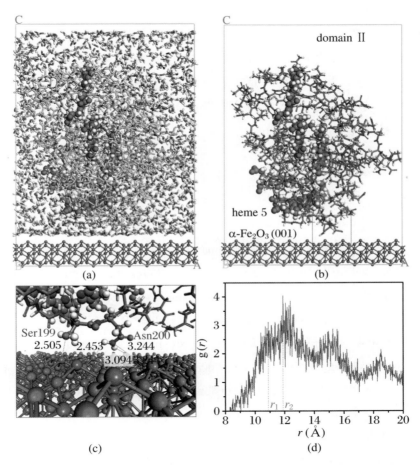

图3.5 外膜蛋白 MtrF 结构域Ⅱ与 α-Fe$_2$O$_3$(001)晶面电子转移体系的分子动力学模拟:水溶液中弛豫后的构型,高亮部分为结构域Ⅱ的肽链,其余为电子转移活性中心血红素的排布结构(a);去除水分子后的相对结构图(b);(b)中蓝框部分放大后的结构,结构域Ⅱ氨基酸官能团至 α-Fe$_2$O$_3$(001)晶面的距离,其中 Ser199 中的羟基 O 与晶面 Fe 之间的距离为 2.505 Å(c);活性中心 heme 与晶面之间的径向分布函数(d)

内的。实验测试结果显示,实际体系中有羟基以 O—Fe 配位的方式存在于 (001)晶面,而配位键的长度应小于两个原子的半径之和(1.97 Å),此时, α-Fe_2O_3(001)晶面上的吸附羟基足以与 heme 5 附近的氨基酸残基形成氢键作用。因此,氢键作用影响了界面附近肽链的结构及血红素 5 相对于其他血红素和电子受体晶面的取向,且缩短了 heme 5 与 α-Fe_2O_3(001)晶面的电子转移距离,为红外测试结果提供了分子水平的依据。为了进一步分析外膜蛋白活性位点与晶面之间的电子转移距离,通过点-点(site-site)之间的径向分布函数(RDFs)提供了出现概率最大的电子转移距离(最高峰)与晶面上接受电子的位点数量(峰个数)。在整个体系弛豫的过程中,电子转移终端活性位点 heme 5 最有可能出现在离晶面 r_1(10.9 Å)的位置处,且 α-Fe_2O_3(001)晶面存在多个接受电子转移的位点[图 3.5(d)],意味着两者之间存在充分的相互作用。因此,在水溶液中外膜蛋白与 α-Fe_2O_3 等金属氧化物可能存在一定的协同作用,从而影响 EAB 的胞外电子转移能力。

参考文献

[1] Ishii S, Suzuki S, Norden-Krichmar T M, et al. A novel metatranscriptomic approach to identify gene expression dynamics during extracellular electron transfer[J]. Nature Communications, 2013:4.

[2] Rittmann B E, Krajmalnik-Brown R, Halden R U. Pre-genomic, genomic and post-genomic study of microbial communities involved in bioenergy[J]. Nature Reviews Microbiology, 2008, 6 (8):604.

[3] Logan B E, Rabaey K. Conversion of wastes into bioelectricity and chemicals by using microbial electrochemical technologies[J]. Science, 2012, 337 (6095): 686-690.

[4] Weber K A, Achenbach L A, Coates J D. Microorganisms pumping iron: anaerobic microbial iron oxidation and reduction[J]. Nature Reviews Microbiology, 2006, 4(10):752.

[5] Nichols E M, Gallagher J J, Liu C. et al. Hybrid bioinorganic approach to solar-to-chemical conversion[J]. Proceedings of the National Academy of Sciences, 2015, 112(37):11461.

[6] Jeong H E, Kim I, Karam P, et al. Bacterial recognition of silicon nanowire arrays[J]. Nano Letters, 2013,13(6):2864-2869.

[7] White G F, Shi Z, Shi L, et al. Rapid electron exchange between surface-exposed bacterial cytochromes and Fe(Ⅲ) minerals[J]. Proceedings of the National Academy of Sciences of the United States of America, 2013, 110(16):6346-6351.

[8] Malvankar N S, Vargas M, Nevin K P, et al. Tunable metallic-like conductivity in microbial nanowire networks[J]. Nature Nanotechnology, 2011, 6(9):573.

[9] Chen J J, Chen W, He H, et al. Manipulation of microbial extracellular electron transfer by changing molecular structure of phenazine-type redox mediators[J]. Environmental Science & Technology, 2012,47(2):1033-1039.

[10] Pfeffer C, Larsen S, Song J, et al. Filamentous bacteria transport electrons over centimetre distances[J]. Nature, 2012,491(7423):218-221.

[11] Nielsen L P, Risgaard-Petersen N, Fossing H, et al. Electric currents couple spatially separated biogeochemical processes in marine sediment[J]. Nature, 2010,463(7284):1071-1074.

[12] Pirbadian S, Barchinger S E, Leung K M, et al. *Shewanella oneidensis* MR-1 nanowires are outer membrane and periplasmic extensions of the extracellular electron transport components[J]. Proceedings of the National Academy of Sciences of the United States of America, 2014,111(35):12883-12888.

[13] Leys D, Meyer T E, Tsapin A S, et al. Crystal structures at atomic resolution reveal the novel concept of 'electron-harvesting' as a role for the small tetraheme cytochrome c[J]. The Journal of Biological Chemistry, 2002, 277(38):35703-35711.

[14] Richardson D J, Butt J N, Fredrickson J K, et al. The 'porin-cytochrome' model for microbe-to-mineral electron transfer[J]. Molecular Microbiology, 2012,85(2):201-212.

[15] Shaik S. Biomimetic chemistry: Iron opens up to high activity[J]. Nature Chemistry, 2010, 2(5):347.

[16] Wang F, Gu Y, O'Brien J P, et al. Structure of microbial nanowires reveals stacked hemes that transport electrons over micrometers[J]. Cell, 2019, 177(2):361-369.

[17] Lovley D R. Live wires: direct extracellular electron exchange for bioenergy and the bioremediation of energy-related contamination[J]. Energy & Environmental Science, 2011,4(12):4896-4906.

[18] Allen C C, Westall F, Schelble R T. Importance of a martian hematite site for astrobiology[J]. Astrobiology, 2001,1(1):111-123.

[19] Lohaus C, Klein A, Jaegermann W. Limitation of Fermi level shifts by polaron defect states in hematite photoelectrodes[J]. Nature Communications, 2018,9(1):4309.

[20] Jiang Z, Liu Q, Dekkers M J, et al. Control of earth-like magnetic fields on the transformation of ferrihydrite to hematite and goethite[J]. Scientific Reports, 2016,6:30395.

[21] Sheng A, Liu F, Shi L, et al. Aggregation kinetics of hematite particles in the presence of outer membrane cytochrome OmcA of *Shewanella oneidenesis* MR-1[J]. Environmental Science & Technology, 2016, 50(20): 11016-11024.

[22] Clarke T A, Edwards M J, Gates A J, et al. Structure of a bacterial cell surface decaheme electron conduit[J]. Proceedings of the National Academy of Sciences of the United States of America, 2011,108(23):9384-9389.

[23] Byun H S, Pirbadian S, Nakano A, et al. Kinetic monte carlo simulations and molecular conductance measurements of the bacterial decaheme cytochrome MtrF[J]. Chemelectrochem, 2014,1(11):1932-1939.

[24] Watanabe H C, Yamashita Y, Ishikita H. Electron transfer pathways in a multiheme cytochrome MtrF[J]. Proceedings of the National Academy of Sciences of the United States of America, 2017,114(11):2916-2921.

[25] Breuer M, Zarzycki P, Blumberger J. et al. Thermodynamics of electron flow in the bacterial deca-heme cytochrome MtrF[J]. Journal of the American Chemical Society, 2012,134(24):9868-9871.

[26] Kerisit S, Rosso K M, Dupuis M, et al. Molecular computational investigation of electron-transfer kinetics across cytochrome-iron oxide interfaces[J]. Journal of Physical Chemistry C, 2007,111(30):11363-11375.

[27] Essmann U, Perera L, Berkowitz M L, et al. A smooth particle mesh Ewald method[J]. The Journal of Physical Chemistry, 1995,103(19):8577-8593.

[28] Andersen H C. Molecular dynamics simulations at constant pressure and/or temperature[J]. The Journal of Physical Chemistry, 1980, 72(4): 2384-2393.

[29] Johs A, Shi L, Droubay T, et al. Characterization of the decaheme c-type cytochrome OmcA in solution and on hematite surfaces by small angle X-ray

scattering and neutron reflectometry[J]. Biophysical Journal, 2010, 98(12): 3035-3043.

[30] Xiong Y J, Shi L, Chen B W, et al. High-affinity binding and direct electron transfer to solid metals by the *Shewanella oneidensis* MR-1 outer membrane c-type cytochrome OmcA[J]. Journal of the American Chemical Society, 2006, 128(43): 13978-13979.

[31] Eggleston C M, Voros J, Shi L, et al. Binding and direct electrochemistry of OmcA, an outer-membrane cytochrome from an iron reducing bacterium, with oxide electrodes: A candidate biofuel cell system[J]. Inorganica Chimica Acta, 2008, 361(3): 769-777.

[32] Speare J O, Rush III T S. IR spectra of cytochrome c denatured with deuterated guanidine hydrochloride show increase in β sheet[J]. Biopolymers, 2003, 72(3): 193-204.

[33] Dong A, Huang P, Caughey W S. Protein secondary structures in water from second-derivative amide I infrared spectra[J]. Biochemistry, 1990, 29(13): 3303-3308.

[34] Balakrishnan G, Hu Y, Oyerinde O F, et al. conformational switch to β-sheet structure in cytochrome c leads to heme exposure. Implications for cardiolipin peroxidation and apoptosis[J]. Journal of the American Chemical Society, 2007, 129(3): 504-505.

[35] Lower B H, Lins R D, Oestreicher Z, et al. In vitro evolution of a peptide with a hematite binding motif that may constitute a natural metal-oxide binding archetype[J]. Environmental Science & Technology, 2008, 42(10): 3821-3827.

[36] Edwards M J, Baiden N A, Johs A, et al. The X-ray crystal structure of *Shewanella oneidensis* OmcA reveals new insight at the microbe-mineral interface[J]. FEBS Letters, 2014, 588(10): 1886-1890.

[37] Farahmandjou M, Soflaee F. Synthesis and characterization of α-Fe_2O_3 nanoparticles by simple co-precipitation method[J]. Physical Chemistry Chemical Physics, 2015, 3(3): 191-199.

第 4 章

EAB 外膜细胞色素 c 与氧化石墨烯的界面电子传递机制

4.1 胞外电子传递链与石墨烯材料相互作用的可能性

微生物燃料电池(microbial fuel cell，MFC) EAB 回收废水有机物中的化学能[1-3]，在这个能量转化中，EAB/电极界面的电子转移起着关键作用[4-6]。EET 通过连续的多个氧化还原反应实现电子的跨膜转移，最终将电子传递给胞外电子受体，其机制分为直接传递和间接传递。直接 EET 主要通过金属氧化还原蛋白[7]或者导电纳米鞭毛来实现[8]；间接 EET 主要通过核黄素等氧化还原媒介分子来实现电子在 EAB 和电子受体之间的转移。在上述 EET 途径中，含有多个多血红素的 OM c-Cyts 是必不可少的：① 它可以直接与电子受体物理接触；② 它可以起到桥梁的作用，将电子从上游蛋白传递到下游氧化还原媒介。有研究报道，通过电化学手段来调控 OM c-Cyts 与氧化还原媒介如核黄素的相互作用，可提高 MFC 的产电[9-11]。在 MFC 体系中，生物/无机界面(EAB/电极)的有效电子传递对 MFC 在环境修复和能量生产的应用至关重要。

同时，石墨烯因其优异的导电性、稳定性以及生物相容性等性质而成为一种有前景的二维材料，目前已被广泛应用于包括 MFC 在内的能量转化和储存领域。例如，石墨烯巨大的比表面积可增强电极对细胞的黏附，从而使 MFC 的产电提高了 25 倍[12]。石墨烯 sp^2 杂化碳原子产生的共轭结构赋予电极良好的导电性，能够减少能量损耗，促进 EAB 和电极界面的电子转移[13]。除此之外，以 *Shewanella* 为代表的 EAB 在生理条件下能把电子传递给氧化石墨烯(graphene oxide, GO)，原位生成石墨烯，并形成导电的生物无机网络，促进电子的转移[12,14]。通过蛋白敲除实验揭示了直接与 GO 接触的 OM c-Cyts(OmcA，MtrC 等)在还原时扮演的重要角色[15-16]。

尽管如此，现在对从 OM c-Cyts 到 GO 的电子转移机制仍然知之甚少。为了能够调控 EET 并从废水有机物中回收更多的能量，更好地解析生物/无机界面的 EET 显得十分重要。迄今为止，各领域的研究者通过分子建模、体内、体外实验对电子从 OM c-Cyts 到铁锰矿物的转移行为做了大量的研究[17-23]：借助于荧光相关光谱[21]、中子反射技术[24]、原子力显微镜[25]，观察到 OM c-Cyts 在赤铁矿表面的吸附结合行为会改变蛋白结构，形成利于电子传递的构型，其中

OmcA 对赤铁矿的亲和力强于 MtrC[25]。OmcA 通过 heme 10 附近的保守肽链 Ser/Thr-Pro-Ser/Thr 与羟基化赤铁矿表面形成氢键[26],进而得到 0.11 s^{-1} 的表观 EET 速率[21]。通常来说,以可溶性金属复合物作电子受体的电子转移速率要快于金属氧化物固体[27]。电子从 c-Cyts 到铁氧化物的分子模拟结果表明,电子转移速率易受电子供体-受体的距离、卟啉环的构型以及表面 termination 等因素的影响[17]。因此,EET 速率在很大程度上受电子受体理化性质的影响。

在本章工作中,GO 被选作胞外电子受体,为研究其可能的 EET 分子机制,OmcA 被提取纯化,并在不同条件下进行圆二色谱(circular dichroism,CD)、红外(attenuated total reflectance Fourier transform infrared spectroscopy,ATR-FTIR)以及二维相关分析(two-dimensional correlation analysis,2D COS)表征,期望能够实时监测 OmcA 将电子转移给 GO 之前两者相互作用过程中的结构变化。该研究的意义一方面在于可以拓展、完善对 EET 过程的理解;另一方面启发研究人员可以通过调控相互作用过程来调节 EET 促进产电。可以结合基因工程和蛋白质工程对 EAB 或者外膜蛋白进行改造,甚至可以实现纳米材料生物合成的定向选择进化[28-30]。

4.2 细胞色素的提纯、氧化石墨烯制备与表征

4.2.1

OmcA 的表达和提纯

将 *Shewanella* MR-1 3332 菌种转接到无菌 LB 培养基中,在 30 ℃下培养至 OD600 为 0.6 左右,加入终浓度为 1 mmol·L^{-1} 的阿拉伯糖进行诱导表达。14 h 后 8000 g 离心 30 min 分离细菌和上清。Ni 柱(Ni Sepharose GE Healthcare

Life Sciences,USA)用至少 3 倍柱体积的 buffer B(20 mmol·L^{-1} Hepes,pH 7.8,10%甘油,10 mmol·L^{-1} 咪唑)预平衡。将离心后的上清液和预平衡的 Ni 柱混合,4 ℃下缓速搅拌 2 h。重新填装 Ni 柱并用 5 倍柱体积的洗脱液 buffer C (20 mmol·L^{-1} Hepes, pH 7.8,10%甘油,250 mmol·L^{-1} 咪唑)洗脱蛋白。目标蛋白组分通过超滤管浓缩并去除咪唑,−80 ℃冷冻保存备用。

4.2.2
蛋白膜循环伏安法

热解石墨(pyrolytic electrode,PG)电极(武汉高仕睿联公司)用 Al_2O_3 打磨,在水中超声清洗 3 次。将表面洁净的 PG 电极浸泡在 4 ℃的 OmcA 溶液中(20 mmol·L^{-1} Hepes,pH 7.4),24 h 后拿出轻轻润洗即可得 OmcA 修饰的电极(OmcA/PG)。在三电极体系下用上海辰华工作站 660C 进行循环伏安扫描 (cyclic voltammetry,CV):Ag/AgCl 为参比电极,Pt 丝为对电极,OmcA/PG 为工作电极,电势窗口为 −0.8~0 V,扫速为 100 mV·s^{-1},电解质为 50 mmol·L^{-1} Hepes(pH 7.4)。CV 均在厌氧条件下进行,温度为 25 ℃。

4.2.3
基于驻流分光光度计的动力学实验

测试前所有溶液均通氮气除氧。纯化的 OmcA 通过逐滴滴加连二亚硫酸钠来还原,同时使用 UV-Vis 实时监测 OmcA 的还原程度。采用配有二极管阵列检测器的驻流分光光度计(Stopped-Flow spectrophotometry,SFS;SX20,英国应用光物理公司)来监测电子转移动力学反应。其死时间是 20 ms。每次测试前,用厌氧的缓冲盐(Hepes,pH 7.5)冲洗管路以保证严格的厌氧条件。预实验结果表明没有底物(GO)时,OmcA 不能被氧化,证明了 SFS 在实验期间能维持良好的厌氧环境。等体积的 OmcA 和 GO 溶液经压缩氮气驱动后,在 2 mm 光程的光谱池混合,每隔一定时间监测 550 nm 的吸收变化(低浓度 GO 时每 2 min 采集一次数据,高浓度时每 0.05 s 采集一次数据)。反应温度为 25 ℃。采集的数据点通过拟合得到反应速率常数。

4.2.4
OmcA 的二级结构表征

OmcA 的二级结构用 CD(中国科大生命科学实验中心)来表征,扫描范围为 190～250 nm,扫速为 100 nm·min^{-1},光谱池为 1 mm 光程的石英比色皿。样品为 50 mg·L^{-1} 蛋白溶液,缓冲液为 2 mmol·L^{-1} Hepes(pH 7.0)。电压小于 400 mV 时方可采集样品信号。扣除缓冲盐背景后,每个样品扫描 3 次取平均值,原始谱图用 Jascow 32 软件进行定量分析。

4.2.5
OmcA/GO 相互作用过程的红外监测及二维相关分析

ATR-FTIR 用带有碲化汞镉检测器的 Vertex 70 光谱仪(德国布鲁克公司)记录,检测器使用前需用液氮冷却至 77 K,红外晶体为 ZnSe。将 800 μL GO 溶液(2 g·L^{-1})滴涂在 ZnSe 晶体表面,自然干燥。GO 膜在缓冲盐中平衡 2 h,采集光谱作为背景,理论上是包括 ZnSe、GO 膜、缓冲盐和 H_2O 4 种成分的混合信号。将缓冲盐替换为 15 μmol·L^{-1} 的 OmcA 溶液(0.6 mL),每 3 min 采集一次红外光谱,扫描范围为 1800～1000 cm^{-1},扫速为 4 cm^{-1}。所获得的红外光谱经基线校正、平滑、去背景后用 2D COS 分析(时间为外部变量),分析软件为 2D Shige (Kwansei-Gakuin 大学,日本)。根据同步相关峰 $\Phi(\nu_1,\nu_2)$ 和异步相关峰 $\Psi(\nu_1,\nu_2)$ 的正负来判定键 ν_1 和 ν_2 的变化顺序:当 $\Phi(\nu_1,\nu_2)$ 和 $\Psi(\nu_1,\nu_2)$ 同时为正或同时为负时,键 ν_1 先于 ν_2 发生变化;当 $\Phi(\nu_1,\nu_2)$ 和 $\Psi(\nu_1,\nu_2)$ 正负号相反时,键 ν_2 先于 ν_1 发生变化;当 $\Psi(\nu_1,\nu_2)$ 为 0 时,键 ν_1 和 ν_2 同时发生变化[31]。

4.3 OmcA 蛋白与 GO 之间的电子转移过程解析

4.3.1 OmcA 和 GO 的光谱和电化学特性

氧化态 OmcA 的紫外光谱在 410 nm 处表现出 Soret 吸收带（γ 带），在 525 nm 处表现出 Q 吸收带，展现了轴向两个组氨酸配位的低自旋血红素的光谱特征[32]。还原后，γ 带红移到 420 nm，Q 带分裂成更加尖锐的 α 带和 β 带[图 4.1(a)]。这些紫外光谱特征表明提纯的 OmcA 含有 6 配位且低自旋的血红素[33]。OmcA 的 A_{410}/A_{280}（表征血红素蛋白纯度的指数）比值为 4.8，说明蛋白的纯度很高[34]，这一结果与 SDS-PAGE 的结论一致。在 SDS-PAGE 上只能清晰地观察到分子量为 85 kDa 的条带[图 4.1(b)]，说明 OmcA 的纯度在 95% 以上[35]。在 CV 测试里，OmcA 表现出一对氧化还原峰：出现在 −0.3 V 的明显的还原峰和出现在 −0.2 V 的弱的氧化峰[图 4.1(c)]归属于蛋白结构中 10 个血红素的连续氧化还原[36]。

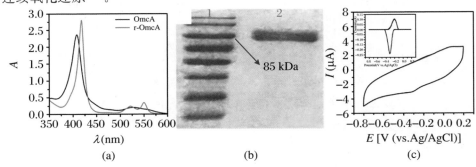

图 4.1　外膜蛋白 OmcA 的性质：氧化态 OmcA（黑色）和还原态 OmcA（r-OmcA，红色）的紫外可见光谱(a)；OmcA 的 SDS-PAGE，通道 1 是标记物，通道 2 是 OmcA 样品(b)；OmcA 在 PG 表面的 CV 曲线，缓冲液为 20 mmol·L^{-1} Hepes（pH 7.8），扫速为 100 mV·s^{-1}，插图是将背景电流扣除后的 CV 曲线(c)

GO 一方面以 PG 为工作电极时能在 -0.7 V 甚至更低的电位下被还原[37]，另一方面也能被 OM c-Cyts 还原，然而电化学活性细菌 OM c-Cyts 的电势在 -0.5 V 到 -0.2 V 之间[36]，因此，理论上电子从 OM c-Cyts 到 GO 的转移是热力学不利的。为探究这一有意思的现象，对两者的混合体系分别进行了紫外、Raman 等光谱表征和 CV 表征。将 GO 滴加到还原态 OmcA 溶液之后，OmcA 的 γ 带从 420 nm 跃迁回 410 nm，伴随着 α 和 β 带的消失以及 Q 带的出现，将 GO 的吸收扣除之后，最终光谱和氧化态 OmcA 的光谱一致[图 4.2(a)]，该结果说明电子从还原态 OmcA 发生转移。Raman 光谱可非破坏性地表征碳材料，因此，在本研究中被用于检测 GO 的结构变化。GO 表现出碳材料特征的 G 带（~1600 cm^{-1}）和 D 带（~1355 cm^{-1}）[图 4.2(b)]，其中 G 带来源于 sp^2 杂化的石墨化碳原子，D 带归结于无序态碳原子[38]。和还原态 OmcA 混合后，G 带迁移至~1585 cm^{-1}，意味着缺陷碳原子恢复了六边形网络结构[39]。除此之外，D 带和 G 带的比值略有上升，意味着 GO 缺陷度升高[40]，可能是由于 OmcA 在 GO 表面的吸附造成的。GO 的 Raman 变化说明 GO 被还原态 OmcA 的电子还原。

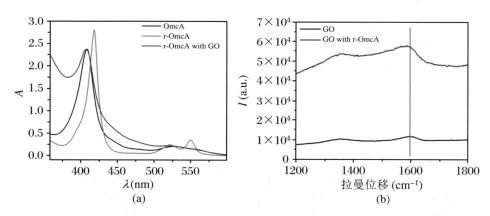

图 4.2　外膜蛋白 OmcA 与 GO 混合后光谱变化：不同状态 OmcA 的紫外可见光谱(a)。黑色：氧化态 OmcA；红色：还原态 OmcA(r-OmcA)；蓝色：r-OmcA 与 0.5 mL 2 g·L^{-1} GO 混合；不同状态 GO 的 Raman 光谱(b)。黑色：纯 GO；蓝色：r-OmcA 与 0.5 mL 2 g/L GO 混合后

4.3.2
OmcA/GO 界面电子传递的动力学过程

将 OmcA 修饰在 PG 电极表面,进行蛋白膜伏安分析。当 GO 存在时,还原电流随着 GO 浓度增加而提高[图 4.3(a)],而裸 PG 电极的电流在相同条件下保持恒定,不随着 GO 的浓度变化而变化[图 4.3(b)]。该对比表明电子经 OmcA 转运后才能传递给 GO,证明 OmcA 与 GO 接触后发生了某些变化才能实现电子的转移。通过对电流进一步定量分析发现,电流随着 GO 浓度增加而逐渐趋于稳定[图 4.3(c)],符合米氏方程,表现出酶促反应的特征[41]。

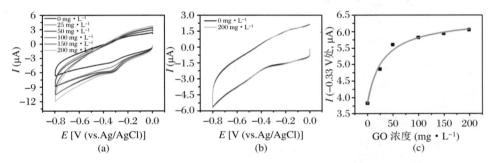

图 4.3 GO 在不同电极上的 CV 响应:不同浓度 GO 在 OmcA/PG 电极上的 CV 信号(a);不同浓度 GO 在裸 PG 电极上的 CV 信号(b);两种电极条件的扫速均为 100 mV·s^{-1},-0.33 V 处电流与 GO 浓度的米氏方程拟合(c)

由于伏安法是一个动态过程,而米氏方程的基础是一个稳态条件,因此,电化学方法并不能精确描述电子转移动力学过程。鉴于此,采用了 SFS 来研究厌氧条件下 OmcA/GO 界面的电子转移动力学。

将 60 mg·L^{-1} GO 与 OmcA 以 1∶1 体积混合之后,α 带和 β 带吸收减小,γ 带从 420 nm 蓝移到 410 nm[图 4.4(a)],说明蛋白被氧化,电子转移给 GO。进一步拟合发现 550 nm 吸收表现出准一级反应特征,电子转移速率为 0.37 min^{-1}。如图 4.4(b)和(c)所示,当 OmcA 浓度一定时,反应速率常数与 GO 浓度呈正线性关系。根据米氏方程,当酶浓度一定时,反应速率有一个上限[42]。如图 4.4(d)所示,当 GO 浓度为 90 mg·L^{-1} 时,反应速率为 0.53 s^{-1},浓度为 100 mg·L^{-1} 时反应速率为 0.56 s^{-1},两者非常接近,可以佐证 OmcA 与 GO 之间的电子传递符合米氏方程,遵循酶促反应特征。有研究报道,OmcA 与赤铁矿的电子传递速率为 0.11 s^{-1}[21],MtrC 与赤铁矿的电子传递速率为 0.26 s^{-1}[43]。相较而言,这些速率常数是可比的,然而 OmcA、MtrC 与赤铁矿的电子转移遵循一级反应动力学。当电子受体为针铁矿以及铁的配合物时,电子转移遵循二级反应动力学[27]。这些结果说明 EET 在很大程度上与电子受体的种类和存在形式相关。

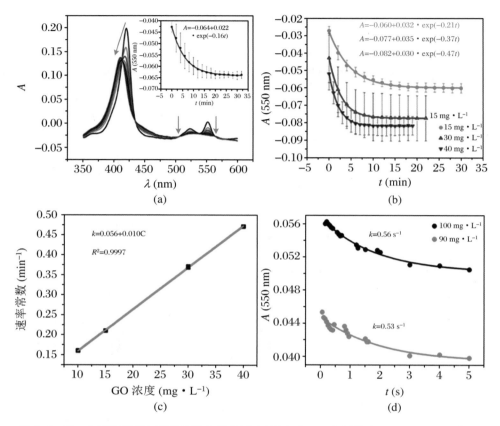

图 4.4 OmcA/GO 界面电子转移动力学分析:还原态 OmcA 与等体积 GO(20 mg·L^{-1})混合后随时间变化的紫外光谱(a);采样间隔为 2 min,红色箭头表示对应吸收峰的变化方向,插图表示对 550 nm 吸收峰随时间变化曲线的拟合,不同 GO 浓度下还原态蛋白在 550 nm 吸收峰随时间变化曲线的拟合(b);不同浓度下准一级电子转移速率常数与 GO 浓度的线性拟合(c);100 mg·L^{-1} 和 90 mg·L^{-1} GO 浓度下电子转移速率常数的对比(d)

这些结果证明,当以电极或者还原剂为电子供体时,即使 OmcA 的氧化还原电势比 GO 还原的起始电势要高,OmcA 在体外仍可以将电子传递给 GO。与之相似,OmcA 同样可以将电子转移给赤铁矿,即使赤铁矿的氧化还原电势为 −430 mV(pH 7.0)[44],该电子转移主要归因于 OmcA 的 C 段肽链(Ser/Thr-Pro-Ser/Thr)与赤铁矿之间形成了氢键,降低了反应能垒[25-26]。据此推测,OmcA 与 GO 之间可能形成了某种相互作用,如氢键,从而促进了电子转移的发生。

4.3.3
OmcA 二级结构的变化

为进一步研究电子转移机制,采用了 CD 来研究蛋白的二级结构及其变化(图 4.5)。OmcA 在 205 nm 和 220 nm 处有两个负峰,意味着蛋白存在 α 螺旋结构。经定量分析发现 OmcA 二级结构包括 α 螺旋、β 折叠、β 转角以及无序结构,各组分含量如表 2.1 所示。根据这些组分含量,可将 OmcA 划分为 α+β 蛋白,与文献报道一致[45]:OmcA 的结构域Ⅰ和结构域Ⅲ主要是 β 结构,结构域Ⅱ和结构域Ⅳ富含 α 结构[46]。尽管如此,OmcA 含有 40% 的无序结构,说明 OmcA 的结构比较松散。当 OmcA 被还原后,β 转角转换为 α 螺旋、β 折叠以及无序结构(表 4.1),该结构变化同样被小角 X 射线散射所证明,说明当蛋白从氧化态变成还原态时,蛋白整体长度缩短了 7 Å[24]。氧化态 OmcA 和 GO 混合后,依然保持着原来的二级结构,对于还原态 OmcA,混合后其 β 转角结构会大幅度减少并转化为 α 螺旋和无序结构。这些结构变化可能是由于活性中心位置的 heme 平面发生重构引起的,从而使得 heme 更利于与底物结合,促进电子传

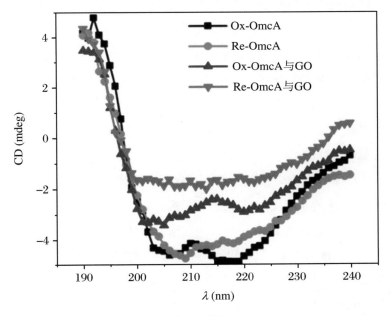

图 4.5　OmcA 在不同状态下的 CD 谱图

　　Ox-OmcA:氧化态 OmcA;Re-OmcA:还原态 OmcA;Ox-OmcA 与 GO:与 GO 混合后的氧化态 OmcA;Re-OmcA 与 GO:与 GO 混合后的还原态 OmcA

递[47]。考虑到蛋白的初始结构以及结构变化，推测 OmcA 通过结构域Ⅰ或者结构域Ⅲ与 GO 结合。由于 OmcA 可通过结构域Ⅱ的 heme 5 形成二倍体[46]，因此，结构域Ⅰ或者Ⅲ可通过分子间作用力与 GO 结合。

表 4.1　OmcA 在不同状态下的二级结构各组分含量

	α 螺旋	β 转角	β 折叠	无序结构
Ox-OmcA	0.114	0.49	0	0.396
Re-OmcA	0.179	0.252	0.115	0.454
Ox-OmcA 与 GO	0.142	0.4	0	0.458
Re-OmcA 与 GO	0.325	0.059	0	0.617

4.3.4

OmcA/GO 界面电子传递的分子机制

为了解 OmcA/GO 界面的可能作用机制，分析了结构域Ⅰ和Ⅲ在 heme 附近的氨基酸。位于 heme 5 附近结构域Ⅱ上的酪氨酸残基 374 去质子化后与另一个 OmcA 分子的 Tyr374 残基形成氢键[5, 46]。在这种构象下，结构域Ⅰ（heme 2 附近）或者结构域Ⅲ（heme 7 附近）上的氨基酸可形成一个暴露于溶剂环境的袋形空间，用于结合底物，传递电子[图 4.6(a)]。例如，距离 heme 20 Å 且位于结构域Ⅰ上的氨基酸残基（苯丙氨酸 141、谷氨酸 124、脯氨酸 125、苏氨酸 127、酪氨酸 301）可能提供与 GO 的结合位点[图 4.6(b)]。谷氨酸 124 的羧基或者酪氨酸 301 的羟基可能与 GO 的含氧官能团（羟基、环氧基）形成氢键。在结构域Ⅲ，暴露的官能团，包括羟基（酪氨酸 436、酪氨酸 460、苏氨酸 595）、羧基（天冬氨酸 462 和天冬氨酸 596）、胍基（精氨酸 459）以及氨基（赖氨酸 594）参与了 GO 表面氢键的形成[图 4.6(c)]。因此，氢键作为 OmcA 和 GO 之间的主要作用力，强化了两者之间的相互作用。此外，苯丙氨酸和酪氨酸的苯环结构可部分与 GO 的 sp^2 网络共轭，可进一步缩短电子转移距离。因此，结构分析的结果表明电子的传递是通过结构Ⅰ或Ⅲ与 GO 的相互作用（氢键）实现的。然而，在细胞外膜上，OmcA 的构型需要慎重考虑，因为它能够与 MtrC 形成一个稳定的复合体被固定在细胞外膜上[32]。

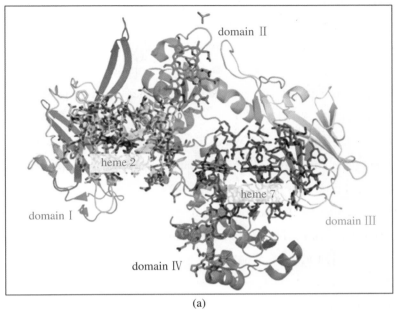

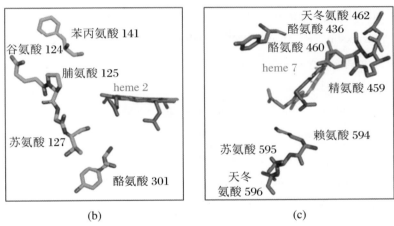

图 4.6 OmcA 的晶体结构和 GO 可能的结合位点：OmcA 的晶体结构（a）（PDB：4LMH），含有 4 个结构域和 10 个血红素；heme 2 附近（小于 10 Å）位于结构域Ⅰ上的带有功能基团的氨基酸残基（b）；heme 7 附近（小于 10 Å）位于结构域Ⅲ上的带有功能基团的氨基酸残基（c）

参考文献

［1］ Logan B E，Rabaey K. Conversion of wastes into bioelectricity and chemicals by using microbial electrochemical technologies［J］. Science，2012，337

(6095): 686-690.

[2] He Z. Development of microbial fuel cells needs to go beyond 'power density' [J]. ACS Energy Letters, 2017, 2 (3): 700-702.

[3] Li W W, Yu H Q, He Z. Towards sustainable wastewater treatment by using microbial fuel cells centered technologies [J]. Energy & Environmental Science, 2014, 7 (3): 911-924.

[4] Hasan K, Grattieri M, Wang T, et al. Enhanced bioelectrocatalysis of *Shewanella oneidensis* MR-1 by a naphthoquinone redox polymer [J]. ACS Energy Letters, 2017, 2 (9): 1947-1951.

[5] Shi L, Dong H, Reguera G, et al. Extracellular electron transfer mechanisms between microorganisms and minerals [J]. Nature Reviews Microbiology, 2016, 14 (10): 651-662.

[6] Zhao S, Li Y, Yin H, et al. Three-dimensional graphene/Pt nanoparticle composites as freestanding anode for enhancing performance of microbial fuel cells [J]. Science Advances, 2015, 1 (10): e1500372.

[7] Fredrickson J K, Romine M F, Beliaev A S, et al. Towards environmental systems biology of *Shewanella* [J]. Nature Reviews Microbiology, 2008, 6 (8): 592-603.

[8] Subramanian P, Pirbadian S, El-Naggar M Y, et al. Ultrastructure *Shewanella oneidensis*/em> MR-1 nanowires revealed by electron cryotomography [J]. Proceedings of the National Academy of Sciences of the United States of America, 2018, 115 (14): E3246.

[9] Okamoto A, Hashimoto K, Nealson K H, et al. Rate enhancement of bacterial extracellular electron transport involves bound flavin semiquinones [J]. Proceedings of the National Academy of Sciences of the United States of America, 2013, 110 (19): 7856-7861.

[10] Li F, Li Y X, Cao Y X, et al. Modular engineering to increase intracellular NAD(H/+) promotes rate of extracellular electron transfer of *Shewanella oneidensis* [J]. Nature Communications, 2018, 9 (1): 3637.

[11] Okamoto A, Saito K, Inoue K, et al. Uptake of self-secreted flavins as bound cofactors for extracellular electron transfer in *Geobacter* species [J]. Energy & Environmental Science, 2014, 7 (4): 1357-1361.

[12] Yong Y C, Yu Y Y, Zhang X, et al. Highly active bidirectional electron transfer by a self-assembled electroactive reduced graphene oxide hybridized

biofilm [J]. Angewandte Chemie International Edition, 2014, 53 (17): 4480-4483.

[13] Xie X, Yu G, Liu N, et al. Graphene-sponges as high-performance low-cost anodes for microbial fuel cells [J]. Energy & Environmental Science, 2012, 5 (5): 6862-6866.

[14] Shi T, Hou X, Guo S, et al. Nanohole-boosted electron transport between nanomaterials and bacteria as a concept for nano-bio interactions [J]. Nature Communications, 2021, 12 (1): 493.

[15] Jiao Y, Qian F, Li Y, et al. Deciphering the electron transport pathway for graphene oxide reduction by *Shewanella oneidensis* MR-1 [J]. Journal of Bacteriology, 2011, 193 (14): 3662-3665.

[16] Yang C, Aslan H, Zhang P, et al. Carbon dots-fed Shewanella oneidensis MR-1 for bioelectricity enhancement [J]. Nature Communications, 2020, 11 (1): 1379.

[17] Yu Y Y, Wang Y Z, Fang Z, et al. Single cell electron collectors for highly efficient wiring-up electronic abiotic/biotic interfaces [J]. Nature Communications, 2020, 11 (1): 4087.

[18] Shi L, Dong H, Reguera G, et al. Extracellular electron transfer mechanisms between microorganisms and minerals [J]. Nature Reviews Microbiology, 2016, 14 (10): 651-662.

[19] Breuer M, Zarzycki P, Blumberger J, et al. Thermodynamics of electron flow in the bacterial deca-heme cytochrome MtrF [J]. Journal of the American Chemical Society, 2012, 134 (24): 9868-9871.

[20] Watanabe H C, Yamashita Y, Ishikita H. Electron transfer pathways in a multiheme cytochrome MtrF [J]. Proceedings of the National Academy of Sciences of the United States of America, 2017, 114 (11): 2916-2921.

[21] Xiong Y J, Shi L, Chen B W, et al. High-affinity binding and direct electron transfer to solid metals by the *Shewanella oneidensis* MR-1 outer membrane c-type cytochrome OmcA [J]. Journal of the American Chemical Society, 2006, 128 (43): 13978-13979.

[22] White G F, Shi Z, Shi L, et al. Rapid electron exchange between surface-exposed bacterial cytochromes and Fe (III) minerals [J]. Proceedings of the National Academy of Sciences of the United States of America, 2013, 110 (16): 6346-6351.

[23] Breuer M, Zarzycki P, Shi L, et al. Molecular structure and free energy landscape for electron transport in the decahaem cytochrome MtrF [J]. Biochemical Society Transactions., 2012, 40: 1198-1203.

[24] Johs A, Shi L, Droubay T, et al. Characterization of the decaheme c-type cytochrome OmcA in solution and on hematite surfaces by small angle X-ray scattering and neutron reflectometry [J]. Biophys. J., 2010, 98 (12): 3035-3043.

[25] Lower B H, Shi L, Yongsunthon R, et al. Specific bonds between an iron oxide surface and outer membrane cytochromes MtrC and OmcA from *Shewanella oneidensis* MR-1 [J]. Journal of Bacteriology, 2007, 189 (13): 4944-4952.

[26] Lower B H, Lins R D, Oestreicher Z, et al. In vitro evolution of a peptide with a hematite binding motif that may constitute a natural metal-oxide binding archetype [J]. Environmental Science & Technology, 2008, 42 (10): 3821-3827.

[27] Wang Z M, Liu C X, Wang X L, et al. Kinetics of reduction of Fe(Ⅲ) complexes by outer membrane cytochromes MtrC and OmcA of *Shewanella oneidensis* MR-1 [J]. Applied and Environmental Microbiology, 2008, 74 (21): 6746-6755.

[28] Tian L J, Li W W, Zhu T T, et al. Directed biofabrication of nanoparticles through regulating extracellular electron transfer [J]. Journal of the American Chemical Society, 2017, 139 (35): 12149-12152.

[29] Wargacki A J, Wörner T P, Van de Waterbeemd M, et al. Complete and cooperative in vitro assembly of computationally designed self-assembling protein nanomaterials [J]. Nature Communications, 2021, 12 (1): 883.

[30] Tu Y M, Song W, Ren T, et al. Rapid fabrication of precise high-throughput filters from membrane protein nanosheets [J]. Nature Materials, 2020, 19 (3): 347-354.

[31] Chen W, Qian C, Liu X Y, et al. Two-dimensional correlation spectroscopic analysis on the interaction between humic acids and TiO_2 nanoparticles [J]. Environmental Science & Technology, 2014, 48 (19): 11119-11126.

[32] Shi L, Chen B, Wang Z, et al. Isolation of a high affinity functional protein complex between OmcA and MtrC: two outer membrane decaheme c-type cytochromes of Shewanella oneidensis MR-1 [J]. Journal of Bacteriology,

2006, 188 (13): 4705-4714.

[33] Qian X, Mester T, Morgado L, et al. Biochemical characterization of purified OmcS, a c-type cytochrome required for insoluble Fe (Ⅲ) reduction in Geobacter sulfurreducens [J]. Biochemical and BioPhysical Research Communications, 2011, 1807 (4): 404-412.

[34] Anderson J L R, Armstrong C T, Kodali G, et al. Constructing a man-made c-type cytochrome maquette in vivo: electron transfer, oxygen transport and conversion to a photoactive light harvesting maquette [J]. Chemical Science, 2014, 5 (2): 507-514.

[35] Ross D E, Brantley S L, Tien M. Kinetic characterization of OmcA and MtrC, terminal reductases involved in respiratory electron transfer for dissimilatory iron reduction in *Shewanella oneidensis* MR-1 [J]. Applied and Environmental Microbiology, 2009, 75 (16): 5218-5226.

[36] Firer-Sherwood M, Pulcu G S, Elliott S J. Electrochemical interrogations of the Mtr cytochromes from *Shewanella*: opening a potential window [J]. The Journal of Biological Chemistry, 2008, 13 (6): 849-854.

[37] Guo H L, Wang X F, Qian Q Y, et al. A green approach to the synthesis of graphene nanosheets [J]. ACS Nano, 2009, 3 (9): 2653-2659.

[38] Kudin K N, Ozbas B, Schniepp H C, et al. Raman spectra of graphite oxide and functionalized graphene sheets [J]. Nano Letters, 2008, 8 (1): 36-41.

[39] Ramesha G K, Sampath S. Electrochemical reduction of oriented graphene oxide films: an in situ Raman spectroelectrochemical study [J]. Journal of Physical Chemistry Letters, 2009, 113 (19): 7985-7989.

[40] Stankovich S, Dikin D A, Piner R D, et al. Synthesis of graphene-based nanosheets via chemical reduction of exfoliated graphite oxide [J]. Carbon, 2007, 45 (7): 1558-1565.

[41] Belchik S M, Kennedy D W, Dohnalkova A C, et al. Extracellular reduction of hexavalent chromium by cytochromes MtrC and OmcA of *Shewanella oneidensis* MR-1 [J]. Applied and Environmental Microbiology, 2011, 77 (12): 4035-4041.

[42] Ritchie R J, Prvan T. Current statistical methods for estimating the K-m and V-max of Michaelis-Menten kinetics [J]. Biochemical Education, 1996, 24 (4): 196-206.

[43] Shi L, Richardson D J, Wang Z, et al. The roles of outer membrane cyto-

chromes of *Shewanella* and *Geobacter* in extracellular electron transfer [J]. Environmental Microbiology Reports, 2009, 1 (4): 220-227.

[44] Shi Z, Zachara J M, Wang Z M, et al. Reductive dissolution of goethite and hematite by reduced flavins [J]. Geochimica et Cosmochimica Acta, 2013, 121: 139-154.

[45] Bodemer G J, Antholine W A, Basova L V, et al. The effect of detergents and lipids on the properties of the outer-membrane protein OmcA from *Shewanella oneidensis* [J]. The Journal of Biological Chemistry, 2010, 15 (5): 749-758.

[46] Edwards M J, Baiden N A, Johs A, et al. The X-ray crystal structure of *Shewanella oneidensis* OmcA reveals new insight at the microbe-mineral interface [J]. FEBS Letters, 2014, 588 (10): 1886-1890.

[47] Patila M, Pavlidis I V, Diamanti E K, et al. Enhancement of cytochrome c catalytic behaviour by affecting the heme environment using functionalized carbon-based nanomaterials [J]. Process Biochemistry, 2013, 48 (7): 1010-1017.

第 5 章

生物电化学系统阳极碳材料表面修饰对微生物 EET 的调控

5.1
生物电化学系统阳极材料的特征

生物电化学系统(BES)能将有机废物中的化学能转化为电能。EAB可以将电子转移到金属氧化物、电极等胞外的电子受体,在环境修复、生物地球化学循环及BES中发挥了重要作用。为了提高微生物协助的能量回收,必须有效控制电子从细胞到石墨纳米片电极的胞外传递效率,其中包含着能量的流动。因此,深刻理解微生物EET是十分必要的[9-10]。目前,已有大量的研究工作探索EET的机制,并已报道了几条EET路径[11]。上一章中已经提到,EET路径包括通过与外膜蛋白直接接触实现[12]、利用可溶性具有氧化还原性质的电子穿梭体实现[13-14]、通过细胞菌毛(pili)实现[15]。在直接电子传递(DET)过程中,电子通过外膜蛋白细胞色素c(OM c-Cyts)直接从胞内传递到电极[16-17]。为了引导从胞内代谢到胞外电极的电子流动,研究生物-化学界面的相互作用以及各种DET决定因素的分析是至关重要的。关于 *Shewanella* 外膜蛋白的结构以及 *Geobacter* 的pili上的细胞色素c已在第2.2.5节做了详细介绍。

虽然已有研究提出了若干关于DET的假设[16],但对于通过OM c-Cyts传递电子机制的基本生物过程的了解依然有限。在自然系统中,由于环境的复杂性和多变性,难以捕捉到界面的电子传递过程。在这种情况下,分子模拟则可以提供有用的信息以阐明分子与表面的相互作用[19],在分子水平上更细致分析微生物胞外电子传递到电极表面的过程,可以通过优化电极表面以更有效地实现电子转移[20-21]。因此,包括分子动力学模拟(MD)和量子化学计算在内的分子模拟方法,能够有效研究电子从OM c-Cyts转移到胞外固体电极的机制,评价EET对环境条件改变的响应,并在分子尺度上理解和调控生物系统的EET过程。

考虑到DET过程会随电子受体的不同而发生改变,这里选择最常用的石墨纳米片电极作为胞外电子受体。由于较大的表面积、良好的导电性、低廉的价格、良好的生物相容性以及表面改性的灵活性,石墨对于许多BES来说是一种颇具吸引力的电极材料[22]。石墨的每一层由sp^2杂化的碳原子组成,与石墨烯性质相同。达到准平衡状态时,氧化石墨烯的氧/碳比例稳定(O/C=0.4),主要含氧官能团为环氧基和羟基[23],它们可以与电极表面的OM c-Cyts直接作用而影响DET过程。

在本章中，综合利用 MD 模拟和量子化学计算，在分子尺度上探索 Cyt-石墨界面的 DET 机制，分析加速 DET 过程的电极材料的特征，同时进行了电化学实验来验证计算结果。该研究结果能够在分子水平上研究 EAB 的 OM c-Cyts 与石墨电极 DET，有助于更好地理解和工程化微生物与石墨或其他电极材料的 DET 过程，以设计并优化高效的 BES。

5.2 阳极材料表面 EET 过程研究方法

5.2.1 电化学分析测试

循环伏安法（CV）是非常通用的电化学工具，广泛应用于识别氧化还原对、描述电化学过程和研究反应动力学。循环伏安曲线上出现的特定峰值电位和电流差对于理解机制和结构-性质关系是非常重要的。本研究利用 CHI660 电化学工作站，在三电极体系下进行循环伏安实验（CV）。石墨棒为工作电极，Pt 丝为对电极，Ag/AgCl 为参比电极。石墨电极（d = 10 mm）先用粒度为 0.3 μm 和 0.05 μm 的氧化铝粉抛光，然后依次在去离子水和乙醇中超声清洗。称取 6.4 mg c-Cyt（from equine heart，Sigma-Aldrich Co.，USA）溶于 5 mL PBS（pH 6.95，50 mmol·L^{-1} $K_2HP_4 + KH_2PO_4$），然后浸泡于冰水中。CV 扫面的范围是 $-0.4 \sim 0.6$ V（vs. Ag/AgCl）。无 c-Cyt 的 PBS 溶液在相同条件下也进行 CV 测试作为空白对照。

5.2.2 细胞色素-石墨纳米片的电子转移模拟

在蛋白质数据库（PDB）中 *Shewanella oneidensis* MR-1 的 STC 的代码是 1M1R。添加 H 原子来饱和石墨层边界上的悬空键，如图 5.1(a)所示。图 5.1(b)中，STC 含 91 个氨基酸和 4 个 c 型的 heme 辅基，heme 1 和 4 相互垂直，而 heme 2 和 3 相互平行。MD 盒子中含 2 个硫酸根离子和 15 个 Na^+ 以中和蛋白的电荷。Fe 原子的轴向配体都是双组氨酸（bisHis）残基[图 5.1(b)]。垂直方向的配位键连着组氨酸残基的 N 原子。模型中在石墨表面是水溶液体系：包括 1930 个水分子、STC 以及 15 个 Na^+。通过移除肽链以分析氨基酸残基对电子从 heme 到石墨的转移的影响。另外，在石墨表面引入环氧基和羟基，作为主要的官能团，使石墨表面维持稳定的氧/碳比例（O/C=0.4）[23-24]，研究含氧基团对电子传递的影响[图 5.1(a)]。水溶液中细胞色素-石墨（STC-G）和细胞色素-

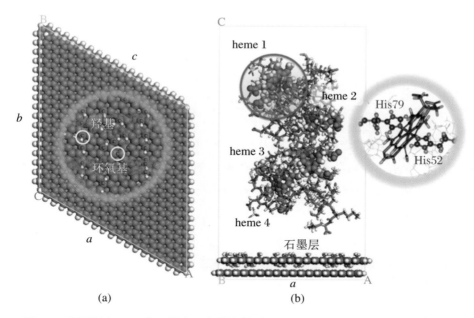

图 5.1　石墨层和 STC 在石墨表面上的构型：在石墨结构的第一层修饰了环氧基和羟基作为含氧官能团，石墨结构是 18×18×10 超晶胞：$a=b=44.28$ Å，$c=68.00$ Å(a)；*Shewanella oneidensis* MR-1 的 STC 分子构型，结构中添加了 H 以及区分了不同键的类型，局部放大的 heme 1 周围的结构，双组氨酸配位的卟啉环结构(b)

氧化石墨(STC-GOH)的 MD 盒子先进行能量最小化步骤，然后分别在 298 K 的 NPT 系综、NVT 系综和 NVE 系综中弛豫 50 ps、50 ps 和 1500 ps。相关计算方法已在 2.2.6 章节中进行了详细的说明。

5.2.3
量子力学/分子力学(QM/MM)计算

在 QM/MM 方法中，系统被分为两个区域，中心化学活性区域用 QM 计算，外部区域用力场 MM[25]计算。QM 区域的选择通常是考虑到化学反应：化学参数作为最小尺寸的 QM 区域，可以用此方法来检测 QM/MM 结果的灵敏度[26]。在血红素与石墨表面电子传递的情况下，最小尺寸的 QM 区域包含卟啉铁和接受电子的两层石墨结构。双组氨酸配体引入血红素基团中来解析氨基酸残基对电子传递的影响，并且考虑了石墨表面的含氧官能团。计算方法已在 2.2.4 节中进行了详细说明。

5.2.4
过渡态的研究

计算反应路径的最简单方法是从鞍点开始沿着负梯度方向进行计算。这种最速下降法(steepest descent approach)得到的结果是最小能量路径(MEP)或质量加权坐标系中的一个内禀反应坐标(IRC)。线性同步转变(LST)/二次同步度越(QST)过渡态搜索方法以两个端点反应物(R)和产物(P)作为输入内容，定位反应路径中的具有最高能量势垒的极大值，也就是过渡态。LST 算法在 R($P-Fe^{2+}$)和 P($P-Fe^{3+}$)之间进行几何插值，从而产生一条反应路径。当 LST 算法通过插值法在起点和终点中形成插值点后，QST 算法还要在这三个点之间进一步插值，因此需要第三个"中间"点。QST 算法从中间点开始进行极大值搜索，施加约束使得到的几何结构与初始结构具有相同的路径坐标。QST 优化过程中会在施加约束的过程中对过渡态有一个上限。因此，优化的 LST 算法和约束最小化 QST 算法可以进行结合，LST/QST 通过执行 LST 优化计算开始，其中 QST 计算是为了获得过渡态的近似构型，而共轭梯度(CG)技术则被用于精修鞍点的几何结构。

5.3 阳极材料表面修饰对 EET 的调控机制

5.3.1 细胞色素 c 与石墨(G,GO)结构的电子传递

在等温(298 K)、等压(0.1 MPa)条件下,对含有 4 个 heme 辅基的 Cyt-石墨(STC-G)系统进行 MD 模拟。STC 与含有 10 个 heme 辅基的 Cyt 具有相同的氧化还原活性中心——heme,又名卟啉铁(P-Fe)[27]。为了确定石墨表面最优的 STC 电子传递构型[图 5.1(a)],需要进行一系列的 MD 模拟。

在模拟盒子中,最初 STC 分子的质量中心(COM)位于石墨纳米片表面上 30 Å 的位置[图 5.1(b)]。处于界面的 Cyt 和 heme 的构型直接影响电子转移速率[28]。heme 1 和 4 暴露于溶剂中,均是六配位低自旋的 heme,并分别与 His52, 79 和 His19, 65 相连。由于 STC 的 heme 4 离石墨表面很近,因此,它是最有效的电子传递位点,这已被实验研究结果证实[29]。通过 OM c-Cyt 实现 DET 需要细菌细胞与石墨表面之间的物理接触(附着)[30]。

STC-G 体系的平衡状态如图 5.2(a)和(c)所示,表明在大部分情况下,细胞色素与电极表面的接触面积很小,并且吸附的强度取决于氨基酸残基的数目。为了分析多肽链的影响,用 MD 模拟研究了只有 heme 存在时与石墨表面的电子传递,即游离态 heme-G 体系(图 5.3)。此外,在实际的电子转移系统的碳电极表面存在着含氧官能团,如图 5.2(b)及(d)所示,可以用电化学方法对石墨表面第一层进行修饰,实验可用于分析石墨表面官能团对 DET 的影响。

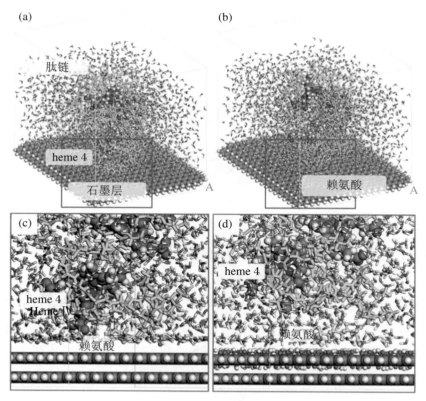

图 5.2 电子传递系统的平衡结构:水溶液中的系统(a);水溶液中的系统(b);STC 与未修饰石墨界面的结构特征(c);STC 与含氧官能团修饰的石墨界面的结构特征;高亮标出的是肽链(d)

微观构象的波动通常是通过 site-site 径向分布函数表示(RDFs),它揭示了细胞色素接近石墨表面第一层时的构象。图 5.4 对比了 STC-G,游离态 heme-G,STC-GOH 三种体系的 site-site RDFs。峰值 $g(r)$ 代表 STC 与表面接触时 heme 4 的 Fe 原子与石墨层的表面原子之间出现概率最大的距离。

体系游离态 heme-G 的峰值($r_2 = 5.1$ Å)比 STC-G 体系的峰值($r_1 = 12.5$ Å)小,表明肽链增加了暴露于溶剂中的血红素与石墨表面之间的距离。STC-GOH 体系的石墨层表面有碳原子和羟基与环氧基中的氧原子,Fe—C 之间 RDF 具有最大的峰值($r_3 = 13.8$ Å),而 Fe—O 之间的 RDF 最大峰值出现在 $r_4 = 9.4$ Å,小于 STC-G 体系 r_1 的值。该结果表明具有极性和亲水性的含氧官能团能缩短电子转移距离,从而加快 EET。STC-GOH 体系中第一峰值为 $r_3' = 11.3$ Å,是 Fe 原子与碳原子之间最短的距离。可以根据 RDFs 提供的 $g(r)$ 曲线中峰的数量来判断电极表面可利用的电子接受位点的数目。如图 5.4 所示,STC-GOH 体系具有更多的电子接受位点,表明含氧官能团提高了石墨表面的转移电子的能

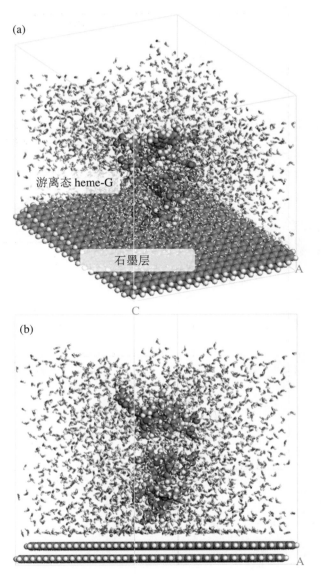

图 5.3 游离态 heme-G 系统的平衡结构：溶剂化的游离态 heme-G 平衡结构的 3D 视图(a)；前视图(b)

力，这是因为含氧官能团具有特殊的电子性质。STC-G 和 STC-GOH 体系平衡的分子构象说明位于 c-Cyts 表面的赖氨酸(Lys)残基能与石墨相互作用[图 5.2(c)和(d)]。赖氨酸残基能与石墨表面形成弱的氢键，缩短了 Cyt 活性中心与石墨表面的距离，从而加快 DET。上述结果表明从 Cyt 到石墨电极的 DET 过程中，Cyt 动力学、构象变化以及石墨表面结构都在细胞色素与石墨电极的电子传递中发挥了重要作用。

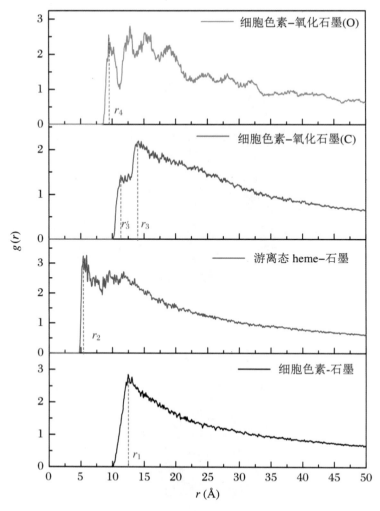

图 5.4　电子传递系统中最靠近石墨层的 heme 的 Fe 原子与石墨最顶层 C 原子之间的 RDFs，系统则包括了 Fe 原子分别与石墨层上 C 原子和 O 原子之间的 RDFs

5.3.2
电子转移中卟啉环结构的变化

作为 STC 的氧化还原活性中心，卟啉环是由 4 个修饰的吡咯亚基形成的大杂环。因此，其高度共轭的几何结构的改变能显著影响电子转移过程。表 5.1 列举了卟啉环结构中活性位点的典型几何参数。

表 5.1　初始 P-Fe 以及 porphyrin-G,porphyrin-biHis-G 和 porphyrin-biHis-GOH 系统中电子传递前后卟啉环的几何结构参数

System	$l(Fe—N_1)$(Å)	$l(Fe—N_2)$(Å)	$l(Fe—N_3)$(Å)	$l(Fe—N_4)$(Å)	$\theta(N_1—Fe—N_3)$(°)	$\theta(N_2—Fe—N_4)$(°)
Initial P-Fe	1.986	1.983	1.979	1.979	176.290	176.850
P-Fe^{2+}（porphyrin-G）	2.062	2.060	2.046	2.051	172.902	176.461
P-Fe^{3+}（porphyrin-G）	2.062	2.060	2.046	2.051	177.194	173.747
P-Fe^{2+}（porphyrin-biHis-G）	2.154	2.039	1.968	2.082	178.549	177.302
P-Fe^{3+}（porphyrin-biHis-G）	2.138	2.023	1.989	2.078	178.067	177.442
P-Fe^{2+}（porphyrin-biHis-GOH）	2.135	2.013	1.994	2.118	176.146	177.720
P-Fe^{3+}（porphyrin-biHis-GOH）	2.111	2.013	1.995	2.122	176.583	177.868

对比 porphyrin-G 复合体中卟啉铁(图 5.5)以及起始卟啉环[图 5.1(b)]的结构参数可以发现,在电子转移过程中,当卟啉铁靠近石墨表面的第一层时,卟啉铁的结构发生改变;当卟啉环接近石墨表面时,卟啉环高度共轭的结构被打破,以激活电子转移过程。在这样的分子趋向中,卟啉铁与石墨之间的距离如图 5.5(d)所示,该距离会因为石墨表面的亲水性含氧官能团而减小,从而加快电子转移。分子轨道能级受卟啉环构象波动的影响,从而加快 Cyt-石墨界面的电子传递。

卟啉环中心的 Fe 原子配位 4 个 N 原子,该部分是共轭卟啉铁的活性位点。因此,其高度共轭结构的改变能显著影响电子转移过程。表 5.1 列出了卟啉环活性位点的几何参数,这些参数是用 PW91/DNP 基组计算的。对比 porphyrin-G 复合体系中卟啉铁和起始卟啉环的结构参数,可以发现在电子转移过程中当卟啉铁接近石墨表面第一层时,卟啉铁的结构会发生改变。三个体系 porphyrin-G, porphyrin-biHis-G 和 porphyrin-biHis-GOH 中 Fe—N 键(Fe—N_1,Fe—N_2,Fe—N_3 和 Fe—N_4)的长度比起始卟啉铁中的 Fe—N 键要长,且氧化态 Fe—N 的长度不同于还原态的 Fe—N 的长度。在 DET 过程中,不仅 Fe—N 的键长会变,其 N_1—Fe—N_3 和 N_2—Fe—N_4 的键角也会发生变化。特别是在 porphyrin-G 体系中,游离态 heme 分子的键角改变更大。这些结果说明当卟啉环接近石墨表面时,其高度共轭的分子结构被破坏,从而诱发电子转移过程。

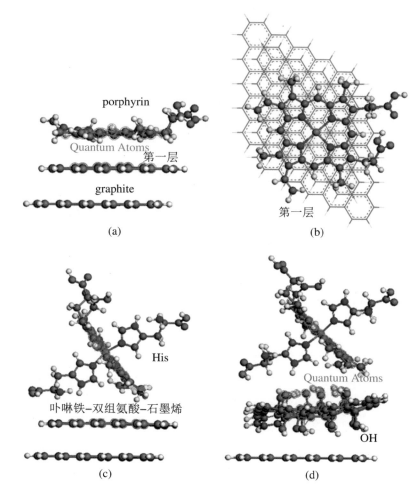

图 5.5　电子传递系统的最低能量结构：porphyrin-G 还原体系的前视图(a)，包括石墨层和 P-Fe^{2+}；porphyrin-G 还原体系俯视图(b)；porphyrin-biHis-G 还原体系(c)，包含石墨层与双组氨酸配位的 P-Fe^{2+}；porphyrin-bi-His-GOH 还原体系的第一层共价键键合了环氧基和羟基(d)

　　为了分析不同的相互作用对 DET 的影响，研究了卟啉环接近石墨表面时若干可能的取向。图 5.5 显示的是从 QM/MM 计算中所得到的利于电子转移到石墨表面的最佳结构。在 porphyrin-G 体系中，卟啉环平行于石墨表面[图 5.5(a)和(b)]。然而，对于 porphyrin-biHis-G 体系，轴向组氨酸残基的位阻效应导致卟啉环（还原态和氧化态）与石墨表面呈 45°[图 5.5(c)和(d)]。卟啉环的取向与分子动力学模拟中所展示的结构相一致[图 5.2(c)和(d)]。至于 porphyrin-biHis-GOH 体系，部分石墨表面的 π-π 共轭结构被共价结合在碳原子上的环氧基和羟基所破坏。值得注意的是，在电子转移发生之前，porphyrin-G 体系中最短的 Fe—C 距离（还原态时为 3.392 Å）是 4.0 Å，比 porphyrin-biHis-G

体系的距离(还原态时为 7.383 Å)要小。相比之下,处于还原态时,porphyrin-biHis-GOH 体系中最短 Fe—C 距离为 7.679 Å,而最短 Fe—O 距离为 6.371 Å。这表明该取向下氧化还原中心的距离会因为石墨表面的亲水含氧基团而减小,从而加快电子传递。这些距离比分子动力学模拟中 RDF 的结果(图 5.4)要小,是因为在 QM/MM 计算中未考虑近程水分子的影响。此外,据报道轴向配体起到稳定金属中心的作用,从而维持甚至是加强了自然环境下 Fe(Ⅱ)/Fe(Ⅲ) 之间的氧化还原循环[32]。在水体系下有共轭配体分子通过配位键结合在卟啉环分子的 Fe 原子上,在微生物体系中,这些物质可形成单一的低聚卟啉分子导线[33,34],吸附在微生物的 MtrABC 电子导线暴露在蛋白外侧的 heme 上,从而促进电子传递(图 5.6)。因此,如果在细胞色素和胞外电子受体之间形成了卟啉分子组成的复合物分子导线体系,形成有效的电子传递通路,将会有效地提高电子传递效率。

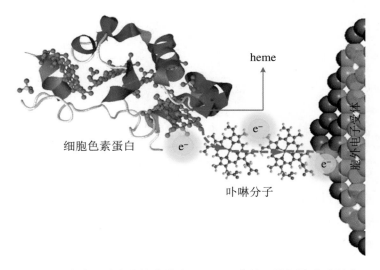

图 5.6 存在于溶液中的卟啉分子可以组成单一的低聚卟啉铁分子导线,链接着微生物的 MtrA、MtrB、MtrC 电子导线暴露在蛋白外侧的 heme 与胞外电子受体促进 EET

5.3.3
卟啉铁与石墨(G,GO)结构电子传递过程中的能量

利用 QM/MM 计算探索了从还原态卟啉铁($P\text{-}Fe^{2+}$)到石墨纳米片表面的 DET 的能量变化。在卟啉环的 Fe 原子上引入轴向配体 bisHis 来研究氨基酸残基对电子传递的影响。在 QM/MM 计算中,总能量是 QM 的能量、MM 的能量

以及 QM/MM 耦合项能量的三者之和。电子转移(表 5.2)总能量变化为正($\Delta E_{Total} = 12.476\,\text{eV}$),说明还原态 P-$Fe^{2+}$-G[图 5.5(a)]比氧化态 P-$Fe^{3+}$-G[图 5.5(b)]更稳定,这意味着游离的血红素(free heme)难以将电子转移到没有官能团修饰的石墨层。体系 porphyrin-biHis-G 的 ΔE_{Total} 为正,但小于游离态 heme 体系,表明两个轴向配体组氨酸残基影响了卟啉铁与石墨层之间的电子传递。然而石墨层含有含氧基团的 porphyrin-biHis-GOH 体系的 ΔE_{Total} 相当小(0.021 eV)。这些结果表明有多肽链氨基酸配位的还原态 heme 更有利于将电子转移到氧化石墨的表面。总能量主要是由 QM 区域贡献的,因此,计算所得到的 ΔE_{Total} 结果说明 Fe 原子和血红素在从 c-Cyt 到石墨表面的 DET 过程中起着十分重要的作用。

表 5.2 直接电子传递系统 porphyrin-G,porphyrin-biHis-G 和 porphyrin-biHis-GOH 的能量变化分析,包括总能量和热力学性质(298.15 K,1 atm)

系统	ΔE_{Total} (eV)	ΔE_{QM} (eV)	ΔE_{MM} (eV)	$\Delta G^{\ominus}$ (eV)	$\Delta H^{\ominus}$ (eV)	$\Delta S^{\ominus}$ $10^{-3}(\text{eV}\cdot\text{K}^{-1})$
porphyrin-G	12.476	12.502	-0.025	12.439	12.449	0.033
porphyrin-biHis-G	10.522	10.908	-0.386	10.820	10.933	0.377
porphyrin-biHis-GOH	0.021	0.052	-0.031	0.079	0.056	-0.075

利用 QM/MM 方法计算了在标准状态下,电子从还原态卟啉铁(P-Fe^{2+})传递到石墨纳米片表面的热力学性质。在卟啉环的铁原子上轴向引入 bisHis 来解释氨基酸残基对电子传递的影响。体系 porphyrin-biHis-GOH[图 5.5(d)]的总能量变化(ΔE_{Total})表明卟啉铁轴向配位多肽链中的氨基酸时,暴露于溶剂中还原态的 heme 更有利于将电子转移到氧化态的石墨表面(表 5.2)。电子转移能量的计算结果所显示的趋势,与分子动力学模拟中 RDFs 的电子转移的有效距离相一致,也说明当石墨表面具有含氧官能团、卟啉铁轴向配位组氨酸时更有利于电子传递。

表 5.2 列出了在室温和标准大气压条件下,电子转移反应的标准 Gibbs 自由能($\Delta G^{\ominus}$)、焓变($\Delta H^{\ominus}$)和熵变($\Delta S^{\ominus}$)。$\Delta G^{\ominus}$ 大于 0 表示游离态 heme 不能自发地将电子转移到受体。然而,当 *Shewanella* 与非水溶性的三价铁化合物胞外电子受体相互作用时,OmcA 和 MtrC 的还原电势分别为 $-0.32\sim-0.24$ V 和 -0.1 V(vs. NHE),表明还原过程是自发的,这可以用标准氧化还原半反应电势($E^{\ominus}$ vs. NHE)来解释,公式如下[37]:

$$E^{\ominus} = \frac{-\Delta G^{\ominus}}{nF} - E_{H}^{\ominus} \quad (5.1)$$

式中，n 是反应式左边的电子数，F 是法拉第常数，等于 23.06 kcal·(mol·V)$^{-1}$，$E_H^\ominus$ 是 NHE 的标准还原电势，等于 4.28 V。

因此，当 $E^\ominus$ 大于 -4.28 V 时，还原反应的 $\Delta G^\ominus$ 小于 0，说明根据报道的实验结果，还原过程是自发进行的，所以其逆过程（即氧化过程），就是将电子导出细胞色素 c 的过程是非自发的，计算结果与报道的实验结果一致。当石墨层表面具有含氧官能团时，虽然 $\Delta G^\ominus$ 仍然大于 0，但只需要克服一个很小的能量（0.079 eV）就可以将电子传递给石墨层，明显小于 porphyrin-G 和 porphyrin-biHis-G 体系，说明在石墨表面的含氧官能团可以加速 DET 过程，这与之前的 RDF 计算结果以及总能量的计算结果趋势一致。

5.3.4

卟啉铁与外膜细胞色素 c 的电化学特性

c-Cyt 的氧化还原电势通过 CV 实验来测定，以验证计算的结果（图 5.7）。空白 PBS 的 CV 曲线没有明显的氧化还原峰，含有 c-Cyt 的体系中氧化峰和还原峰的位置大概在 0.12 V 和 -0.23 V（vs. Ag/AgCl），是马心 c-Cyt 的电化学响应信号。

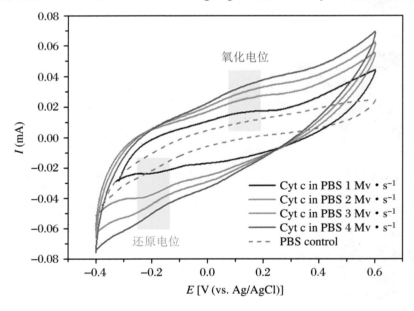

图 5.7　不同扫速下，细胞色素 c 在石墨电极上 PBS 中的 CVs，其中 PBS 溶液是 50 mmol·L^{-1} K$_2$HP$_4$ + KH$_2$PO$_4$（pH 6.95），石墨电极的直径是 10 mm，相同条件下测试 PBS 溶液作为对照

因此,实验所得到的马心 c-Cyt 的氧化还原电势与已有的研究进一步证明该反应的 $\Delta G^{\ominus}$ 小于 0,其逆反应,即马心 c-Cyt 的氧化反应的 $\Delta G^{\ominus}$ 大于 0。这也和计算得到的结果相一致。计算与实验结果数值上的差异源于 QM/MM 模型中应用的是卟啉铁分子而不是整个蛋白结构,但得到的趋势是一致的。这些结果证实了该模型适用于研究从 OM c-Cyt 到固态电子受体的电子传递。

5.3.5 电子转移的能隙

根据前线轨道理论,卟啉(电子供体)的最高占据分子轨道(HOMO)与石墨(电子受体)的最低未占据分子轨道(LUMO)之间最容易发生反应。这是因为同其他能级相比,此两者之间的能隙最小。因此,电子应是从卟啉的 HOMO 转移到石墨的 LUMO。连接在 Fe 原子上的两个轴向组氨酸残基能降低卟啉的 HOMO 和石墨的 LUMO 之间的能隙(E_g),从 $E_{g,1} = 2.40$ eV 降到 $E_{g,2} = 1.86$ eV(图 5.8)。分子轨道能级受卟啉环构象波动的影响。因此,血红素周围多肽链上的氨基酸残基能降低了能隙(E_g),从而使细胞色素-石墨界面的电子传递所需的能量降低。

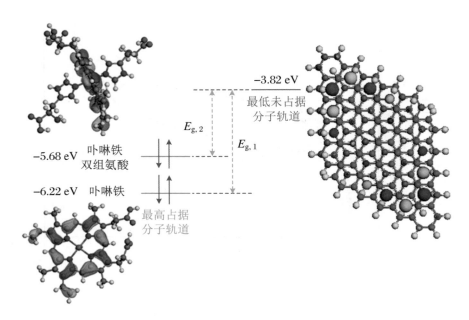

图 5.8 电子传递体系的前线分子轨道解析,porphyrin-biHis-G 体系的 HOMO-LUMO 能隙小于 porphyrin-G 体系

5.3.6

电子转移的动力学

动力学分析表明卟啉环接近石墨层时的构象在很大程度上影响能垒值（E_a，图 5.9 和表 5.3）、电子转移速率，甚至可能引起能垒值的由正值变为负

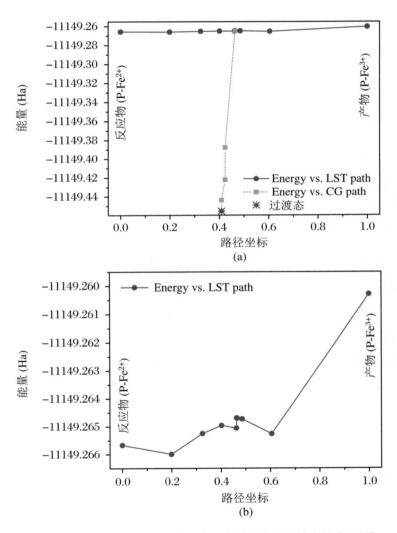

图 5.9　porphyrin-biHis-GOH 体系电子传递过程的过渡态搜索：过渡态搜索中能量与广义反应坐标的关系(a)；对(a)中的路径细节显示(b)

值[39]。对于 porphyrin-biHis-GOH 体系，从 heme 到石墨层的电子转移可能受轴向组氨酸残基的空间效应的影响，但是 $E_a (= E_{TS} - E_R = -5.17 \text{ eV})$ 仍为负值，这是由于过渡状态(TS)的总能量比反应物(R)的总能量要低。基元反应呈现出负的 E_a 值是典型的无势垒(barrierless)反应[40]，其反应的发生依赖于电子受体表面势阱对电子的捕获[41]。因此，这里负的 E_a 值表明界面电子转移的发生不需要克服能垒。其速率常数可通过下列方程计算：

$$k_{ET} = A\exp\left(-\frac{E_a}{RT}\right) = \frac{k_B T}{h} E (c^{\ominus})^{1-n} \exp\left(\frac{\Delta^{\neq} S^{\ominus}}{R}\right) \exp\left(-\frac{E_a}{RT}\right) \quad (5.2)$$

式中，k_{ET} 是电子转移速率系数 $[\text{mol}^{1-n} \cdot (\text{L}^{1-n} \cdot \text{s})^{-1}]$，$A$ 是指前因子 $[\text{mol}^{1-n} \cdot (\text{L}^{1-n} \cdot \text{s})^{-1}]$，$k_B$ 是玻尔兹曼常数 $[1.381 \times 10^{-23} \text{ m}^2 \cdot \text{kg} \cdot (\text{s}^2 \cdot \text{K})^{-1}]$，$h$ 是普朗克常数 $(6.626 \times 10^{-34} \text{ m}^2 \cdot \text{kg} \cdot \text{s}^{-1})$，$c^{\ominus}$ 是标准摩尔浓度 $(1 \text{ mol} \cdot \text{L}^{-1})$，$n$ 是反应级数，$\Delta^{\neq} S^{\ominus}$ 是活化熵 $[\text{J} \cdot (\text{mol} \cdot \text{K})^{-1}]$。过渡态的热力学数据如表 5.3 所示。

表 5.3 卟啉铁到石墨层电子传递的动力学性质，包括活化能(E_a)，标准条件下的过渡态热力学性质($\Delta^{\neq} G^{\ominus}$，$\Delta^{\neq} H^{\ominus}$，$\Delta^{\neq} S^{\ominus}$)和速率常数($k_{ET}$)

系统	E_a (eV)	$\Delta^{\neq} G^{\ominus}$ (eV)	$\Delta^{\neq} H^{\ominus}$ (eV)	$\Delta^{\neq} S^{\ominus}$ (10^{-3} eV·K^{-1})	k_{ET} [mol^{1-n}·(L^{1-n}·s)$^{-1}$]
porphyrin-biHis-GOH	-5.17	-0.701	-0.859	-0.529	9.10×10^{97}

结合热力学和动力学分析结果，表明 EET 过程只需要由细菌代谢底物产生一个很低的热力学驱动力就可以发生，实验研究的文献中也有类似的报道[42-43]。

对于 porphyrin-biHis-GOH 体系，电子传递过程很快就能达到平衡。因此当 OM c-Cyt 直接接触电极表面时，电子从卟啉铁传递到阳极材料表面并不是限速步骤，表观电子转移速率主要取决于该体系中 Cyt 的构象，卟啉环接近石墨表面时的分子取向以及电子转移的距离。

参考文献

[1] Chen Y X, Ji W H, Yan K, et al. Fuel cell-based self-powered electrochemical sensors for biochemical detection [J]. Nano Energy, 2019, 61: 173-193.

[2] Harnisch F, Urban C. Electrobiorefineries: Unlocking the synergy of electrochemical and microbial conversions [J]. Angewandte Chemie-Interna-

tional Edition, 2018, 57(32): 10016-10023.

[3] Jiang Y, May H D, Lu L, et al. Carbon dioxide and organic waste valorization by microbial electrosynthesis and electro-fermentation [J]. Water Research, 2019, 149: 42-55.

[4] Wang H M, Park J D, Ren Z J. Practical energy harvesting for microbial fuel cells: A review [J]. Environmental Science & Technology, 2015, 49(6): 3267-3277.

[5] Zhao C E, Gai P P, Song R B, et al. Nanostructured material-based biofuel cells: recent advances and future prospects [J]. Chemical Society Reviews, 2017, 46(5): 1545-1564.

[6] Flexer V, Jourdin L. Purposely designed hierarchical porous electrodes for high rate microbial electrosynthesis of acetate from carbon dioxide [J]. Accounts of Chemical Research, 2020, 53(2): 311-321.

[7] Logan B E, Rossi R, Ragab A A, et al. Electroactive microorganisms in bioelectrochemical systems [J]. Nature Reviews Microbiology, 2019, 17(5): 307-319.

[8] Pankratova G, Hederstedt L, Gorton L. Extracellular electron transfer features of Gram-positive bacteria [J]. Analytica Chimica Acta, 2019, 1076: 32-47.

[9] Chabert V, Babel L, Fueeg M P, et al. Kinetics and mechanism of mineral respiration: How iron hemes synchronize electron transfer rates [J]. Angewandte Chemie International Edition, 2020, 59: 1-7.

[10] Heidary N, Kornienko N, Kalathil S, et al. Disparity of cytochrome utilization in anodic and cathodic extracellular electron transfer pathways of *Geobacter sulfurreducens* biofilms [J]. Journal of the American Chemical Society, 2020, 142(11): 5194-5203.

[11] Torres C I, Marcus A K, Lee H S, et al. A kinetic perspective on extracellular electron transfer by anode-respiring bacteria [J]. FEMS Microbiology Reviews, 2010, 34(1): 3-17.

[12] Jiang X Y, Burger B, Gajdos F, et al. Kinetics of trifurcated electron flow in the decaheme bacterial proteins MtrC and MtrF [J]. Proceedings of the National Academy of Sciences of the United States of America, 2019, 116(9): 3425-3430.

[13] Light S H, Su L, Rivera-Lugo R, et al. A flavin-based extracellular electron

transfer mechanism in diverse Gram-positive bacteria [J]. Nature, 2018, 562(7725): 140-144.

[14] Wu Y, Luo X, Qin B, et al. Enhanced current production by exogenous electron mediators via synergy of promoting biofilm formation and the electron shuttling process [J]. Environmental Science & Technology, 2020, 54 (12): 7217-7225.

[15] Ru X, Zhang P, Beratan D N. Assessing possible mechanisms of micrometer-scale electron transfer in heme-free *Geobacter sulfurreducens* Pili [J]. Journal of Physical Chemistry B, 2019, 123(24): 5035-5047.

[16] White G F, Shi Z, Shi L, et al. Rapid electron exchange between surface-exposed bacterial cytochromes and Fe(Ⅲ) minerals [J]. Proceedings of the National Academy of Sciences of the United States of America, 2013, 110 (16): 6346-6351.

[17] Busalmen J P, Esteve-Nunez A, Berna A, et al. C-type cytochromes wire electricity-producing bacteria to electrodes [J]. Angewandte Chemie International Edition, 2008, 47(26): 4874-4877.

[18] Liu H, Newton G J, Nakamura R, et al. Electrochemical characterization of a single electricity-producing bacterial cell of *Shewanella* by using optical tweezers [J]. Angewandte Chemie-International Edition, 2010, 49(37): 6596-6599.

[19] Li J, Geng C, Weiske T, et al. Revisiting the intriguing electronic features of the BeOBeC carbyne and some isomers: A quantum-chemical assessment [J]. Angewandte Chemie (International ed in English), 2020, 10.1002/anie.202007990.

[20] Li C, Cheng S. Functional group surface modifications for enhancing the formation and performance of exoelectrogenic biofilms on the anode of a bioelectrochemical system [J]. Critical Reviews in Biotechnology, 2019, 39 (8): 1015-1030.

[21] Yang C, Aslan H, Zhang P, et al. Carbon dots-fed *Shewanella oneidensis* MR-1 for bioelectricity enhancement [J]. Nature Communications, 2020, 11(1): 1379-1379.

[22] Logan B E. Exoelectrogenic bacteria that power microbial fuel cells [J]. Nature reviews Microbiology, 2009, 7(5): 375-381.

[23] Tu Y, Lv M, Xiu P, et al. Destructive extraction of phospholipids from

Escherichia coli membranes by graphene nanosheets [J]. Nature Nanotechnology, 2013, 8(8): 594-601.

[24] Gao W, Alemany L B, Ci L, et al. New insights into the structure and reduction of graphite oxide [J]. Nature Chemistry, 2009, 1(5): 403-408.

[25] Senn H M, Thiel W. QM/MM Methods for Biological Systems [C]//Reiher M. Atomistic approaches in modern biology: From quantum chemistry to molecular simulations. Berlin: Springer Berlin Heidelberg, 2007: 173-290.

[26] Sherwood P, Vries A, Guest M, et al. QUASI: A general purpose implementation of the QM/MM approach and its application to problems in catalysis [J]. Journal of Molecular Structure: THEOCHEM, 2003, 632: 1-28.

[27] Shaik S. Biomimetic chemistry iron opens up to high activity [J]. Nature Chemistry, 2010, 2(5): 347-349.

[28] Garcia-Meseguer R, Marti S, Javier Ruiz-Pernia J, et al. Studying the role of protein dynamics in an $S(N)_2$ enzyme reaction using free-energy surfaces and solvent coordinates [J]. Nature Chemistry, 2013, 5(7): 566-571.

[29] Harada E, Kumagai J, Ozawa K, et al. A directional electron transfer regulator based on heme-chain architecture in the small tetraheme cytochrome c from *Shewanella oneidensis* [J]. FEBS Letters, 2002, 532(3): 333-337.

[30] Pham T H, Aelterman P, Verstraete W. Bioanode performance in bioelectrochemical systems: recent improvements and prospects [J]. Trends in Biotechnology, 2009, 27(3): 168-178.

[31] Roberts J G, Moody B P, Mccarty G S, et al. Specific oxygen-containing functional groups on the carbon surface underlie an enhanced sensitivity to dopamine at electrochemically pretreated carbon fiber microelectrodes [J]. Langmuir, 2010, 26(11): 9116-9122.

[32] Kopf S H, Henny C, Newman D K. Ligand-enhanced abiotic iron oxidation and the effects of chemical versus biological iron cycling in anoxic environments [J]. Environmental Science & Technology, 2013, 47(6): 2602-2611.

[33] Breuer M, Zarzycki P, Blumberger J, et al. Thermodynamics of electron flow in the bacterial deca-heme cytochrome MtrF [J]. Journal of the American Chemical Society, 2012, 134(24): 9868-9871.

[34] Sedghi G, Garcia-Suarez V M, Esdaile L J, et al. Long-range electron tunnelling in oligo-porphyrin molecular wires [J]. Nature Nanotechnology, 2011, 6(8): 517-523.

[35] Marsili E, Baron D B, Shikhare I D, et al. *Shewanella* secretes flavins that mediate extracellular electron transfer [J]. Proceedings of the National Academy of Sciences of the United States of America, 2008, 105(10): 3968-3973.

[36] Field S J, Dobbin P S, Cheesman M R, et al. Purification and magneto-optical spectroscopic characterization of cytoplasmic membrane and outer membrane multiheme c-type cytochromes from Shewanella frigidimarina NCIMB400 [J]. The Journal of Biological Chemistry, 2000, 275 (12): 8515-8522.

[37] Ertem M Z, Konezny S J, Araujo C M, et al. Functional Role of Pyridinium during Aqueous Electrochemical Reduction of CO_2 on Pt(111) [J]. Journal of Physical Chemistry Letters, 2013, 4(5): 745-748.

[38] Millo D, Harnisch F, Patil S A, et al. In situ spectroelectrochemical investigation of electrocatalytic microbial biofilms by surface-enhanced resonance Raman spectroscopy [J]. Angewandte Chemie-International Edition, 2011, 50(11): 2625-2627.

[39] Silverstein T P. Falling Enzyme activity as temperature rises: Negative activation energy or denaturation? [J]. Journal of Chemical Education, 2012, 89(9): 1097-1099.

[40] Alvarez-Idaboy J R, Mora-Diez N, Vivier-Bunge A. A quantum chemical and classical transition state theory explanation of negative activation energies in OH addition to substituted ethenes [J]. Journal of the American Chemical Society, 2000, 122(15): 3715-3720.

[41] Mozurkewich M, Benson S W. Negative activation energies and curved Arrhenius plots. 1. Theory of reactions over potential wells [J]. The Journal of Physical Chemistry, 1984, 88(25): 6429-6435.

[42] Summers Z M, Gralnick J A, Bond D R. Cultivation of an obligate Fe(II)-oxidizing *Lithoautotrophic* bacterium using electrodes [J]. Mbio, 2013, 4(1): e00420-12.

[43] Beard D A, Qian H. Relationship between thermodynamic driving force and one-way fluxes in reversible processes [J]. PLoS One, 2007, 2(1): e144.

第 6 章

EAB 外膜蛋白活性中心与媒介分子核黄素的电子传递机制

6.1
外膜蛋白细胞色素 c 活性中心的特征

　　胞外呼吸及其 EET 过程广泛存在于生物体系中[1]。具有胞外呼吸功能的微生物称为胞外呼吸菌。研究发现，一些胞外呼吸菌如 *Shewanella oneidensis* MR-1 和 *Geobacter* 硫还原菌等，它们的胞外电子转移过程可能存在三种方式：① 通过直接的细胞-金属氧化物（电子受体）直接接触的直接电子传递方式[2]；② 细胞外氧化还原电子穿梭体[3]还原金属（如 Fe 和 Mn）氧化物的间接电子传递方式[2,4]；③ 细菌纤毛组织电子传递方式[5-7]。近几年，也有研究表明在纤毛电子传递中，细菌纤毛可能来源于细菌自身外膜的延伸，故也可以归类于直接传递。在不同的菌类中，胞外电子传递过程均依赖于多血红素 c 型细胞色素蛋白（Cyt c）。MtrC 和 OmcA 都是 *Shewanella* 菌种的外膜 Cyt c，且均已被证明是 EET 过程在外膜体系上的核心电子传递蛋白[8-10]。Cyt c 中的 heme[heme(Ⅱ)]作为电子转移中的电子源，每个蛋白中有 10 个 heme(Ⅱ)以特定的顺序排列，电子定向流向最终的 heme 5[heme(Ⅱ)5][11-13]。heme(Ⅱ)是一种大的杂环有机环（卟啉环）构成的辅助因子，其中心为亚铁离子（Fe^{2+}）[14-16]。电子从 heme(Ⅱ)5 直接或通过电子穿梭体（也称为氧化还原介质）传递至胞外受体。氧化还原介质是可以被可逆地氧化和还原的有机分子，因此，在 EET 中作为电子载体[17-19]。核黄素（RF）是从希瓦氏菌属分泌的非常重要的氧化还原介质。RF 作为黄素类电子传递体，是黄素腺嘌呤二核苷酸（FAD）和黄素单核苷酸（FMN）合成的主要前驱体[21]，由 N-杂环（异咯嗪环）氧化还原活性中心组成，并作为辅因子参与多种电子转移反应[22-23]。RF 与对应的 FAD 和 FMN，一般通过一个或两个电子转移过程参与氧化还原反应[22-23]。至今，细菌中 Cyt c 活性中心 heme(Ⅱ)和电子穿梭体（RF）之间的反应途径、细胞外氧化还原介质的电子转移过程尚未完全阐明。为此，需要以光谱电化学方法来深入了解这种间接电子传递方式。

　　光谱电化学分析是物质聚焦光谱与反应导向电化学的结合，可以提供光谱或电化学方法单独无法给出的其他重要信息[24]。例如，可以立即探测中间体或氧化还原反应产物形成的物质[25]。其中，光谱手段多样，电化学方式也层出不穷，不同种类的光谱和电化学的结合更是可从多角度互相验证分析，光谱电化学方法已被用于探索类醌化合物的特征。在本章工作中设计了一款同时适用于电化学、紫外-可见光谱电化学、荧光光谱电化学的检测池，利用该池与多种实验手段探讨血红素与 RF 间的电子传递过程,模拟电子传递给血红素 hemin(Ⅲ)[heme(Ⅱ)未接受到电子时的氧化状态]后,接收到电子的 hemin(Ⅲ)与 RF 间的电子转移方式,两种物质的可能结构变化及中间产物,采用电化学 CV、DPV 方式,探讨 RF 和 hemin(Ⅲ)的氧化还原性质,结合紫外-可见光谱电化学和荧光光谱电化学方法,了解两种物质电子传递过程及变化,采用电子顺磁共振(EPR)法检测自由基和以二维相关分析检测电子传递过程中两物质结构变化的先后次序。

6.2 氧化还原活性有机分子电子转移的研究方法

6.2.1 电化学实验

1. 实验材料

血红素 hemin(Ⅲ)和核黄素 RF 购自美国 Sigma 公司,空气接触状态下均为氧化状态。分析级石墨粉(颗粒大小:300 目)和固体石蜡(融化温度:46~48 ℃)购自中国上海化学公司,石墨与石蜡两种材料用来制备所有实验中的工作电极——固体碳糊电极(CPE)。实验所用试剂为 0.1 mol·L^{-1} 磷酸盐缓冲液(PBS,1 L PBS 配方:NaCl,8 g;KCl,0.2 g;Na$_2$HPO$_4$,1.44 g;KH$_2$PO$_4$,0.24 g;H$_2$O,1 L),PBS 经 1 mol·L^{-1} HCl 溶液或 NaOH 溶液调节至 pH=7.0,hemin(Ⅲ)和 RF 的溶液浓度均为 50 μmol·L^{-1}。实验用水均为在全玻璃蒸馏装置中制备的双蒸馏水,所有样品溶液制备之后和实验之前都使用高纯度 N$_2$ 鼓吹 20～25 min 进行脱氧。所有实验都在温度 $T = 22 \pm 2\ ℃$ 下完成。

2. 光谱电化学检测池设计

常规的电化学研究是以电信号为激励和检测手段得到各种微观信息的统计结果的，难以直观、准确地反映电极和溶液界面的反应过程。通过把谱学方法（紫外-可见光、荧光、拉曼和红外光谱）应用于电化学原位测试，可实时跟踪、认识电化学过程。当改变电极电位时，电极和电解液界面区的结构和性质随之变化，相应地采集到的光学谱图也随之变化。因此，通过同步采集光谱信息和电化学信号，为研究电极过程及电极表面性质提供分子水平上的信息。如图 6.1 所示，本工作构建了一种专门用于 UV-Vis 光谱电化学和荧光光谱电化学检测的 1 mm 薄层检测池（Thin Layer Cell，TL-检测池）。TL-检测池的主要组成部分为 1 cm 石英比色皿，市售的为两个聚苯乙烯板（其中一块板内嵌石英玻璃）与电化学三电极系统。TL-检测池的两板均为 60 mm×10 mm×5 mm（长×宽×厚）的聚苯乙烯板。如图 6.1 所示，聚苯乙烯板（右）内侧设计有一个 10 mm×6 mm×2.5 mm 几何体积的中空空间，供工作电极（碳糊电极 CPE）的制作放入。右板内挖出一条凹槽，供钛丝 Ti 放置，连接碳糊电极和外部空间，以便电化学工作站的工作电极端接入。聚苯乙烯板（左）内侧，正对着碳糊电极的部位，全部置换为可用于荧光光谱检测的石英玻璃（10 mm×10 mm×4 mm），并设计出两个 2 mm 深的腔连接外侧空间以便电化学工作站的对电极和参比电极端接入，一个腔供对电极（Pt 电极）和氮气鼓吹，另一个腔供参比电极（Ag/AgCl）放置。实验所用的参比电极是特制的毛细管小参比电极，管体为直径为 2 mm 的毛细管，内

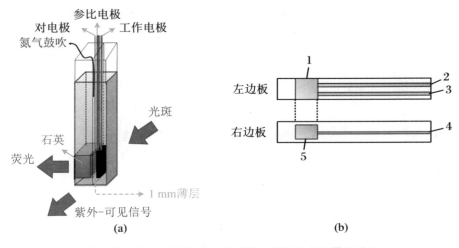

图 6.1　光谱电化学池的设计：立体图(a)；两块聚苯乙烯板内部凹槽设计(b)

部放置饱和 KCl 溶液和 Ag/AgCl 线。将两板合并放入 1 cm 比色皿内,正好空出 1 mm 薄层空间供待测溶液放置(10 mm×10 mm×1 mm,样品体积:0.1 mL)。

3. 电化学实验准备

电化学测试使用上海辰华的 CHI660E 电化学工作站,接入三电极体系:工作电极:CPE 碳糊电极;对电极:Pt 丝电极;参比电极:毛细管 Ag/AgCl 参比电极。

电化学检测(循环伏安法 CV、微分脉冲伏安法 DPV 等)均在 TL-检测池中进行。碳糊电极的制作方法如下:

(1) 将分析级石墨粉放入无水乙醇中超声。

(2) 20 min,静置 1 h,除去上层乙醇溶液(包含可能杂质),剩余石墨放烘箱中干燥(60 ℃,2 h)。

(3) 将干燥后的石墨粉和预熔蜡在玻璃表面皿中充分混合(质量比为 5.02 : 1.92),然后将熔融状态的碳糊放入聚苯乙烯右板的中空空间,等待冷却。

(4) 冷却后在粗砂纸上进行 CPE 初步打磨,接着在金相砂纸中继续打磨至平坦,CPE 表面积为 0.6 cm^2。

(5) 将初步制作的碳糊电极进行电化学活化,活化条件:活化液:1 mol·L^{-1} KCl 溶液,pH=7.0;活化电势范围:−1.0~1.0 V(vs. Ag/AgCl);扫描速度:100 mV·s^{-1};扫描圈数:60 圈;扫描间隔:0.001 V;静止时间:2 s。扫描出的 CV 图如图 6.2 所示。

(6) 扫描出的电化学信号若达到 10^{-3} A(微安级别),则该打磨的 CPE 电极性能较好,即可使用。如若未达微安级别,则需重复(3)、(4)步操作至性能达标方可。

组装好 TL-检测池后,通过注射器将已氮气鼓吹 20 min 的 0.5 mL RF、hemin(Ⅲ)、或 RF-hemin(Ⅲ)混合溶液压入薄层区域(会有些许溢出,注意不能产生气泡),保证溶液完全进入薄层和放置电极的腔体,放入三电极体系,连接 CHI660E 电化学工作站,将氮吹管放在待测液表面并通入氮气。连接完成后根据需求进行电化学 CV 或 DPV 实验。

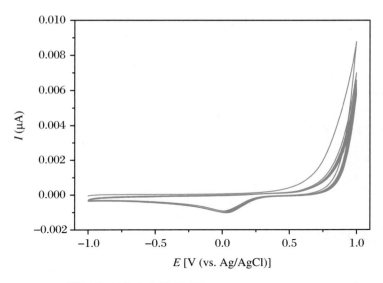

图 6.2　碳糊电极活化电化学 CV 图

6.2.2 紫外可见光谱电化学

紫外可见分光光度法是根据物质分子对波长在 200~760 nm 范围内的电磁波的吸收特性所建立起来的一种定性、定量和结构分析方法。它具有灵敏度高、选择性好、适用浓度范围广等特点,因此,将其与小型化电化学测试装置相结合,用于实时鉴别电化学反应过程的中间体种类和检测其浓度变化。如图 6.1 所示,紫外可见光谱的光路为直线光路,将放入已氮气鼓吹 20 min 的 0.5 mL 待测液和三电极体系接入电化学工作站的 TL-检测池,放入紫外可见分光光度计,将氮吹管放在待测液表面并通入氮气,保证测试光度计光路通畅,保证测试电化学工作站电极连接顺利。测试完成后根据需求同时进行紫外可见光谱和电化学实验。

6.2.3 二维相关光谱分析

二维光谱技术在谱学分析上已有广泛的应用,可实现对核磁、红外、拉曼、紫外和荧光等谱图的数学分析。二维相关光谱的原理：将光谱信号扩展到二维以提高光谱分辨率，通过简化含有许多重叠峰的复杂光谱，选择目标光谱信号以研究分子内和分子间的相互作用。本实验中采用的二维相关分析是基于紫外-可见光谱电化学数据进行分析，计算和生成 2D 轮廓图的软件是由日本 Kwansei-Gakuin 大学 Shigeaki Morita 和 Yukihiro Ozaki 教授编程的 2D-IR 相关分析软件 2D-shige version 1.3。

6.2.4

荧光光谱电化学

荧光法作为一种光谱检测技术，广泛应用于水中有机质成分分析，特别是可以进行低能量跃迁的含有芳香环结构或不饱和脂肪链的物质。目前，已发展了各种荧光光谱研究技术，比如固定激发光波长的荧光发射光潜、固定发射光与激发光波长差的同步荧光光谱及三维荧光光谱等。由于荧光光谱技术具有灵敏度高，选择性好，且不破坏样品结构的优点，非常适合与电化学技术联用。荧光光谱检测装置为美国 Perkin-Elmer 公司 LS 55 荧光分光光度计，其光源为脉冲氙气激发光源。如图 6.1 所示，荧光光谱的光路为直角光路，将放入已氮气鼓吹 20 min 的 0.5 mL 待测液和三电极体系接入电化学工作站的 TL-检测池，放入荧光分光光度计，放入氮吹管在待测液表面并通入氮气，保证测试光度计光路通畅，保证测试电化学工作站电极连接顺利。测试完成后根据需求同时进行荧光光谱和电化学实验，为检测核黄素的信号，荧光光谱实验激发光均是 $\lambda_{ex} = 444$ nm，实验激发光和发射光狭缝均为 5 nm，扫描速度为 600 nm·min^{-1}。

6.2.5

电子顺磁共振 EPR

电子顺磁共振（EPR）技术是一项检测具有未成对电子样品的波谱方法。根据泡利原理：每个分子轨道上已成对的电子自旋运动产生的磁矩是相互抵消的，只有存在未成对电子的物质才具有永久磁矩；单电子在外磁场中呈现顺磁性。EPR 技术不易受到正在进行的化学反应和物理过程的干扰，能获得有意义的物质结构信息和动态信息。电子顺磁共振 EPR 光谱检测装置为日本 JEOL 公司的 JES-FA200EPR 光谱仪。在 TL-检测池中分别电解已氮气鼓吹 20 min 的 0.5 mL RF、hemin(Ⅲ) 或 RF-hemin(Ⅲ) 混合溶液,电解 100 min 后迅速取出,在氮气环境下将电解溶液转移到石英扁平检测池中,并快速用光谱仪记录连续波 EPR 光谱。仪器设置:微波频率,9.6985 GHz;磁场调制幅:1 mT;调制频率:100 kHz;微波功率:2 mW;调制幅度:0.2 G;时间常数:300 ms。

6.3
核黄素与卟啉铁分子之间的电子转移

6.3.1
电化学结果分析

在图 6.3(a) 和图 6.3(c) 中分别显示了不同扫描速率的 PBS 中 50 μmol·L^{-1} RF 和 50 μmol·L^{-1} hemin(Ⅲ) 的 CV 图。由图 6.3(a) 可知,在 PBS 缓冲液中,RF 的氧化还原过程是可逆的。RF 的循环伏安图的阳极峰电流 I_{pa} 和阴极峰电流 I_{pc} 的比例范围为 0.921~1.024(近似为"1"),且这个比例与扫描速率、开关电位(E_λ)和溶液扩散系数均无关。这表明 RF 降低的产品相当稳定,过程是可逆的。

hemin(Ⅲ) 的氧化还原过程在 PBS 缓冲液中不是严格可逆的。hemin(Ⅲ) 的循环伏安图中 I_{pa} 和 I_{pc} 的比例在 0.674~0.848 的范围内。尽管该比例对于扫描速率、E_λ 和扩散系数也是无关的,但与 RF 的氧化还原过程相比,hemin(Ⅲ) 的氧化还原不是完全可逆的过程。

RF 可以通过可逆的单电子还原途径还原到 $RF_{rad}{}^{\cdot-}$ 或者通过双电子传递还原到 $RF_{red}{}^{2-}$。根据以下等式可以基于循环伏安图中的峰值电位差(ΔE_p)计算出 RF 在 PBS 缓冲液中的转移电子数(n):

$$\Delta E_p = E_{pa} - E_{pc} \tag{6.1}$$

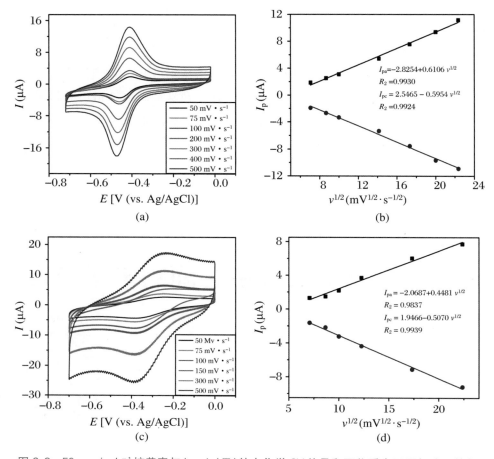

图 6.3 50 μmol·L⁻¹核黄素与 hemin(Ⅲ)的电化学 CV 信号和两物质在不同扫速下的与扫速的平方根关系图:RF 不同扫速下 CV 图(a);RF 的 I_{pa} 和 I_{pc} 与扫速的平方根关系图(b);hemin(Ⅲ)不同扫速下 CV 图(c);hemin(Ⅲ)的 I_{pa} 和 I_{pc} 与扫速的平方根关系图(d)

$$\Delta E_p = \frac{2.3\,RT}{nF} \tag{6.2}$$

其中,E_{pa} 是阳极峰值电位,E_{pc} 是阴极峰值电位,R 是热力学常数 $R=8.314$ J·(mol·K)⁻¹,T 是热力学温度,F 是法拉第常数($F=96485$ ℃·mol⁻¹)。

在不同的扫描速率下,n 的计算值在 0.991～1.083 的范围内变化(近似为"1")。因此,在水溶液中,RF 在相对于参比电极 Ag/AgCl 的 -0.70～0 V 的扫描范围内发生一个电子的还原过程。

因为 RF 和 hemin(Ⅲ)的峰值电位与扫描速率无关,将 RF 和 hemin(Ⅲ)在不同扫速下的 I_{pa} 和 I_{pc} 与扫速的平方根作图,得到图 6.3(b)和图 6.3(d),两种物质的 I_{pa} 和 I_{pc} 与扫速的平方根成线性相关。根据 Randles-Sevčik 方程[26],有

$$I_\mathrm{p} = 0.4463 nFAc_\mathrm{o}^* \left(\frac{nF}{RT}\right)^{1/2} D_\mathrm{o}^{1/2} v^{1/2} \tag{6.3}$$

其中，n 是转移的电子数，A 是电极的面积，D_o 是待测物质的扩散系数，c_o^* 是溶液中待测物质的浓度，v 是电化学扫描速度。

由方程公式可知，I_p 和扫速的平方根呈线性关系，RF 和 hemin(Ⅲ)的氧化还原过程是溶液扩散控制。RF 和 hemin(Ⅲ)的阴极峰电位(E_pc)分别为 -0.477 V 和 -0.384 V(vs. Ag/AgCl)。

在 RF 和 hemin(Ⅲ)[RF-hemin(Ⅲ)]的混合溶液中，可以通过 CV 和 DPV 分别观察它们的还原峰。图 6.4 是混合溶液在不同扫描速度下的 CV 图，可以看到，大部分扫速下，虽然氧化过程(图中上部曲线)在曲线中难以分辨两种物质的阳极峰值电位，但在还原过程中(图中下部曲线)可以比较清晰地分辨两种物质有着不同的阴极峰值电位。将还原过程单独列出，用反应更为灵敏的 DPV 扫描两种物质混合溶液，得到图 6.5。在图 6.5 中，黑色曲线为混合液扫描出的 DPV 结果，曲线经高斯分峰得到红色和蓝色两条曲线，分别对应 RF 和 hemin(Ⅲ)，RF 和 hemin(Ⅲ)的阴极峰电位[E_pc：-0.489 V 和 -0.379 V(vs. Ag/AgCl)]与两种物质单独电化学 CV 扫描的结果(-0.477 V 和 -0.384 V)相似，但又不完全相同。说明两种物质的电子传递过程的确存在，且影响了两种物质的还原过程，但因为在水溶液中，这个中间过程很难使用电化学方法监控得到具体信息，所以使用光谱电化学方法可更清楚地检测中间过程。图 6.5 可清楚证实 RF 和 hemin(Ⅲ)之间的 E_pc 差异，也正是由于该差异，可在两者的混合溶液中以光谱电化学方法原位实时监测 hemin(Ⅲ)到 heme(Ⅱ)的电化学还原过程，且该过程不影响 RF。此外，heme(Ⅱ)产生后，使用光谱电化学方法也可在混合溶液中跟踪从 heme(Ⅱ)到 RF 的电子转移过程。

6.3.2
UV-Vis 光谱电化学及二维相关结果分析

基于上述图 6.4(CV)和图 6.5(DPV)结果，可以将电化学还原电位确定为 -0.415 V(vs. Ag/AgCl)，即当水溶液含有 hemin(Ⅲ)和 RF 两种组分时，在该电位下，hemin(Ⅲ)可以电化学还原，而 RF 不能。首先，将单一组分溶液放入 TL-检测池中并给予一段时间的 -0.145 V 恒电位进行紫外可见 UV-Vis 光谱电化学实验，得到图 6.6(a)和(b)，分别对应为 RF 和 hemin(Ⅲ)。图 6.6(c)是实验得到的电化学还原 hemin(Ⅲ)的还原产物与化学还原 hemin(Ⅲ)的还原产物 heme(Ⅱ)的 UV-Vis 谱图的对比。

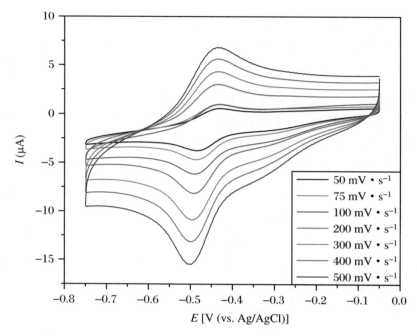

图 6.4　RF-hemin(Ⅲ)混合体系在不同扫速下的 CV 信号

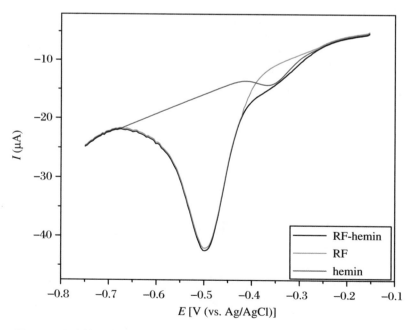

图 6.5　混合体系在不同扫速下的信号

如图 6.6(a)所示，RF 在 -0.145 V 恒电位 90 min 内光谱几乎保持不变，所以没有 RF 被电还原。但对于 hemin(Ⅲ)[图 6.6(b)],在 -0.145 V 恒电位 90 min 内,370 nm 处的宽峰(卟啉环特有的 Soret 带)峰值不断降低,存在着一定的还原。但与图 6.6(c)的对比可知,化学还原 hemin(Ⅲ)到最终产物 heme(Ⅱ),紫外可见光谱中有着 385 nm 处的 Soret 带和 560 nm 处的 Q 带,而电化学 hemin(Ⅲ)的过程不存在 Soret 带的红移和 Q 带新峰的产生,这表明电化学还原 hemin(Ⅲ)没有直接还原成 heme(Ⅱ)。根据实验结果,假定 hemin(Ⅲ)被电化学还原成中间体 hemx(Ⅱ),它是像 heme(Ⅱ)一样接受了电子,但仍然维持 hemin(Ⅲ)化学结构的一种物质。

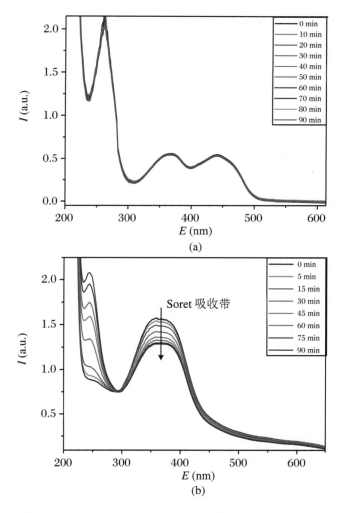

图 6.6 -0.415 V (vs. Ag/AgCl)电位下 RF 单独体系的紫外可见光谱电化学(a);-0.415 V (vs. Ag/AgCl)电位下 hemin(Ⅲ)单独体系的紫外可见光谱电化学(b);电还原和化学还原 hemin(Ⅲ)的紫外信号对比(c)

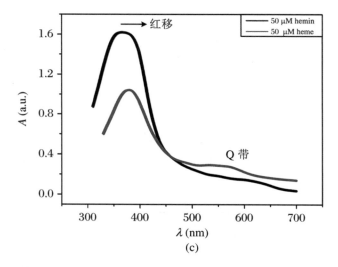

图6.6 －0.415 V (vs. Ag/AgCl)电位下RF单独体系的紫外可见光谱电化学(a)；－0.415 V (vs. Ag/AgCl)电位下hemin(Ⅲ)单独体系的紫外可见光谱电化学(b)；电还原和化学还原hemin(Ⅲ)的紫外信号对比(c)(续)

此外，图6.7中比较了RF-hemin(Ⅲ)混合物(未加电还原)、RF和hemin(Ⅲ)的三种UV-Vis光谱。施加电解电位之前，在RF-hemin(Ⅲ)混合物体系中，

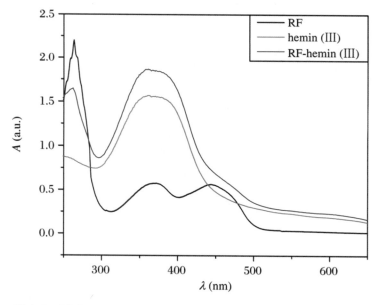

图6.7 RF、hemin(Ⅲ)、RF-hemin(Ⅲ)混合体系，三者的紫外可见光谱图

没有新的峰产生,没有峰位的红/蓝移现象,也没有峰强上的增色/减色效应。虽然混合体系中核黄素 465 nm 处的标志峰不算明显,但仍然可以看出峰的存在。结果表明,RF 与 hemin(Ⅲ)混合后的紫外-可见光谱只是两物质光谱的简单叠加,没有发生结构上的变化。

在 TL-检测池中放入 RF-hemin(Ⅲ)混合溶液,给予体系 -0.415 V(vs. Ag/AgCl)恒电位后,混合液的紫外可见光谱电化学结果示于图 6.8 中。光谱峰的变化对应 RF 和 hemin(Ⅲ)的不同状态。为详细地说明 hemin(Ⅲ)和 RF 之间的电子转移情况,将 RF_{ox}/RF_{red} 和 hemin(Ⅲ)/heme(Ⅱ)特征峰的变化列表 6.1。

表 6.1 RF_{ox}/RF_{red} 和 hemin(Ⅲ)/heme(Ⅱ)的特征峰变化

峰位置	所属物质	变化趋势	对变化的可能解释
300 nm 峰	RF	不断增加	RF 不断被还原
360~390 nm 宽峰	RF 和 hemin(Ⅲ)	1. 0~5 min:370 nm 峰不断减少; 2. 5~30 min:峰不断红移至 385 nm; 3. 30~90 min:385 nm 峰不断减少	1. 主要:hemin(Ⅲ) + e⁻ ⇌ hemx(Ⅱ)[RF 的存在加速了 hemin(Ⅲ)的电子传递给 hemx(Ⅱ)];次要:RF 的 π-π* 跃迁特征峰 372 nm 峰减少,RF 不断还原; 2. 主要:hemin(Ⅲ)或 hemx(Ⅱ)的折叠结构消失,所以 heme(Ⅱ)不断产生;次要:$RF_{ox} + e^- \rightleftharpoons RF_{rad}\cdot^{-[27]}$。 3. heme(Ⅱ)不断生成且 $RF_{rad}\cdot^-$ 产生
465 nm 小峰	RF	不断减少,30 min 后最终消失	RF 的 π-π* 跃迁特征峰 465 nm 峰不断减少并消失,RF 不断被还原[28]
560 nm 峰	hemin(Ⅲ)	30 min 后峰出现,随时间不断增加(基线不断抬升)	heme(Ⅱ)不断产生(Q 带)

由图 6.8 和表 6.1 总结,且对比图 6.6(b)中单一 hemin(Ⅲ)的还原,同样是 5 min 恒电位还原过程,因为有 RF 的加入,370 nm 处宽峰的峰强减弱速度变快,表明与没有 RF 添加的溶液相比,混合物中的 hemin(Ⅲ)被更快地还原为 hemx(Ⅱ)。接着,混合溶液的 UV-Vis 光谱在恒电位 30 min 后出现 heme(Ⅱ)的特征峰 Q 带(560 nm)。这些结果表明 RF 的存在可以加速且进一步促进 hemin(Ⅲ)的还原。还原速率的加速可能归因于 hemin(Ⅲ)中的卟啉环和 RF 的异咯嗪环之间的共轭效应。

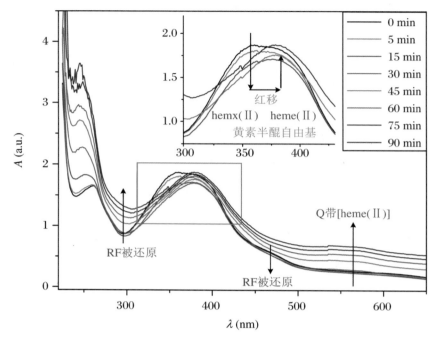

图6.8 −0.415 V (vs. Ag/AgCl) 电位下 RF-hemin(Ⅲ) 混合体系的紫外-可见光谱电化学信号

在实验探究 RF-hemin(Ⅲ) 混合溶液在两者之间电子传递的几个反应步骤（提供−0.415 V 恒电位）时，对比图6.6(a)和图6.8发现，最重要的一步则是实验证明当有 hemin(Ⅲ) 存在并不断还原产生 hemx(Ⅱ) 或 heme(Ⅱ) 时，RF 被还原。当电子连续流入混合物时，观察到在 372 nm 处的吸收峰峰值不断降低、465 nm 处的吸收峰峰值不断减少至消失，与此同时 300 nm 和 385 nm 处的吸收峰不断增加。这些变化对应氧化状态的 RF(RF_{ox}) 转化为其还原形式（黄素半醌物质：RF_{red})[15]。在 hemx(Ⅱ) 或 heme(Ⅱ) 持续不断产生后，385 nm 处的吸收峰峰值增加，可能对应连续形成一种黄素半醌自由基：$RF_{rad} \cdot ^-$[31]。以上结果表明转化为 RF_{rad} 的电子来源于 hemx(Ⅱ) 或 heme(Ⅱ)。

峰值变化次序可以从使用时间作为干扰条件轴的 RF-hemin(Ⅲ) 混合溶液的原位 UV-Vis 光谱的二维相关分析（图6.9）获得。表6.2显示了二维相关光谱的同步谱和异步谱中交叉峰的信息。通过分析两个谱图中主要的五对交叉峰，根据 Noda 规则[32]，它们的顺序如下：首先在 430 nm 处的峰值变化，这是 RF 和 hemin(Ⅲ) 光谱的重叠，然后 RF 的 465 nm 峰值（RF 的 π-π* 跃迁），接着 560 nm 和 300 nm 峰同时改变，分别对应 heme(Ⅱ) 产生和 RF 还原。

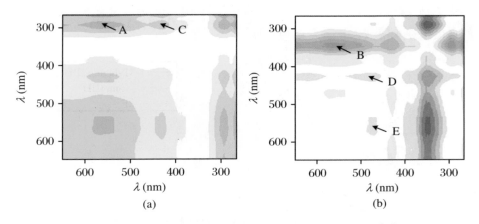

图 6.9 紫外-可见光谱电化学的二维相关光谱分析图：同步谱(a)；异步谱(b)

表 6.2 二维相关光谱中的交叉峰

编号	交叉峰(nm)	同步谱	异步谱	峰先后次序(nm)
A	(560, 300)	> 0	$= 0$	560 = 300
B	(560, 360)	$= 0$	> 0	不相关
C	(430, 300)	> 0	> 0	430 > 300
D	(465, 430)	> 0	< 0	465 < 430
E	(465, 560)	> 0	< 0	465

6.3.3
电子顺磁共振 EPR 结果分析

为了证实 RF-hemin(Ⅲ)在加电情况下，两者间发生了电子传递，RF 被还原，选择 EPR 的方法验证 RF 还原的中间产物自由基的存在。分别在 TL-检测池中加入 RF、hemin(Ⅲ)和 RF-hemin(Ⅲ)三种待测液，给予体系 -0.415 V (vs. Ag/AgCl)恒电位后，EPR 光谱结果示于图 6.10 中。实验证明，单一组分的体系没有观察到 RF 或 hemin(Ⅲ)的 EPR 信号(图 6.10)。然而，混合溶液体系在被电化学还原后，RF hemin(Ⅲ)中检测出 g 值为 2.0037 的 EPR 信号。在该体系中，只有 RF 或 hemin(Ⅲ)两种组分，且实验证明 hemin(Ⅲ)还原产物无 EPR 信号(无自由基产生)，所以该 EPR 信号最有可能为 RF 的还原产物信号，从信号的 g 值和特征谱线形状也都对应着 RF 自由基信号。参照文献得出，RF 和 hemin(Ⅲ)电子传递产生的 EPR 信号很有可能为自由基 $RF_{rad}\cdot$ 和 $RF_{red}H^-$，且

主要产物是抗磁 $RF_{red}H$[33-34]。EPR 信号超过 10 min 仍持续存在,表明检测到的自由基在水溶液体系里寿命相对较长。

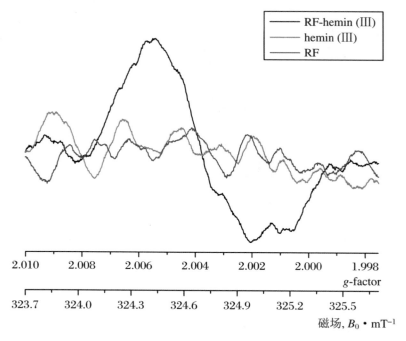

图 6.10 连续波 X 光波段 EPR 光谱中的 RF、hemin(Ⅲ)和 RF-hemin(Ⅲ)混合物在 TL-检测池中在 -0.415 V(vs. Ag/AgCl)恒电位下还原信号

6.3.4
荧光光谱电化学结果分析

由于 hemin(Ⅲ),heme(Ⅱ)和 RF_{red} 均没有荧光信号,而 RF_{ox} 本身就具有荧光信号,因此,用荧光光谱电化学的方法可以很好地监测 RF-hemin(Ⅲ)混合溶液中 RF_{ox} 的还原过程。首先对于单独 RF 组分,给予 -0.415 V(vs. Ag/AgCl)恒电位电化学信号,在 λ_{ex} = 444 nm 的激发光下,λ_{em} = 535 nm 处的荧光发射强度荧光光谱信号并没有变化(图 6.11),说明单独组分在 -0.415 V 时的确未被还原。接着对 RF-hemin(Ⅲ)混合溶液进行相同的实验,结果如图 6.12 所示,λ_{em} = 535 nm 处的荧光发射强度在最初 20 min 内急剧下降,接着经过 85 min 的还原后,RF 荧光信号从最开始 899 a.u. 下降至最终只有 54 a.u.,且在最后的

15 min 内几乎保持不变,足以说明 RF 在混合溶液体系中的还原。溶液最终的低荧光强度不再降低,表明 RF 几乎完全被还原。荧光光谱电化学的实验结果进一步验证了是 hemin(Ⅲ)还原产生的 heme(Ⅱ)和 hemx(Ⅱ)促进了 RF 的还原,说明两者发生电子传递。

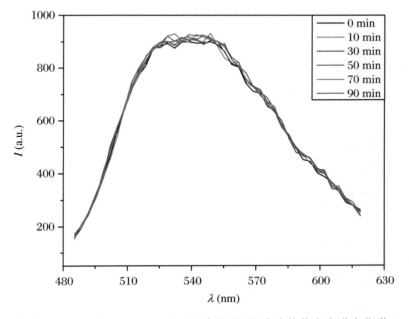

图 6.11　−0.415 V (vs. Ag/AgCl)电位下 RF 溶液的荧光光谱电化学

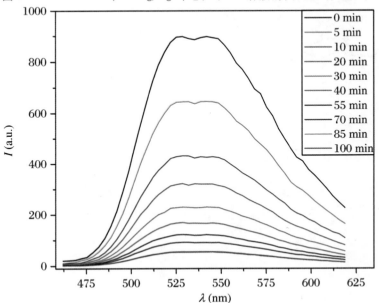

图 6.12　−0.415 V (vs. Ag/AgCl)电位下 RF-hemin(Ⅲ)混合体系的荧光光谱电化学信号

6.3.5
RF-hemin(Ⅲ)间的电子传递机制

为模拟外膜蛋白体系中的电子转移情况,通过上述所有的实验结果,可以总结出从 RF-hemin(Ⅲ)在给予电信号后的电子转移机制,两物质的结构与共轭如图 6.13 所示,电子传递机制如图 6.14 所示。首先,hemin(Ⅲ)(图 6.13(a))在给予电子流的情况下被还原成 hemx(Ⅱ)[式(6.1)],该物质为血红素还原的中间产物,它具有 hemin(Ⅲ)的结构,但带一个电荷。仅提供电子流时,无法让 hemx(Ⅱ)的结构改变为 heme(Ⅱ)[图 6.13(b)],所以需要 RF 的存在。因为 RF 和 hemx(Ⅱ)/hemin(Ⅲ)发生共轭[图 6.13(c)(上视角)和图 6.13(d)(前视角)],改变了 RF 和 hemx(Ⅱ)/hemin(Ⅲ)的电子离域状态,提高了电子转移能力,且促进了 hemx(Ⅱ)的进一步结构,成为完全还原态的 heme(Ⅱ)。与此同时,RF 也和 hemx(Ⅱ)或之后结构变化的 heme(Ⅱ)发生电子传递,RF 被第一

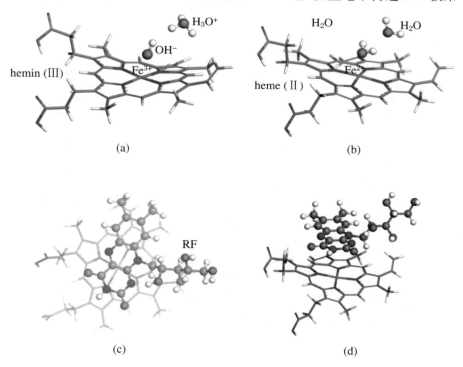

图 6.13　hemin(Ⅲ)的结构(a);heme(Ⅱ)的结构(b);RF 和 heme 结合状态(上视角)(c);RF 和 heme 结合状态(前视角)(d)

步还原产生 $RF_{rad}\cdot^-$[式(6.5)]，此时反应产生的 hemin(Ⅲ)仍然会被实验提供的电子流不断还原。连续波 X 波段 EPR 的结果显示混合物中可能存在两个自由基，且荧光光谱电化学也表明 RF 几乎被完全还原，所以可以进一步证明产生的 $RF_{rad}\cdot^-$自由基继续在体系中发生反应。$RF_{rad}\cdot^-$可以由两个不同的途径中与电子或质子发生反应。而在水溶液环境中，$RF_{rad}\cdot^-$更倾向于和质子发生质子加成产生 $RF_{rad}H\cdot$[式(6.6)]。接着，$RF_{rad}H\cdot$也可进一步被 hemx(Ⅱ)或 heme(Ⅱ)还原产生 $RF_{red}H^-$[式(6.7)]。最后发生 $RF_{red}H^-$质子化步骤以形成 $RF_{red}H_2$ 在水溶液中相对稳定存在[式(6.8)]。

$$hemin(Ⅲ) + e^- \rightleftharpoons hemx(Ⅱ) \tag{6.4}$$

$$RF_{ox} + hemx(Ⅱ)/heme(Ⅱ) \rightleftharpoons RF_{rad}\cdot^- + hemin(Ⅲ) \tag{6.5}$$

$$RF_{rad}\cdot^- + H_3O^+ \rightleftharpoons RF_{rad}H\cdot + H_2O \tag{6.6}$$

$$RF_{rad}H\cdot + hemx(Ⅱ)/heme(Ⅱ) \rightleftharpoons RF_{red}H^- + hemin(Ⅲ) \tag{6.7}$$

$$RF_{red}H^- + H_3O^+ \rightleftharpoons RF_{red}H_2 + H_2O \tag{6.8}$$

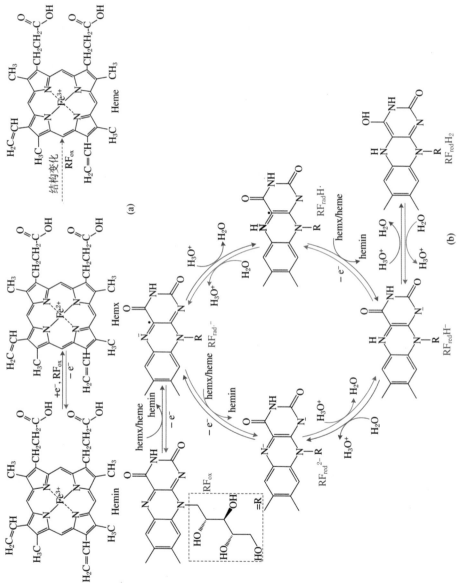

图 6.14 hemin(Ⅲ)的电子传递与结构变化(a); RF 与 hemin(Ⅲ)的电子传递机制(b)

参考文献

[1] Pankratova G, Hederstedt L, Gorton L. Extracellular electron transfer features of Gram-positive bacteria [J]. Analytica Chimica Acta, 2019, 1076: 32-47.

[2] Tokunou Y, Hashimoto K, Okamoto A. Acceleration of extracellular electron transfer by alternative redox-Active molecules to riboflavin for outer-membrane cytochrome c of *Shewanella oneidensis* MR-1 [J]. Journal of Physical Chemistry C, 2016, 120(29): 16168-16173.

[3] Huang B, Gao S, Xu Z, et al. The Functional mechanisms and application of electron shuttles in extracellular electron transfer [J]. Current Microbiology, 2018, 75(1): 99-106.

[4] Lies D P, Hernandez M E, Kappler A, et al. *Shewanella oneidensis* MR-1 uses overlapping pathways for iron reduction at a distance and by direct contact under conditions relevant for Biofilms [J]. Applied and Environmental Microbiology, 2005, 71(8): 4414-4426.

[5] Cosert K M, Castro-Forero A, Steidl R J, et al. Bottom-up fabrication of protein nanowires via controlled self-assembly of recombinant *Geobacter* Pilins [J]. Mbio, 2019, 10(6): e02721-19.

[6] Michelson K, Sanford R A, Valocchi A J, et al. Nanowires of *Geobacter sulfurreducens* require redox cofactors to reduce metals in pore spaces too small for cell passage [J]. Environmental Science & Technology, 2017, 51(20): 11660-11668.

[7] Tan Y, Adhikari R Y, Malvankar N S, et al. Expressing the *Geobacter metallireducens* PilA in Geobacter *sulfurreducens* yields Pili with exceptional conductivity [J]. Mbio, 2017, 8(1): e02203-16.

[8] Sheng A, Liu F, Shi L, et al. Aggregation kinetics of hematite particles in the presence of outer membrane cytochrome OmcA of *Shewanella oneidenesis* MR-1 [J]. Environmental Science & Technology, 2016, 50(20): 11016-11024.

[9] Han J C, Chen G J, Qin L P, et al. Metal respiratory pathway-independent Cr isotope fractionation during Cr(Ⅵ) reduction by *Shewanella oneidensis*

MR-1 [J]. Environmental Science & Technology Letters, 2017, 4(11): 500-504.

[10] Breuer M, Rosso K M, Blumberger J. Electron flow in multiheme bacterial cytochromes is a balancing act between heme electronic interaction and redox potentials [J]. Proceedings of the National Academy of Sciences of the United States of America, 2014, 111(2): 611-616.

[11] Hartshorne R S, Jepson B N, Clarke T A, et al. Characterization of *Shewanella oneidensis* MtrC: a cell-surface decaheme cytochrome involved in respiratory electron transport to extracellular electron acceptors [J]. Journal of Biological Inorganic Chemistry, 2007, 12(7): 1083-1094.

[12] Hopfield J J. Electron transfer between biological molecules by thermally activated tunneling [J]. Proceedings of the National Academy of Sciences of the USA, 1974, 71(9): 3640-3644.

[13] Poulos T L. Structural biology of heme monooxygenases [J]. Biochemical and BioPhysical Research Communications, 2005, 338(1): 337-345.

[14] Loew G H, Harris D L. Role of the heme active site and protein environment in structure, spectra, and function of the cytochrome p450s [J]. Chemical Reviews, 2000, 100(2): 407-420.

[15] Edwards Marcus j, Hall A, Shi L, et al. The crystal structure of the extracellular 11-heme cytochrome UndA reveals a conserved 10-heme motif and defined binding site for soluble iron chelates [J]. Structure, 2012, 20(7): 1275-1284.

[16] Van Der Zee F R, Cervantes F J. Impact and application of electron shuttles on the redox (bio)transformation of contaminants: A review [J]. Biotechnology Advances, 2009, 27(3): 256-277.

[17] Grattieri M, Rhodes Z, Hickey D P, et al. Understanding Biophotocurrent generation in photosynthetic purple bacteria [J]. Acs Catalysis, 2019, 9(2): 867-873.

[18] Nie H, Nie M, Wang L, et al. Evidences of extracellular abiotic degradation of hexadecane through free radical mechanism induced by the secreted phenazine compounds of P-aeruginosa NY3 [J]. Water Research, 2018, 139: 434-441.

[19] Wang W L, Min Y, Yu S S, et al. Probing electron transfer between hemin and riboflavin using a combination of analytical approaches and theoretical

calculations [J]. Physical Chemistry Chemical Physics, 2017, 19(48): 32580-32588.

[20] Li J, Tang Q, Li Y, et al. Rediverting Electron flux with an engineered CRISPR-ddAsCpf1 system to enhance the pollutant degradation capacity of *Shewanella oneidensis* [J]. Environmental Science & Technology, 2020, 54(6): 3599-3608.

[21] Martin C B, Tsao M-L, Hadad C M, et al. The reaction of triplet flavin with indole. A study of the cascade of reactive intermediates using density functional theory and time resolved infrared spectroscopy [J]. Journal of the American Chemical Society, 2002, 124(24): 7226-7234.

[22] Kaim W, Fiedler J. Spectroelectrochemistry: the best of two worlds [J]. Chemical Society Reviews, 2009, 38(12): 3373-3382.

[23] Liu L, Zeng L, Wu L, et al. Label-free surface-enhanced infrared spectroelectrochemistry studies the interaction of cytochrome c with cardiolipin-containing membranes [J]. Journal of Physical Chemistry C, 2015, 119(8): 3990-3999.

[24] Hui Y H, Chng E L K, Chng C Y L, et al. Hydrogen-bonding interactions between water and the one- and two-electron-reduced forms of vitamin K-1: Applying quinone electrochemistry to determine the moisture content of non-aqueous solvents [J]. Journal of the American Chemical Society, 2009, 131(4): 1523-1534.

[25] Wang B, Yu S, Shannon C. Reduction of 4-nitrothiophenol on Ag/Au bimetallic alloy surfaces studied using bipolar Raman spectroelectrochemistry [J]. Chemelectrochem, 2020, 7(10): 2236-2241.

[26] Sokolova R, Degano I, Ramesova S, et al. The oxidation mechanism of the antioxidant quercetin in nonaqueous media [J]. Electrochimica Acta, 2011, 56(21): 7421-7427.

[27] Chen W, Chen J J, Lu R, et al. Redox reaction characteristics of riboflavin: A fluorescence spectroelectrochemical analysis and density functional theory calculation [J]. Bioelectrochemistry, 2014, 98: 103-108.

[28] He J B, Ma G H, Chen J C, et al. Voltammetry and spectroelectrochemistry of solid indigo dispersed in carbon paste [J]. Electrochimica Acta, 2010, 55(17): 4845-4850.

[29] He J B, Wang Y, Deng N, et al. Study of the adsorption and oxidation of

antioxidant rutin by cyclic voltammetry-voltabsorptometry [J]. Bioelectrochemistry (Amsterdam, Netherlands), 2007, 71(2): 157-163.

[30] Xiao M D, Bo Y, Hai Y Z, et al. Generalized two-dimensional correlation spectroscopy [J]. Science in China Series B: Chemistry, 2004, 47(3): 257-266.

[31] Niemz A, Imbriglio J, Rotello V M. Model systems for flavoenzyme activity: One- and two-electron reduction of flavins in aprotic hydrophobic environments [J]. Journal of the American Chemical Society, 1997, 119(5): 887-892.

[32] Wang X A, Xiang K W, Nie Y J, et al. Intermediate state and weak intermolecular interactions of alpha-trans-1, 4-Polyisoprene during the gradual cooling crystallization process investigated by In situ FTIR and two-dimensional infrared correlation spectroscopy [J]. Macromolecular Research, 2013, 21(5): 493-501.

[33] El-Naggar M Y, Wanger G, Leung K M, et al. Electrical transport along bacterial nanowires from Shewanella oneidensis MR-1 [J]. Proceedings of the National Academy of Sciences of the United States of America, 2010, 107(42): 18127-18131.

[34] De La Escosura-Muniz A, Ambrosi A, Merkoci A. Electrochemical analysis with nanoparticle-based biosystems [J]. Trac-Trends in Analytical Chemistry, 2008, 27(7): 568-584.

第 7 章

氧化还原媒介分子结构对微生物 EET 路径的调控

7.1
氧化还原媒介分子的质子耦合电子传递反应

吩嗪是一类氧化还原媒介分子,可不断在微生物细胞和胞外氧化物之间来回穿梭,并在微生物的氧化还原过程中起到重要的作用[1-2]。这类物质在胞内氧化还原循环中可还原氧气分子,使活性氧不断积累[3-4],并可促进环境中矿物的还原[5]。天然存在和人工合成的吩嗪派生物对细菌相互作用和代谢的影响已有报道[6]。在微生物燃料电池(MFC)中,吩嗪能够使阳极上的多层生物膜具有很高的导电性,从而增强产电效果[7]。

吩嗪的氧化还原反应是质子耦合电子转移反应(PCET)[8-9],这种反应在许多化学和生物过程中发生,如酶促反应和光合作用[10]。再者,腐殖质[11-12]及其类似物蒽醌-2,6-二磺酸盐(2,6-AQDS)[13]均含有氧化还原活性的醌类基团。这些复合物在 Fe(Ⅲ)还原的 PCET 过程中可以作为外源电子穿梭体(ES)。ES 的电子传导能力主要取决于其主要的可还原基团,如对苯醌、邻苯醌和腐殖质中一些含氮官能团和含硫官能团等非醌类基团[14-15]。由于腐殖质含有种类非常广泛的可还原基团,其在还原电势很宽的范围内均能与质子耦合转移电子。含氮杂环的吩嗪分子具有与腐殖质和 2,6-AQDS 类似的功能,也可作为外源氧化还原介体分子[5]。含氮杂环的吩嗪分子的电子传导能力取决于耦合的质子-电子传递过程,这与三元有机杂环的性质密切相关[4]。但是,关于吩嗪如何参与多种电子传递过程的信息目前非常有限,尤其是吩嗪分子的吸电子基团(EWGs)与供电子基团(EDGs)特征以及功能基团对电子转移的影响至今仍不清楚。

杂环上取代基的不同是吩嗪分子物理和化学性质不同的主要原因。据报道,取代基的特点和位置主要决定了吩嗪分子的氧化还原电势、极性和稳定性[16-17],而且可以极大程度地影响 MFC 中的电子传递和电流产生[18]。因此,阐明吩嗪分子的取代基效应,找出最适合的取代基类型和 ES 的分子结构,对环境的生物修复和能量转换具有重要意义。

为了优选合适的氧化还原媒介分子,通常会进行电化学测试或 MFC 运行试验,但这些实验通常费时、费力。而量子化学计算提供了一种有效的具选择性的高速筛选新型氧化还原媒介分子结构的方法[19]。已有文献报道,采用一种结合量子力学和分子力学的方法可以模拟黄素结合的醌氧化还原酶2的电子和质子的加成反应[20]。具有不同取代基的黄素分子，其能量的电化学能量变化采用M06-L密度函数进行模拟，结果发现官能团对黄素分子的氧化还原电位有很大的影响[21]。这说明量化计算可以成为探索吩嗪取代基对电子传递影响的有效手段。

黄素具有多种化学行为的主要原因是可进行连续的($2e^-/1H^+$)反应和同步的($2e^-/2H^+$)反应[21-22]。因此，吩嗪 PCET 过程的量化计算需要考虑连续发生或同步发生两种情况，而水合质子是另一个需要考虑的重要因素，其结构和性质是水化学研究最基本的方面[23]。在水环境系统中，质子化的水簇化合物包括水合氢离子(H_3O^+)、Zundel 阳离子($H_5O_2^+$)[23]、$H_7O_3^+$[24]、Eigen 阳离子($H_9O_4^+$)[25]和以小水簇化合物为核心的大尺寸阳离子。这些水合质子的存在可能也会对 PCET 反应产生影响，但是这个影响因素在已有的 PCET 反应计算中被忽略。

因此，本章主要讨论的是采用基于密度泛函理论(DFT)的量化计算方法来探索水溶液中多种取代基对吩嗪分子电化学性质的影响。此外，还计算了吩嗪 PCET 反应过程中每一步的自由能和氧化还原电位的变化，详细分析环绕在吩嗪分子周围的水合团簇的影响，了解其在质子耦合还原反应中的作用，同时进行电化学实验以验证计算结果。研究结果可以为设计新的氧化还原媒介分子提供有用的结构信息，降低电子传递过程中的能量损失，提高 MFC 的能量转换效率。

7.2 吩嗪分子衍生物氧化还原电位研究方法

7.2.1 电化学实验

吩嗪类物质是一类具有氧化还原响应性的分子，电化学技术可改变其氧化还原状态,使其分子结构在氢键受体和氢键给体之间切换。此外,对吩嗪分子结构作不同取代基修饰,可以调控其氧化还原响应的电化学窗口。为证实计算结果,挑选了几种取代后的吩嗪分子,并采用循环伏安(CV)测定氧化还原电势。挑选的其中一种吩嗪类物质是中性红(NR),其在六氯-1,3-丁二烯、四氯乙烯和六氯苯的微生物厌氧还原中,与V_{B12}具有相似的反应活性[28]。另外一种物质是平面富电子的杂环二胺——2,3-二氨基吩嗪(DAP)[29-30]。所有溶液均采用优级纯化学试剂配制。NR、K_2HPO_4、KH_2PO_4和KCl均购自中国化学试剂国药控股有限公司。Phenazine-1-hydroxy和DAP购自中国北京J&K科技有限公司。

CV采用CHI852C电化学工作站(CHI Instruments Co.,China)进行记录。一个直径2.0 mm的金片用氧化铝抛光后,用水冲洗并超声处理,然后作为工作电极。另外体系中使用Pt对电极和Ag/AgCl参比电极。CV实验在室温(约25 ℃)下磷酸盐缓冲液(pH=6.95,包含0.05 mol·L^{-1} K_2HPO_4 + KH_2PO_4以及0.1 mol·L^{-1} KCl)中进行。在测试之前,电解液要通入氮气除氧,所有电位数据均相对于标准氢电极(SHE),其中换算公式是 E vs. SHE = E vs. Ag/AgCl + 0.197 V,以便于与文献中的相关研究进行比较。

7.2.2
量化计算

水溶液中含取代基的吩嗪分子结构和质子化水簇化合物均采用DFT方法进行计算。计算中采用$DMol^3$程序[31,32]中广义梯度近似(GGA)[33,34]中的Perdew-Wang 91(PW91)泛函来计算交换相关能量部分,使用全电子方法进行处理。另外还采用了包含有p轨道极化函数的双精度数值基组(DNP)进行计算。每个几何优化周期中的能量的收敛精度是$1×10^{-5}$ Hartree,最大位移和梯度分别是$5×10^{-3}$ Å和$2×10^{-3}$ Hartree·$Å^{-1}$。水介质的溶剂效应用类导体屏蔽模型(COSMO)进行描述[35,36]。COSMO是一种连续介质溶剂模型,其中溶质分子形成一个介电常数为ε的介电连续的空腔,用来代表溶剂。

7.2.3
吩嗪的质子耦合电子传递(PCET)反应

吩嗪的还原可能包含形成 semi-吩嗪自由基的不完整 $1e^-/1H^+$ 过程和形成稳定的二氢吩嗪的完整 $2e^-/2H^+$ 过程(图 7.1)[17],这与黄素的还原过程类似[20,21]。因此,吩嗪具有三种氧化态:完全氧化(Phz)、不稳定的 semi-吩嗪自由基阴离子($Phz \cdot ^-$)和两电子还原后的二价阴离子(Phz^{2-})。每种氧化态都具有相应的质子化形态,如图 7.1 所示。$Phz \cdot ^-$ 可以迅速转化成质子化中性的 semi-吩嗪形态,$PhzH \cdot$(第 2 步)在经过 $1e^-$ 还原后会形成二价阴离子 Phz^{2-}(第 5 步)。后一个反应过程属于质子-电子共同转移途径[37]。阴离子—氢化吩嗪阴离子 $PhzH^-$ 能够通过中性的 semi-吩嗪自由基 $PhzH \cdot$ 还原形成(第 3 步),或通过 Phz^{2-} 的质子化形成(第 6 步)。由于这些分子具有高对称结构,吩嗪分子只具有两个不同的单取代位置,即 R_1 和 R_2。因此,基于理论计算,能够评估 EWGs 和 EDGs 对吩嗪的氧化还原电势的影响。另外,水合质子可能在质子化反应中取代质子,因而也对水合质子参与的反应和转化进行了计算。

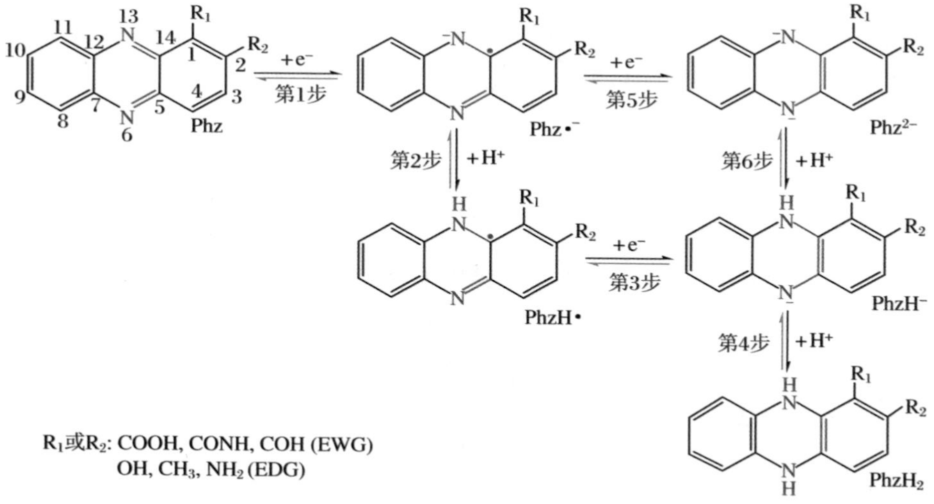

R_1 或 R_2: COOH, CONH, COH (EWG)
OH, CH_3, NH_2 (EDG)

图 7.1 水溶液中吩嗪的 PCET 反应机制

7.3 吩嗪分子取代基对电化学性质的影响

7.3.1 电荷分布和几何结构分析

吩嗪分子结构中的氮原子(N_{13}，N_6)是活性位点，因此，电荷分布和三元环的几何结构变化会对电子传递过程产生重要的影响。表7.1列出了在PW91/DNP基组水平上计算出的典型几何参数和吩嗪中活性位点的电荷分布。

表7.1 吩嗪三元环最低能量结构中的电荷分布和中心环的几何结构分析

吩嗪	N_{13} (e)	N_6 (e)	$l(N_{13}-C_{12})$ (Å)	$l(N_{13}-C_{14})$ (Å)	$\theta(C_5-N_6-C_7)$ (°)	$\theta(C_{12}-N_{13}-C_{14})$ (°)
1-COOH	−0.328	−0.368	1.346	1.342	117.357	117.530
1-CONH	−0.337	−0.372	1.346	1.343	117.290	117.387
1-COH	−0.311	−0.374	1.348	1.340	117.242	117.510
1-OH	−0.404	−0.369	1.347	1.338	117.562	117.261
1-CH$_3$	−0.384	−0.381	1.347	1.344	117.367	117.599
1-NH$_2$	−0.382	−0.377	1.349	1.338	117.710	117.676
2-COOH	−0.363	−0.364	1.344	1.348	117.019	116.934
2-CONH	−0.367	−0.369	1.345	1.347	116.975	116.974
2-COH	−0.370	−0.370	1.343	1.349	117.108	117.007
2-OH	−0.370	−0.366	1.348	1.347	117.180	117.159
2-CH$_3$	−0.382	−0.378	1.348	1.346	117.078	117.137
2-NH$_2$	−0.378	−0.374	1.349	1.352	117.184	117.247
NR	−0.377	−0.370	1.353	1.356	117.071	117.145
DAP	−0.372	−0.372	1.356	1.344	117.213	117.213

电荷分布结果显示取代基导致了N_{13}和N_6原子上电荷分布出现差别。R_1上的吸电子基团，如—COOH，—CONH和—COH，将电子吸引远离氧化还原活性中心，并导致了N_{13}原子上电子的减少。如Phz-1-COOH中N_{13}的电荷数为−0.328，而N_6的绝对电荷数要大一些。因此，吩嗪中N_{13}的R_1位置的吸电子

基团的亲质子性较弱并更容易接受电子。相反,供电子基团包括—OH,—CH$_3$ 和—NH$_2$将电子推向氧化还原活性中心,使其具有更强的亲核性并更难获得电子。三环含氮结构的几何对称性被取代基所破坏,在 Phz-1-EWG 中 C—N—C 角 $\theta(C_{12}—N_{13}—C_{14})$ 大于 $\theta(C_5—N_6—C_7)$。然而由于 R_2 位置离氧化还原活性中心较远,处于这个位置的吸电子基团和供电子基团的影响较小。对于 NR 分子,供电子基团—N(CH$_3$)$_2$,—NH$_2$ 和—CH$_3$ 位于 3 个 R_2 位置,对电荷分布和几何结构影响较小。这些结果说明吩嗪分子的物化性质均会受到分子结构的影响,包括电子转移热力学和氧化还原电位。

7.3.2
PCET 反应的热力学性质

吩嗪的还原反应可以描述成耦合的质子-电子转移过程,在这个过程中电子的转移可以在质子转移后迅速发生或两者同时进行。热力学驱动力决定了两种反应的氧化还原电位,并会影响电子转移动力学[8]。

图 7.1 的反应步骤显示,水溶液中 $1e^-/1H^+$ 反应的 Gibbs 自由能的变化可以由第 1 步和第 2 步的和获得:

$$\Delta G^\ominus(1e^-/1H^+) = \Delta G_1^\ominus + \Delta G_2^\ominus \tag{7.1}$$

完整的 $2e^-/2H^+$ 过程的 Gibbs 自由能变化值可以通过以下方程获得:

$$\begin{aligned}\Delta G^\ominus(2e^-/2H^+) &= \Delta G_1^\ominus + \Delta G_2^\ominus + \Delta G_3^\ominus + \Delta G_4^\ominus \\ &= \Delta G^\ominus(1e^-/1H^+) + \Delta G_3^\ominus + \Delta G_4^\ominus\end{aligned} \tag{7.2}$$

中性的一氢化吩嗪的第二个质子加成反应取决于溶液 pH 和 pK_a,$2e^-/1H^+$ 过程是一氢化吩嗪的去质子反应:

$$\begin{aligned}\Delta G^\ominus(2e^-/1H^+) &= \Delta G_1^\ominus + \Delta G_2^\ominus + \Delta G_3^\ominus \\ &(\text{或} = \Delta G_1^\ominus + \Delta G_5^\ominus + \Delta G_6^\ominus)\end{aligned} \tag{7.3}$$

质子化的离子,自由基或分子可以通过加质子步骤的 $\Delta G^\ominus$ 值来计算其 pK_a 值。

$$pK_a = \frac{-\Delta G^\ominus}{2.303RT} \tag{7.4}$$

对于各种取代吩嗪和水合质子,计算了水溶液中三种耦合质子-电子转移过程的自由能变化。结果显示吩嗪的 PCET 反应能够在室温和大气压下自然发生(图 7.2)。电化学实验在中性条件(pH=6.95)和室温下(~25 ℃)进行,这与计

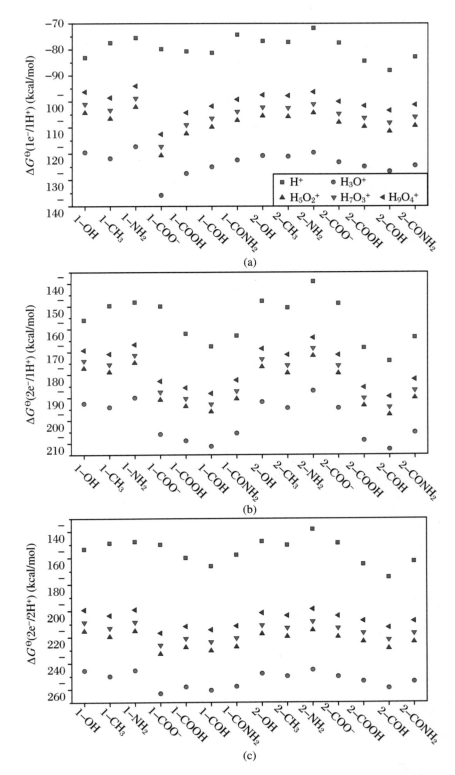

图 7.2 水溶液中吩嗪衍生物与水合质子发生的三种不同的 PCET 反应的标准 Gibbs 自由能变：$1e^-/1H^+$ (a)；$2e^-/1H^+$ (b) 和 $2e^-/2H^+$ (c)

算中的标准条件(298.15 K,100 kPa)保持一致。游离的质子系统的基元反应过程自由能变化如表7.2所示。电子加成后自由能的变化量大于质子加成后的变化量,这说明了整个反应自由能变化主要是由电子加成反应引起的。此外还可进一步观察到两个连续的电子加成反应的能量处于相同的数量级上。

表7.2 水溶液吩嗪衍生物与游离质子进行PCET基元反应的标准Gibbs自由能(kcal/mol)以及$PhzH_2$的pK_a值

吩嗪	$\Delta G_1^\ominus$	$\Delta G_2^\ominus$	$\Delta G_3^\ominus$	$\Delta G_4^\ominus$	$pK_a/PhzH_2$
1-OH	−79.924	−3.192	−72.916	2.689	1.97
1-CH$_3$	−77.566	0.136	−72.263	0.924	0.68
1-NH$_2$	−75.517	0.017	−72.657	0.422	0.31
1-COO$^-$	−80.364	0.553	−70.172	0.106	0.08
1-COOH	−80.272	−0.368	−81.289	2.049	1.50
1-COH	−86.718	5.478	−86.242	1.329	0.97
1-CONH$_2$	−79.740	5.438	−83.084	−0.330	−0.24
2-OH	−77.156	0.449	−70.969	0.247	0.18
2-CH$_3$	−78.023	0.895	−73.414	0.291	0.21
2-NH$_2$	−72.702	0.925	−67.328	0.783	0.57
2-COO$^-$	−76.420	−0.990	−71.279	−0.045	−0.03
2-COOH	−85.759	1.276	−83.702	3.474	2.55
2-COH	−88.860	0.744	−85.891	−0.637	−0.47
2-CONH$_2$	−83.359	0.420	−80.598	1.092	0.80
NR	−66.960	0.737	−62.141	−1.062	−0.78
DAP	−67.816	−0.075	−66.704	1.976	1.45

自由能主要受到吩嗪分子上取代基的位置影响(图7.2)。吸电子基团,如—COOH、—COH和—CONH,使吩嗪分子的亲核性减弱并使其倾向于形成稳定的阴离子或富电子结构[38]。带吸电子基团后的吩嗪分子的自由能变得更负,使得与相同的水合质子反应后更容易接受电子。供电子基团,如—OH、—CH$_3$和—NH$_2$,则具有相反的效果,使吩嗪具有较强的亲核性[38]。所以带有供电子基团的吩嗪分子具有比带吸电子基团的吩嗪分子高的自由能,但对于三种PCET反应都为负值,说明可以自发进行。例如,$\Delta G^\ominus$(1e$^-$/1H$^+$)、$\Delta G^\ominus$(2e$^-$/1H$^+$)和$\Delta G^\ominus$(2e$^-$/2H$^+$)分别是−83.116 kcal·mol^{-1}、−156.033 kcal·mol^{-1}和−153.344 kcal·mol^{-1}。吩嗪分子的R$_1$或R$_2$位置被—COOH取代后,与苯甲酸具有类似的结构。苯甲酸的pK_a值是4.20[39],因此,吩嗪上的羧基在

pH 为 7 时能够轻易地解离形成羧基阴离子,而微生物通常适合在中性条件下生长和发挥活性[40]。

水合质子团簇的种类对 PCET 的热力学性质存在明显的影响。图 7.3 是各种水合质子团簇能量最小化时的结构。在生物系统的水环境中,氢离子以水合结构形式存在。水分子的近程效应可能在质子加成过程中起到了重要的作用,归因于质子转移过程中的弱相互作用力[41]。对于相同的 PCET 反应,游离质子反应过程的自由能高于水合质子团簇反应过程。随着水团簇尺寸变大,其相应的质子-电子转移反应的自由能也逐步增加,但仍是热力学可行的。此外,H_3O^+ 体系和其余的三种水合质子团簇体系进行质子加成作用时的自由能变化存在较为明显的差异($\Delta G_2^{\ominus}$ 和 $\Delta G_4^{\ominus}$)[图 7.2(a)～(c)]。

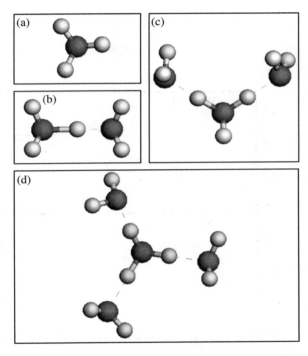

图 7.3 水合质子团簇的能量最小化结构:水合氢离子 H_3O^+ (a);Zundel 阳离子 $H_5O_2^+$ (b);$H_7O_3^+$ (c)和 Eigen 阳离子 $H_9O_4^+$ (d)

随着水合质子团簇尺寸的增大,这种差别会减小。因此,可以推出水合质子结构和取代基都对吩嗪在水溶液中氧化还原反应的热力学性质存在影响。

7.3.3

标准氧化还原电势

通过 DFT 方法结合 COSMO 确定的溶剂效应,计算了水溶液中吩嗪的单电子和双电子还原中的还原电势。半反应的标准电极电位($E^{\ominus}$ vs. SHE)定义为

$$E^{\ominus} = \frac{-\Delta G^{\ominus}}{nF} - E_{H}^{\ominus} \tag{7.5}$$

其中,n 代表反应式左侧的电子数,F 代表法拉第常数[23.06 kcal·(mol·V)$^{-1}$],$E_H^{\ominus}$ 代表标准氢电极的标准还原电位(4.28 V)。

CV 实验测得挑选的吩嗪分子的氧化还原电位(图 7.4),该实验结果直接证实了上述计算结果。

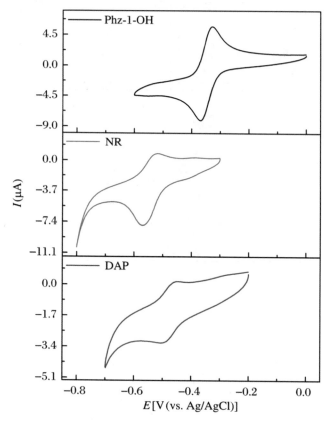

图 7.4　PBS(pH 6.95)中 0.2 mmol·L^{-1} 吩嗪的 CV 图,工作电极为 2 mm 的平板电极,扫速为 100 mV·s^{-1}

同时获得了 pH 为 6.95 的 PBS 溶液中 Phz-1-OH 的不同扫速下的 CV 曲线(图 7.5)。从图中可以看出,阳极电流与阴极电流的比例接近定值,且峰电流与扫速的平方根成线性关系,说明受扩散控制[17,42]。峰电位几乎与扫速无关,这意味着吩嗪衍生物的电极反应是可逆的。这些结果说明表 7.3 中的半波电位,即阳极和阴极峰电位的平均值,能够准确代表在相同实验条件下的形式还原电位。

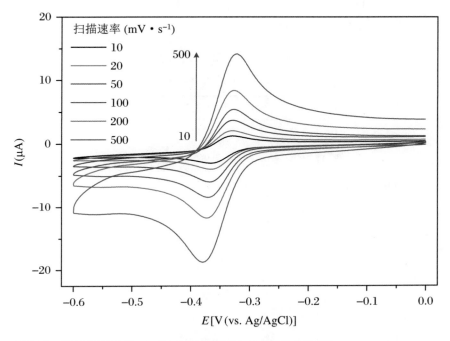

图 7.5 不同扫速下 Phz-1-OH 在 PBS 溶液(pH 6.95)中的 CV

与电子-质子转移反应过程中的自由能一样,氧化还原电位也受官能团和水合质子的影响。此外,第二个质子加成反应的自由能($\Delta G_4^{\ominus}$)说明了这一步对于大多数吩嗪来说是很难发生的(表 7.2)。例如 Phz-1-OH、Phz-1-COO$^-$ 和 DAP 的 $\Delta G_4^{\ominus}$ 值分别是 2.689 kcal·mol^{-1}、0.106 kcal·mol^{-1} 和 1.976 kcal·mol^{-1}。正值表明第二个质子加成反应是在热力学上非自发进行的。虽然第一个质子的加成反应的 $\Delta G_2^{\ominus}$ 与 $\Delta G_4^{\ominus}$ 相似,但随后的电子转移步骤的自由能是负的且能够自发进行。因而反应平衡会发生移动从而驱动前一个基元反应进行。计算结果与测得的实验数据的比较可以判断循环伏安实验中的 PCET 反应类型。生成 PhzH· 的 1e$^-$/1H$^+$ 过程如图 7.1 所示,而由于自由基的高反应活性,实验中难以检测 PhzH· 的形成过程。此外,由于 $\Delta G_4^{\ominus}$(表 7.2)的值已知,二氢吩嗪的 pK_a 值能够由式(7.4)计算出,且由于其 pK_a 小于 7,说明了二氢吩嗪能够在中性环境里大量的去质子化。因此,在这种情况下 2e$^-$/1H$^+$ 产生的一氢化吩嗪阴离子的过程是实际体系中发生的 PCET 反应类型。

表 7.3 中的数据说明了含供电子基团的吩嗪分子的 PCET 反应可能是与水合质子 H_3O^+ 发生了加成反应。例如,计算出的 NR 与 H_3O^+ 进行的 $2e^-/1H^+$ 过程对应的还原电位为 -0.334 V,与实验数据吻合得很好。然而,带有吸电子基团的吩嗪分子应该从 $H_5O_2^+$ 获得质子,如 Phz-1-$CONH_2$ 的计算结果是 -0.154 V,这与 Wang 和 Newman 报道的实验结果[17]非常吻合。实验结果说明了对于 $2e^-/1H^+$ 反应,带供电子基团和吸电子基团的吩嗪分子会分别从 H_3O^+ 和 $H_5O_2^+$ 获得质子进行反应。

表 7.3 水溶液中吩嗪衍生物还原电位(vs. SHE, V)计算值与测量值($E_{exp}^{\ominus}$, pH 7.0)的比较

吩嗪	质子的形成	$E^{\ominus}(2e^-/2H^+)$	$E^{\ominus}(2e^-/1H^+)$	$E_{exp}^{\ominus}$
Phz-1-OH	H^+	-0.955	-0.897	-0.174[17] $-0.152*$
	H_3O^+	1.046	-0.108	
	$H_5O_2^+$	0.387	-0.437	
	$H_7O_3^+$	0.244	-0.509	
	$H_9O_4^+$	0.039	-0.611	
NR	H^+	-1.474	-1.497	-0.325[28] $-0.345*$
	H_3O^+	1.027	-0.334	
	$H_5O_2^+$	0.236	-0.730	
	$H_7O_3^+$	0.094	-0.801	
	$H_9O_4^+$	-0.111	-0.904	
DAP	H^+	-1.405	-1.362	$-0.281*$
	H_3O^+	0.904	-0.265	
	$H_5O_2^+$	0.245	-0.594	
	$H_7O_3^+$	0.102	-0.665	
	$H_9O_4^+$	-0.103	-0.768	
Phz-1-COO^-	H^+	-1.030	-1.028	-0.116[17] -0.177[43]
	H_3O^+	1.426	0.186	
	$H_5O_2^+$	0.767	-0.144	
	$H_7O_3^+$	0.624	-0.215	
	$H_9O_4^+$	0.419	-0.317	

续表

吩嗪	质子的形成	$E^{\ominus}(2e^-/2H^+)$	$E^{\ominus}(2e^-/1H^+)$	$E^{\ominus}_{exp}$
	H^+	−0.863	−0.750	
	H_3O^+	1.310	0.176	
Phz-1-$CONH_2$	$H_5O_2^+$	0.651	−0.154	−0.140[17] −0.115[44]
	$H_7O_3^+$	0.508	−0.225	
	$H_9O_4^+$	0.303	−0.327	

（**来自本章中的 CV 实验，可以根据图 7.4 得到。）

图 7.6 中展示了带有不同取代基吩嗪分子的还原电位。吸电子基团的存在会减小吩嗪环上的电子云密度，并使其更容易稳定富电子结构。吩嗪中的吸电子基团使其更容易获得电子并增加 $2e^-/1H^+$ 过程中的还原电位。相反，供电子基团则具有相反的效果。在 NR 的 R_2 位置上引入—$N(CH_3)_2$ 和—CH_3 等供电子基团具有降低其氧化还原电位的效果（表 7.3）。

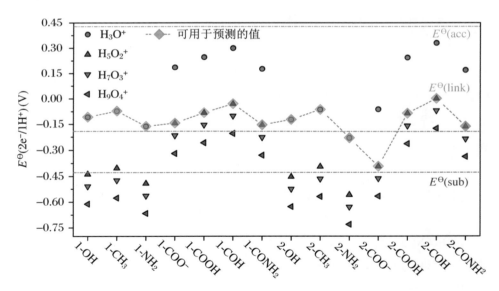

图 7.6 水溶液中吩嗪衍生物电子传递过程中的氧化还原电位，$E^{\ominus}$(sub)，$E^{\ominus}$(link) 和 $E^{\ominus}$(acc) 分别是底物、氧化还原媒介和最终电子受体的氧化还原电位

7.3.4
能量转化效率

氧化还原电位对氧化还原反应的活化和反应均有重要的影响[8]。氧化还原媒介[也被称作细胞与电极之间的链接分子(linking species)]的标准电位 $E^{\ominus}$(link),理想条件下应该接近于主要底物的氧化还原电位 $E^{\ominus}$(sub),或至少显著负于最终受体的 $E^{\ominus}$(acc)[18]。在电子传递途径中,微生物获取的能量可以定义为[18]

$$\Delta G_{biol}^{\ominus} = -nF[E^{\ominus}(\text{link}) - E^{\ominus}(\text{sub})] \tag{7.6}$$

由于微生物的主要消耗底物不同,其对应的 $E^{\ominus}$(sub)可能具有很大差别。例如,CO_2/葡萄糖的 $E^{\ominus}$ 值为 -0.43 V;CO_2/乙酸盐的 $E^{\ominus}$ 值为 -0.28 V;丙酮酸/乳酸的 $E^{\ominus}$ 值为 -0.19 V[45]。可以针对特定的底物系统来挑选合适的氧化还原媒介来降低微生物的能量损耗,如针对乳酸的氧化反应,1-NH_2、1-$CONH_2$ 和 2-$CONH_2$ 取代的吩嗪可以作为合适的外源电子穿梭体来减少能量损失。然而,Phz-2-NH_2 和 Phz-2-COO^- 不能够作为微生物与受体之间的氧化还原介体,因为其 $E^{\ominus}$(link)低于 $E^{\ominus}$(丙酮酸/乳酸)。

对于特定的胞外氧化物(即胞外最终电子受体),输出能量可以通过以下方程计算[18]:

$$\Delta G_{out}^{\ominus} = -nF[E^{\ominus}(\text{acc}) - E^{\ominus}(\text{link})] \tag{7.7}$$

在自然环境条件下,吩嗪呈现出足够低的还原电势[$E^{\ominus}$(link)],使得还原态吩嗪与最终电子受体之间的电子传递是热力学可行的(图 7.6)。例如,$E^{\ominus}$(NO_3^-/NO_2^-)值为 0.43 V,$E^{\ominus}$(Fe^{3+}/Fe^{2+})值为 0.77 V[46]。因此,对于特定的电子流动途径,可以通过选择或设计合适的氧化还原媒介来降低生物能量损耗,提高能量转化效率。

参考文献

[1] Semenec L, Vergara I A, Lalock A E, et al. Adaptive evolution of *Geobacter sulfurreducens* in coculture with pseudomonas aeruginosa [J]. Mbio, 2020, 11(2): e02875-19.

[2] Nie H, Nie M, Wang L, et al. Evidences of extracellular abiotic degradation of hexadecane through free radical mechanism induced by the secreted phenazine compounds of P-aeruginosa NY3 [J]. Water Research, 2018, 139: 434-441.

[3] Hassan H, Fridovich I. Mechanism of the antibiotic of pyocyanin [J]. Journal of Bacteriology, 1980, 141:156-163.

[4] Ahuja E G, Janning P, Mentel M, et al. PhzA/B catalyzes the formation of the tricycle in phenazine biosynthesis [J]. Journal of The American Chemical Society, 2008, 130(50): 17053-17061.

[5] Letourneau M K, Marshall M J, Grant M, et al. Phenazine-1-carboxylic acid-producing bacteria enhance the reactivity of iron minerals in dryland and irrigated wheat rhizospheres [J]. Environmental Science & Technology, 2019, 53(24): 14273-14284.

[6] Simoska O, Sans M, Eberlin L S, et al. Electrochemical monitoring of the impact of polymicrobial infections on *Pseudomonas aeruginosa* and growth dependent medium [J]. Biosensors & Bioelectronics, 2019, 142: 111538.

[7] Glasser N R, Saunders S H, Newman D K. The colorful world of extracellular electron shuttles [C]//Gottesman S. Annual Review of Microbiology, 2017, 71: 731-751.

[8] Weinberg D R, Gagliardi C J, Hull J F, et al. Proton-coupled electron transfer [J]. Chemical Reviews, 2012, 112(7): 4016-4093.

[9] Tan S L J, Webster R D. Electrochemically induced chemically reversible proton-coupled electron transfer reactions of riboflavin (Vitamin B2) [J]. Journal of The American Chemical Society, 2012, 134(13): 5954-5964.

[10] Miyashita O, Okamura M Y, Onuchic J N. Interprotein electron transfer from cytochrome c2 to photosynthetic reaction center: tunneling across an aqueous interface [J]. Proceedings of the National Academy of Sciences of the United States of America, 2005, 102(10): 3558-3563.

[11] He S, Lau M P, Linz A M, et al. Extracellular electron transfer may be an overlooked contribution to pelagic respiration in humic-rich freshwater lakes [J]. Msphere, 2019, 4(1): e00436-18.

[12] Valenzuela E I, Padilla-Loma C, Gomez-Hernandez N, et al. Humic substances mediate anaerobic methane oxidation linked to nitrous oxide reduction in wetland sediments [J]. Frontiers in Microbiology, 2020, 11: 587.

[13] Liu C, Zachara J M, Foster N S, et al. Kinetics of reductive dissolution of hematite by bioreduced anthraquinone-2,6-disulfonate [J]. Environmental Science & Technology, 2007, 41(22): 7730-7735.

[14] Qiao J, Li X, Li F, et al. Humic substances facilitate arsenic reduction and release in flooded paddy soil [J]. Environmental Science & Technology, 2019, 53(9): 5034-5042.

[15] Hernandez-Montoya V, Alvarez L H, Montes-Moran M A, et al. Reduction of quinone and non-quinone redox functional groups in different humic acid samples by *Geobacter sulfurreducens* [J]. Geoderma, 2012, 183: 25-31.

[16] Mavrodi D V, Blankenfeldt W, Thomashow L S. Phenazine compounds in fluorescent Pseudomonas spp. biosynthesis and regulation [J]. Annual Review of Phytopathology, 2006, 44: 417-445.

[17] Wang Y, Newman D K. Redox reactions of phenazine antibiotics with ferric (hydr)oxides and molecular oxygen [J]. Environmental Science & Technology, 2008, 42(7): 2380-2386.

[18] Glasser A M, Collins L, Pearson J L, et al. Overview of electronic nicotine delivery systems: A systematic review [J]. American Journal of Preventive Medicine, 2017, 52 (2): E33-E66.

[19] Marenich A V, Ho J M, Coote M L, et al. Computational electrochemistry: prediction of liquid-phase reduction potentials [J]. Physical Chemistry Chemistry Physical, 2014, 16 (29): 15068-15106.

[20] Rauschnot J C, Yang C, Yang V, et al. Theoretical determination of the redox potentials of NRH: Quinone oxidoreductase 2 using quantum mechanical/molecular mechanical simulations [J]. The Journal of Physical Chemistry B, 2009, 113(23): 8149-8157.

[21] North M A, Bhattacharyya S, Truhlar D G. Improved density functional description of the electrochemistry and structure-property descriptors of substituted flavins [J]. Journal of Physical Chemistry B, 2010, 114 (46): 14907-14915.

[22] Mueller R M, North M A, Hati C Y S, et al. Interplay of flavin's redox states and protein dynamics: An insight from QM/MM simulations of dihydronicotinamide riboside quinone oxidoreductase 2 [J]. Journal of Physical Chemistry B, 2011, 115(13): 3632-3641.

[23] Marx D, Tuckerman M E, Hutter J, et al. The nature of the hydrated

excess proton in water [J]. Nature, 1999, 397(6720): 601-604.

[24] Stoyanov E S, Stoyanova I V, Tham F S, et al. The nature of the hydrated proton H(aq)$^+$ in organic solvents [J]. Journal of the American Chemical Society, 2008, 130(36): 12128-12138.

[25] Miyazaki M, Fujii A, Ebata T, et al. Infrared spectroscopic evidence for protonated water clusters forming nanoscale cages [J]. Science, 2004, 304(5674): 1134-1137.

[26] Zwier T S. Chemistry. The structure of protonated water clusters [J]. Science, 2004, 304(5674): 1119-1120.

[27] Wernet P, Nordlund D, Bergmann U, et al. The structure of the first coordination shell in liquid water [J]. Science, 2004, 304(5673): 995-999.

[28] Mckinlay J B, Zeikus J G. Extracellular iron reduction is mediated in part by neutral red and hydrogenase in Escherichia coli [J]. Applied and Environmental Microbiology, 2004, 70(6): 3467-3474.

[29] Tarcha P J, Chu V P, Whittern D. 2,3-Diaminophenazine is the product from the horseradish peroxidase-catalyzed oxidation of o-phenylenediamine [J]. Analytical Biochemistry, 1987, 165(1): 230-233.

[30] Soulis T, Sastra S, Thallas V, et al. A novel inhibitor of advanced glycation end-product formation inhibits mesenteric vascular hypertrophy in experimental diabetes [J]. Diabetologia, 1999, 42(4): 472-479.

[31] Delley B. An all-electron numerical method for solving the local density functional for polyatomic molecules [J]. The Journal of Chemical Physics, 1990, 92: 508-517.

[32] Delley B. From molecules to solids With the DMol^3 Approach [J]. The Journal of Chemical Physics, 2000, 113: 7756-7764.

[33] Perdew, Chevary, Vosko, et al. Erratum: Atoms, molecules, solids, and surfaces: Applications of the generalized gradient approximation for exchange and correlation [J]. Physical Review B, Condensed Matter, 1993, 48(7): 4978-4978.

[34] Perdew J P, Wang Y. Accurate and simple analytic representation of the electron-gas correlation energy [J]. Physical Review B, 2018, 98(7): 079904.

[35] Klamt A, Jonas V, Bürger T, et al. Refinement and parametrization of COSMO-RS [J]. The Journal of Physical Chemistry A, 1998, 102(26):

5074-5085.

[36] Klamt A, Schüürmann G. COSMO: a new approach to dielectric screening in solvents with explicit expressions for the screening energy and its gradient [J]. Journal of the Chemical Society, Perkin Transactions 2, 1993, 5: 799-805.

[37] Costentin C, Hajj V, Louault C, et al. Concerted proton-electron transfers. consistency between electrochemical kinetics and their homogeneous counterparts [J]. Journal of the American Chemical Society, 2011, 133(47): 19160-19167.

[38] Watanabe K, Manefield M, Lee M, et al. Electron shuttles in biotechnology [J]. Current Opinion in Biotechnology, 2009, 20(6): 633-641.

[39] Jano I, Hardcastle J E, Zhao K, et al. General equation for calculating the dissociation constants of polyprotic acids and bases from measured retention factors in high-performance liquid chromatography [J]. Journal of Chromatography A, 1997, 762(1): 63-72.

[40] He Z, Huang Y, Manohar A K, et al. Effect of electrolyte pH on the rate of the anodic and cathodic reactions in an air-cathode microbial fuel cell [J]. Bioelectro Chemistry, 2008, 74(1): 78-82.

[41] Ludwig R. Water: From clusters to the bulk [J]. Angewandte Chemie International Edition, 2001, 40: 1808-1827.

[42] Aikens D A. Electrochemical methods, fundamentals and applications [J]. Journal of Chemical Education, 1983, 60(1): A25.

[43] Mann S. Studies on identification and redox properties of the pigments produced by *Pseudomonas aureofaciens* and *P. iodina* [J]. Archiv fur Mikrobiologie, 1970, 71(4): 304-318.

[44] Elema B. Oxidation-reduction potentials of chlororaphine [J]. Recueil des Travaux Chimiques des Pays-Bas, 2010, 52:569-583.

[45] Thauer R K, Jungermann K, Decker K. Energy conservation in chemotrophic anaerobic bacteria [J]. Bacteriological Reviews, 1977, 41(1): 100-180.

[46] Hernandez M E, Newman D K. Extracellular electron transfer [J]. Cellular and Molecular Life Sciences, 2001, 58(11): 1562-1571.

第 8 章

内源性氧化还原媒介分子修饰的碳电极对 EAB 电子传递过程的调控

8.1
内源性氧化还原媒介的优势

生物电化学系统(BES)作为一种绿色的能源技术,在处理水中有机污染物的同时可产电、氢及高附加值化学品[1]。前几个章中已提及,它通过电子转移的方式来实现生物组分和非生物组分(电极)的信息交换和能量转移,其优势在于降解有机物时能够产生电能,可同时解决环境污染问题和能源短缺问题[1-2]。微生物燃料电池(microbial fuel cell,MFC)和微生物电解池(microbial electrolysis cell,MEC)作为 BES 的典型代表,存在着商业化的可能。MFC 以阳极微生物为催化剂,电极为电子受体,将有机污染物中的化学能转化为电能;MEC 在外加电压作用下利用微生物将有机质转化为氢气或其他化学品[3]。在 BES 中,决定性步骤是生物/无机界面的 EET[4-5]。作为典型 EAB,*Geobacter* 和 *Shewanella* 能通过 OM c-Cyts[6]、导电鞭毛[7]和电子媒介[8]三种方式实现 EET。对于电子媒介机制,EAB 既可分泌自由态的电子媒介实现长距离电子传递[8-9],又可通过 OM c-Cyts 结合电子媒介实现电子传递[10]。

维生素 RF 作为氧化还原酶的辅因子,能催化许多生物转化和能量转移反应,相比于人工电子媒介,具有生物毒性小等多种优势[9,11]。RF 的功能主要来源于异咯嗪结构[12],其标准氧化还原电势为 −400 mV(vs. Ag/AgCl)[13],低于 OM c-Cyts,利于电子转移。当 RF 和 OM c-Cyts 耦合后,复合物的电势在 −400 mV 到 −60 mV 之间[10,14]。通过更换培养基的方法,发现 RF 以自由态的形式往返电极与 EAB 之间促进电子传递[9],理论计算在分子尺度同样证明了电子媒介在 EET 过程的重要性[15-16]。*Geobacter sulfurreducens* 是 EAB 的模式菌株之一,虽然含有丰富外膜蛋白,但许多研究表明 *Geobacter* 不能利用电子媒介来实现 EET[9],另一方面,目前关于电子媒介的研究所使用的多是自由态的分子。因此,研究 OM c-Cyts 与电子媒介形成复合体后对 EET 的影响具有十分重要的意义。尽管文献报道电极经 RF 修饰后,*Shewanella* 的产电能力高于裸电极,但 RF 作为电极和 OM c-Cyts 之间的分子桥梁,其促进 EET 机制还没有被深入探讨[17]。

因此,该工作制备了生物膜/RF/石墨烯电极,结合电化学表征和 DFT 计算,解析 RF 分子在 EET 过程中的桥梁作用。石墨烯基材料因其优异的导电性而被广泛用作 BES 的阳极材料,用于促进 EET 和能量转换[18]。不仅如此,石墨

烯可通过 sp^2 杂化的二维平面结构与含苯环的分子形成 π-π 共轭,提高 BES 的性能[5,19-20]。本工作中,RF 通过 π-π 作用固定在石墨烯上(rGO),制备了具备良好氧化还原性和高导电性等优点的 RF/rGO 电极,探究了其自身以及附着生物膜的电化学性质,结合分子模型,阐明了 OM c-Cyts-RF-rGO 之间的相互作用以及 EET 机制,为设计有效的 BES 电极提供了新的信息和手段。

8.2 内源性氧化还原媒介提升 EET 过程机制解析方法

8.2.1 电极制备与表征

通过 Hummers 法制备 GO[21],6 μL 0.5 g/L 的 GO 滴加在电极表面,自然干燥后,在三电极体系(工作电极为玻碳电极,参比电极为饱和 KCl 的 Ag/AgCl,对电极为铂丝)下循环伏安扫描 30 圈即可得 rGO/GC 电极,电势窗口为 $-1.4\sim0$ V,扫速为 100 mV·s^{-1},电解质为 0.1 mol·L^{-1} 磷酸盐(pH 7.0)。随后将裸 GC 电极和 rGO/GC 电极在 RF 溶液中进行循环伏安扫描($-0.8\sim0.4$ V,50 mV·s^{-1},30 圈),即可得 RF/GC 电极和 RF/rGO/GC 电极。具体制备过程如图 8.1 所示。电化学阻抗谱(electrochemical impedance spectroscopy,EIS)的运行参数为 0.01 Hz~10 kHz,扰动电压为 5 mV。为表征电极表面亲疏水性,将电极水平置于玻璃片上,通过座滴法来测量电极的水接触角,每个电极测量 10 次后取平均值。

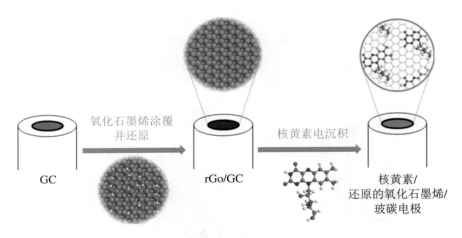

图 8.1 电极制备示意图（VB$_2$ 即核黄素 RF）

8.2.2

Geobacter 培养

矿物盐煮沸除氧之后，用 NaHCO$_3$ 将 pH 调节到 7.0，密封灭菌。向培养基注入 20 mmol·L^{-1} 乙酸钠为底物，50 mmol·L^{-1} 延胡索酸为电子受体，接种 G. *sulfurreducens* DL-1，30 ℃静止培养 2~3 天。矿物盐（1 L 含量）成分为 1.5 g NH$_4$Cl、0.6 g Na$_2$HPO$_4$、0.1 g KCl、10 mL 微量元素母液以及 1 mL 亚硒酸钠-钨酸钠母液。微量元素母液配方为（1 L 含量）：1.50 g 氨三乙酸、3.00 g MgSO$_4$·7H$_2$O、0.50 g MnSO$_4$·H$_2$O、1.00 g NaCl、0.10 g FeSO$_4$·7H$_2$O、0.18 g CoSO$_4$·7H$_2$O、0.10 g CaCl$_2$·2H$_2$O、0.18 g ZnSO$_4$·7 H$_2$O、0.01 g CuSO$_4$·5H$_2$O、0.02 g KAl(SO$_4$)$_2$·12H$_2$O、0.01 g H$_3$BO$_3$、0.01 g Na$_2$MoO$_4$·2 H$_2$O 和 0.025 g NiCl$_2$·6H$_2$O。亚硒酸钠-钨酸钠母液（1 L 含量）为 0.5 g NaOH、3 mg Na$_2$SeO$_3$·5H$_2$O、4 mg Na$_2$WO$_4$·2H$_2$O。

8.2.3

微生物电解池实验

在无菌条件下，将配好的矿物盐培养基注入 MEC 反应器，充分曝氮气除氧，添加 20 mmol·L^{-1}乙酸钠作为底物,接种细菌使最终 OD600 为 0.3。以经典三电极体系组装 MEC[图 8.2(a)],在 30 ℃下施加 0.1 V(vs. Ag/AgCl)电压运行。

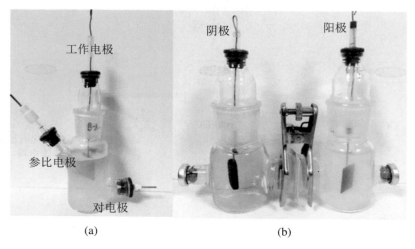

图 8.2　MEC(a)和 MFC 实物图(b)

8.2.4

微生物燃料电池测试

本研究采用双室 MFC 构型[图 8.2(b)],每个腔室体积为 120 mL,中间用质子交换膜(GEFC-10N,GEFC Co.,China)隔离。阳极材料为 RF 和 rGO 修饰的碳纸(RF/CP,rGO/CP,RF/rGO/CP),接种终浓度为 0.3(OD600)的菌液,20 mol·L^{-1}乙酸钠为底物。阴极以碳毡为电极,电解液为含 50 mmol·L^{-1}铁氰化钾溶液的磷酸缓冲盐体系(50 mmol·L^{-1},pH7.0)。阴阳极通过 1000 Ω 的电阻连通,电压通过数据采集卡记录(34970A,Agilent Inc.,USA)。运行结束后,以 1 mV·s^{-1}扫速进行线性循环伏安扫描获得极化曲线。

8.2.5

生物膜电化学分析

将附着生物膜的电极在有底物的培养基中进行 CV 测试,扫速为 5 mV·s^{-1}。

将生物膜进行饥饿处理后,再次进行 CV 测试。

8.2.6
量子力学/分子力学计算

为确定促进微生物/电极相互作用的最佳电极修饰条件,引入相互作用能量(ΔE_{int})这一参数表征 RF 与石墨烯表面的相互作用亲和力以及 RF 的构型。ΔE_{int} 等于两者堆叠体系的总能量($E_{\text{T,RF/GO}}$)减去 GO 表面能 $E_{\text{T,GO}}$ 与 RF 表面能 $E_{\text{T,V}_{\text{B2}}}$ 之和,公式表示如下:

$$\Delta E_{\text{int}} = E_{\text{T,V}_{\text{B2}}/\text{GO}} - (E_{\text{T,GO}} + E_{\text{T,V}_{\text{B2}}}) \tag{8.1}$$

采用了量子力学(quantum mechanics,QM)计算 OM c-Cyts 和电极表面的电子转移,其最小尺寸区域包括铁卟啉,RF 以及作为电子受体的 GO 的 C 原子。在铁卟啉轴向引入两个组氨酸以模拟 OM c-Cyts 的活性中心环境。采用 universal 力场通过 GULP 计算了分子力学区域其他原子的能量[22],密度泛函理论(density functional theory,DFT)计算通过耦合 DMol^3 模块[23]中广义梯度近似[24]中的 Perdew,Burke 和 Ernzerhof 交换相关公式[25]以及包含有 p 轨道极化函数的双精度数值基组(DNP),分析了 RF 在 rGO 表面质子耦合电子转移过程中的热力学性质。

OM c-Cyts 活性中心与 RF-rGO 之间的电子转移只改变它们的带电量,不引起结构的变化,因此,电子转移速率常数(k_{et})可用 Marcus 理论描述[26]:

$$k_{\text{et}} = \frac{2\pi V_{\text{DA}}^2}{h} \sqrt{\frac{\pi}{\lambda k_{\text{B}} T}} \exp\left[-\frac{(\Delta G^o + \lambda)^2}{4\lambda k_{\text{B}} T}\right] \tag{8.2}$$

其中,V_{DA} 代表血红素和 RF/rGO 或者 rGO 在通道配置中交叉点的电子相互作用能,λ 代表重组能,ΔG^o 代表电子转移反应的 Gibbs 自由能,h 代表普朗克常数,k_{B} 是玻尔兹曼常数,T 是温度。

根据式(8.2),为计算电子转移速率常数需要获得参数 λ,V_{DA} 以及 ΔG^o 的值(表 8.1)。λ 可分为内部重组能和外部重组能[27]:

$$\lambda = \lambda_{\text{in}} + \lambda_{\text{out}} \tag{8.3}$$

其中,λ_{in} 是扭曲电子供体-受体化学键所需的能量,可通过基于 V_{B2}/rGO-V_{B2}/rGO 和血红素-血红素自交换反应的四点法计算[28]。因此,λ_{in} 的计算公式如下:

$$\lambda_{\text{in}} = \frac{\lambda_{\text{in,V}_{\text{B2}}/\text{rGO}} + \lambda_{\text{in,heme}}}{2} \tag{8.4}$$

其中，λ_{out}代表因电荷重新分布导致的极化能量，可通过介电连续介质模型计算[29]：

$$\lambda_{out} = \frac{e^2}{4\pi\varepsilon_0}\left(\frac{1}{\varepsilon_{op}} - \frac{1}{\varepsilon_s}\right)\left(\frac{1}{2r_D} + \frac{1}{2r_A} - \frac{1}{R_{DA}}\right) \quad (8.5)$$

其中，r_D 和 r_A 分别是 V_{B2}/rGO 和 rGO 模型的半径，R_{DA} 为电子供体和电子受体质量中心的距离，e 是转移的电量，ε_0、ε_{op} 和 ε_s 分别表示真空介电常数、光介电常数和溶剂介电常数。

表 8.1 RF 在 rGO 表面和水溶液中还原的热力学性质。常压室温下 RF 在 rGO 表面和水溶液中发生单质子单电子反应的 Gibbs 自由能 $\Delta G^{\ominus}$ 和晗变 $\Delta H^{\ominus}$

反应	$\Delta G^{\ominus}$ (eV)	$\Delta H^{\ominus}$ (eV)
$V_{B2,ox}$/rGO + H_3O^+ + $e^- \rightarrow V_{B2,rad}\cdot^-$/rGO + H_2O	−6.02	−6.10
$V_{B2,ox}$ + H_3O^+ + $e^- \rightarrow V_{B2,rad}H\cdot$ + H_2O	−4.95	−4.86

对于阳离子体系的电子转移过程，根据 Koopmans 理论可通过最高占有分子轨道（HOMO）与 HOMO-1 的能隙计算[30]：

$$V_{DA} = \frac{\varepsilon_{HOMO} - \varepsilon_{HOMO-1}}{2} \quad (8.6)$$

根据能斯特方程以及 OM c-Cyts 的氧化还原电势可计算电子转移反应的 ΔG°：

$$\Delta G^{\circ} = -nFE \quad (8.7)$$

其中，n 代表电子数目，F 代表法拉第常数，E 代表氧化还原电势。

分子模拟采用 Materials Studio 软件的 Forcite 模块运行。模型盒子包含 MtrF 的结构域Ⅱ，16 个 RF 分子修饰的 rGO 以及 3000 H_2O 分子。RF 起始状态设定为氧化态以接收来自于结构域Ⅱ血红素中心（Fe^{2+}）的电子，同时引入 16 个 H_3O^+ 以提供质子完成质子耦合电子转移。能量最小化步骤通过耦合最速下降法、共轭梯度法和牛顿法来计算。静电相互作用通过埃瓦尔德求和法模拟[31]，精度为 0.0001 kcal·mol^{-1}。对于其他非键相互作用即范德华力，求和法的范围是 15.5 Å 以内的原子。以结构域Ⅱ/RF/rGO 在水溶液中能量最小化模型为研究对象，分子模拟的时间步长为 1.0 fs 来平衡电子转移模型并获得最终结构。模拟盒子先在 298 K 下的 NVT 系综中弛豫 5 ns，然后在 298 K 的 NVE 系综中弛豫 5 ns。结构域Ⅱ/RF/rGO 的电子转移系统采用了 universal[32] 力场。Andersen algorithm[33] 的碰撞率设为 1.0 以控制模型的温度。

8.3

内源性氧化还原媒介界面电子转移

8.3.1

电极的电化学及表面性质

GO 固定在 GC 电极上进行 CV 扫描,在第一圈时起始还原电位为 -0.7 V,并在 -1.2 V 出现一个不可逆的还原峰[图 8.3(a)],第二圈时,起始还原电位负移,还原电流显著降低直至最终消失,说明 GO 表面的含氧官能团被快速还原,形成 rGO/GC 电极。将裸 GC 电极置于 RF 溶液中进行 CV 扫描,可观察到一对氧化还原峰(-0.149 V,-0.225 V),归属于 RF。随着扫描圈数的增加,氧化还原峰不断增强[图 8.3(b)],说明 RF 逐渐被沉积在电极上。相比于裸 GC 电极,rGO/GC 电极的峰电流更大[图 8.3(c)],说明 rGO/GC 电极表面 RF 浓度更高,这可能是因为 rGO 比表面积更大,能提供更多的吸附位点。以 CP 为基底时,可得到同样的 CV 信号,结合 SEM 表征,可发现大量 rGO 附着在碳纤维表面[图 8.4(a)]。

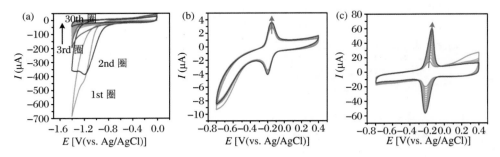

图 8.3 玻碳电极的电化学修饰:GO 在 GC 电极上电化学还原(a);电解液为 0.1 mol·L^{-1} (pH 7.0),扫速为 100 mV·s^{-1},GC 电极在 RF 溶液(pH 3.0)中的 CV 响应(b); rGO/GC 电极在 RF 溶液的 CV 响应(c),扫速为 50 mV·s^{-1}

电极经 RF 修饰后,得出一对明显的氧化还原峰,半波电势为 -0.4 V[图 8.4(b)],多圈扫描后,峰电流保持稳定,说明 RF 被稳定地修饰在电极上[图 8.4(c)]。为研究 RF 在电极表面的动力学行为,进行了不同扫速的 CV 测试,结果

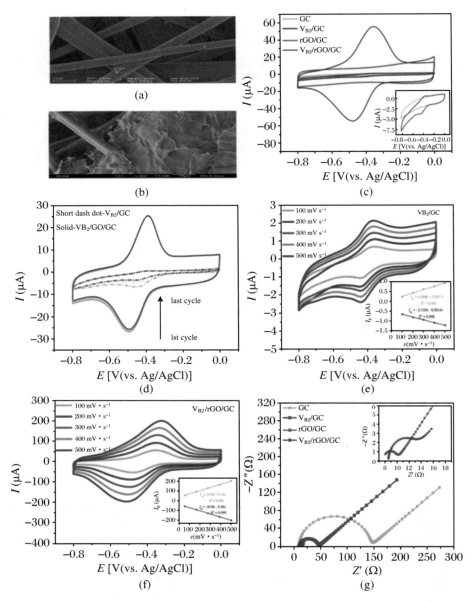

图8.4 修饰电极的性质表征：CP(a)和 rGO/CP(b)电极的 SEM；不同电极在 PBS 缓冲里的 CV 信号(c)，扫速 100 mV·s^{-1}；RF 修饰电极在 PBS 中的多圈 CV 扫描(d)；RF/GC(e)和 RF/rGO/GC(f)电极在不同扫速下的 CV 信号。插图表明峰电流与扫速呈线性关系；不同电极的 EIS(g)；频率为 0.01 Hz~100 kHz，扰动电压为 5 mV

发现氧化峰和还原峰电流与扫速成线性正相关[图 8.4(d)和(f)]，说明 RF 在电极表面的电子传递受限于表面反应速率[34]。进一步根据式(8.8)[35]，计算出 RF 在电极表面的负载量分别为 9.58 nmol·cm^{-2}（RF/rGO/GC）和 4.53×

10^{-2} nmol·cm^{-2}(RF/GC):

$$I_p = \frac{n^2 F^2 A v \Gamma}{4RT} \tag{8.8}$$

其中，v 表示扫速，A 表示电极的表面积，Γ 表示 RF 在电极表面的负载量。

EIS 可表征电极/电解液界面的电荷转移电阻(R_{ct})[36]，参考文献中生物膜体系的等效电路图[37]，模拟得到各电极的 EIS 结果如图 8.4(g)所示。相比于裸 GC 电极，RF/GC 电极的 R_{ct} 极大地减小，说明 RF 可通过其快速可逆的氧化还原反应加速电子传递，由于 rGO 良好的导电能力，rGO/GC 电极表现出极低的转移电阻，RF/rGO/GC 电极的 R_{ct} 略大于 rGO/GC，但远远小于裸 GC。这些结果表明 sp^2 杂化的 rGO 能通过降低电阻的方式促进电极表面的电子转移。

由于含氧官能团减少，rGO/CP 电极的水接触角(120°)与 CP 电极接近(128°)，两者表现出极大疏水性。经 RF 修饰后，水接触角减小(RF/rGO/CP:0°, RF/CP:100°)，说明 RF 增强了电极的亲水性，这可能是因为 RF 通过疏水的异咯嗪环与 rGO 形成 π-π 共轭固定在电极表面[17]，从而将亲水性的核糖醇基暴露，该构型有利于电子从 EAB 到电极的转移。

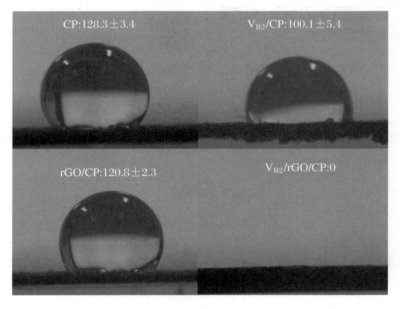

图 8.5　不同电极的水接触角

8.3.2
微生物电解池产电

rGO 和 RF 修饰后,MEC 产电急剧增加[图 8.6(a)]。在启动阶段,由于细胞附着较少,4 个电极体系的产电都很低,相比于裸 GC 电极,rGO 和 RF 修饰后启动时间明显缩短。相比于其他电极,RF/rGO/GC 电极在 118 h 后达到峰值电流($210\ \mu A \cdot cm^{-2}$),并保持了 113 h,该产电能力与其他体系在同一数量级,且 RF/rGO/GC 电极在纯菌体系下的产电高于同体系下的其他电极(表8.2)。

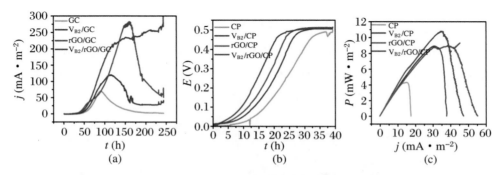

图 8.6 不同电极 MEC 和 MFC 产电对比:MEC 电流密度-时间曲线,施加电压为 0.1 V(a);MFC 电压-时间曲线(b);MFC 的最大功率密度曲线(c)

表 8.2 RF/rGO/GC 电极与其他 MEC 性能对比

微生物	电极	施加电位 [V(vs. NHE)]	电流密度 ($A \cdot m^{-2}$)
城市污水[38]	氨处理的石墨纤维刷	0.6	2.5
厌氧消化池污泥[39]	碳纤维	-0.2	7.5
合成废水[40]	碳毡	1.0	2.7
上流式厌氧污泥[41]	石墨毡	1.0	2.3
城市污水[42]	碳布	0.8	0.9
F. placidus[43]	石墨片	0.7	0.7
G. ahangari[43]	石墨片	0.7	0.6
上流式厌氧污泥[44]	碳毡	0.6	3.0
混菌[45]	活性炭	1.0	5.0
混菌[45]	Fe_3O_4 掺杂的活性炭	1.0	8.2

续表

微生物	电极	施加电位 [V (vs. NHE)]	电流密度 ($A \cdot m^{-2}$)
混菌[45]	FeS 掺杂的活性炭	1.0	10.7
混菌[45]	CaS 掺杂的活性炭	1.0	21.3
混菌[46]	氨处理的石墨纤维刷	-0.9	0.02
Geobacter（本研究）	RF/rGO/GC	0.3	2.5

这些结果说明 rGO 和 RF 的协同作用可促进电子传递。RF 常被用作电子媒介，通过两个电子的氧化还原反应促进电子在 OM c-Cyts 和电子受体之间的转移[9]。不仅如此，RF 还可以和 OmcA 结合，形成复合物，从而通过一个电子的氧化还原过程以更快的速率传递电子[10,14,47]。因此，RF 固定在电极表面后可能进一步与 OM c-Cyts 复合，促进 EET。有研究报道，RF 梯度浓度可引导细胞向不溶性电子受体靠近[48]，因此，通过 π-π 作用固定在电极表面的 RF 可形成高浓度环境，加速生物膜的形成，从而提高产电。除此之外，RF 还可以提高生物膜的附着量[17,49]。综合来看，这些因素导致了 RF/rGO/GC 表现出高产电、高稳定性的优异性能。

8.3.3
微生物燃料电池性能

尽管所有 MFC 在 40 h 内达到稳定电压（500 mV），但它们的启动顺序由快到慢为[图 8.6(b)]：RF/rGO/CP＞rGO/CP＞RF/CP＞CP。在运行时间内，最大功率密度由大到小为 RF/rGO/CP＞RF/CP＞rGO/CP＞CP，趋势与 MEC 的产电一致。这些结果说明 RF/rGO/CP 电极在产电上具有更大的优势，这是因为在该电极表面附着了更多的细胞膜（图 8.7），这可归结于石墨烯的高比面积和 RF 与 OM c-Cyts 的结合有利于细胞膜的形成。各 MFC 的总产电量如表 8.3 所示。相比裸 CP 电极，RF/rGO/CP、rGO/CP 和 RF/CP 电极的电荷量分别增加了 59%、45% 和 28%。总产电量大小顺序为 CP＜RF/CP＜rGO/CP＜RF/rGO/CP。

图 8.7　RF/rGO/CP(a)和 CP(b)电极上生物膜的 SEM

表 8.3　不同阳极 MFC 的总产电量对比

	电荷量(C)	vs. CP(%)	电能(J)	vs. CP(%)
CP	29	—	11	—
RF/CP	37	28	16	45
rGO/CP	42	45	18	64
RF/rGO/CP	46	59	20	42

8.3.4
生物膜的电化学特征

在有底物条件下,生物膜的 CV 表现出典型的 S 形曲线[图 8.8(a)],其中 rGO/GC 电极($12\ \mu A$)和 RF/GC 电极($20\ \mu A$)的催化电流是裸 GC 电极($5\ \mu A$)的三倍左右,RF/rGO/GC 电极的催化电流最大($20\ \mu A$),进一步证明 rGO 和 RF 的协同作用可促进 EET。将 CV 进行一阶导后可观察到在 $-0.35\ \text{V}$ 有一峰[图 8.8(b)],且该峰值随着 rGO 和 VB 的修饰而增强。结合无底物条件下的 CV[图 8.8(c)],判断该峰归属于 OM c-Cyts。相比于裸 GC 电极和 rGO/GC 电极,RF 修饰后使得该峰位置正移,意味着 RF 和 OM c-Cyts 结合,以复合体的形式促进电子转移[10,14,47]。在无底物条件下,生物膜表现出多对氧化还原峰[图 8.8(c)],为进一步分析氧化还原峰归属,对生物膜进行曝 CO 处理,抑制 OM c-Cyts 的氧化还原峰,结果如图 8.8(d)所示。$-0.35\ \text{V}$ 的峰电流被极大抑制,证明该峰归属于 OM c-Cyts。对于 RF/rGO/GC 电极,出现了一对新的氧化还原峰,该峰归属于 RF,相比于曝 CO 之前,该峰位置负移,进一步证明 RF 与 OM c-Cyts 的结合。这些结果表明固定在电极表面的 RF 可与 OM c-Cyts 结合,复合物氧化还原电势正移,有利于 EET 过程。

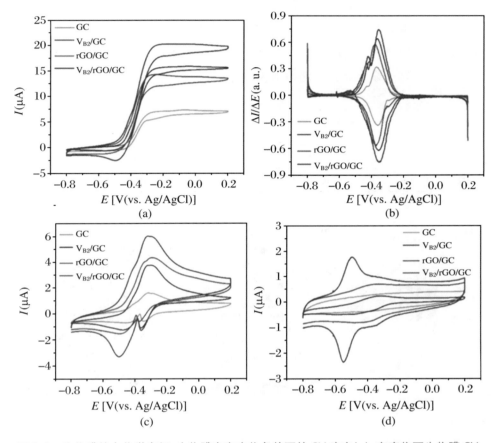

图 8.8　生物膜的电化学表征:生物膜在有底物条件下的 CV 响应(a);有底物下生物膜 CV 信号的一阶导曲线(b);生物膜在无底物条件下的 CV 信号(c);生物膜曝 CO 后的 CV 响应(d);扫速均为 5 mV·s^{-1}

8.3.5

界面电子传递机制的分子解

为提高 BES 阳极的能量转换效率并调节细胞色素-氧化还原活性分子-rGO 基底的协同作用,本书采用了 DFT 计算和分子动力学模拟研究细胞色素与 RF/rGO 界面的电子转移机制。RF 与 rGO 之间相互作用能为负(ΔE_{inter} =

−0.10 eV,表 8.4),说明 rGO 倾向于与 RF 的异咯嗪结构形成共轭[图 8.9 (a)],该平行结构有利于维持 RF 的生物活性并促进 EET。由于共轭结构增强了 π 电子的离域程度,因此,RF 分子可作为能量储存单元。为确定与 rGO 共轭后 RF 分子的氧化还原状态,在标准条件下利用 DFT 计算了 RF 的质子耦合电子转移反应的热力学性质(图 8.10)。RF 可以通过一个电子或者两个电子途径传递电子[9,10,14,47,50],该性质赋予了 RF 携带电子的能力,可从 heme(OM c-Cyts 活性中心)获取一个电子并传递给 rGO。

表 8.4　RF 和 rGO 的总能量以及 RF 与 rGO 的相互作用能(ΔE_{inter})

结合反应	总能量(Ha)			ΔE_{inter}(eV)
	RF	rGO	RF/rGO	
RF+rGO→RF/rGO	−1329.27	−2678.67	−4007.94	−0.10

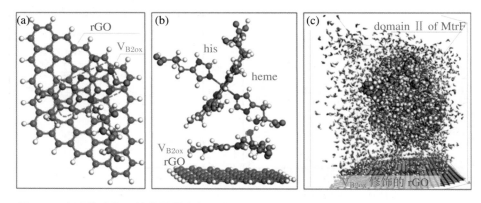

图 8.9　电子转移体系的能量最小化结构:RF 和 rGO 的共轭结构(a)。虚线框里是 RF 的 N 杂化结构异咯嗪,在石墨烯边缘加氢原子使 C 原子饱和;电子供体-受体对(血红素-RF/rGO)的簇模型(b)。轴向有两个组氨酸与 heme 中心铁原子配位;溶剂化结构域Ⅱ/RF/rGO 体系的快照(c)。在平衡状态下 RF 分子依然通过共轭作用固定在 rGO 表面,RF 的核糖醇基朝向电活性蛋白的结构域Ⅱ

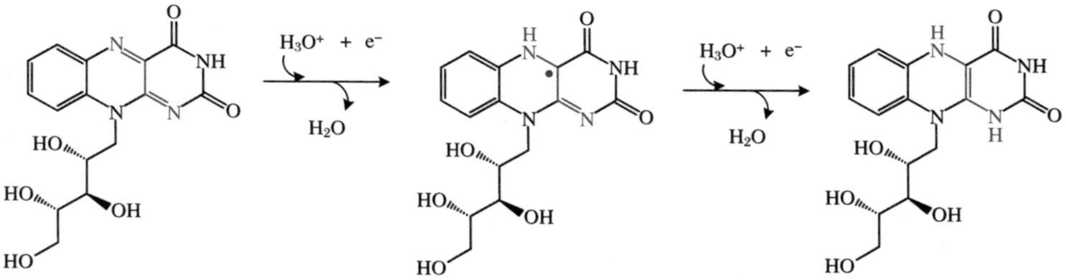

图 8.10　RF 在水溶液中的单质子耦合的单电子还原机制

为研究连接蛋白和无机纳米材料的电子传递链,研究了 RF 单质子耦合的单电子转移反应的热力学性质。RF 在 rGO 表面被还原的标准 Gibbs 自由能($\Delta G°$)为 -6.02 eV,比在水溶液条件的 $\Delta G°$(-4.95 eV)要更负,因此,RF 在热力学上更倾向于保持还原态(半醌态),而且 rGO 能影响 RF 在能量转化时的氧化还原反应。通过 Marcus 理论,计算了 OM c-Cyts 活性中心与 RF/rGO 界面的理论电子转移速率[51]。G. sulfurreducens DL-1 含有很多外膜蛋白,如 OmbB/OmbC,OmaB/OmaC 和 OmcB/OmcC,它们的功能、位置以及活性中心都与 Shewanella oneidensis MR-1 的外膜蛋白类似[52-53]。由于缺乏 G. sulfurreducens DL-1 外膜蛋白的结构信息,采用了 Shewanella oneidensis 的外膜蛋白 MtrF(PDB:3PMQ,图 8.11)作为多血红素细胞色素的蛋白模型[54]。MtrF 作为 MtrC 的同系物,同样含有 4 个结构域和 10 个血红素。采用图 8.9(b)所示的电子供体-受体团簇模型计算了内部重组自由能,其中电子供体是卟啉环中心的 Fe 原子,并在轴向引入两个组氨酸来模拟氨基酸残基对电子传递的影响。

基于 OM c-Cyts 的氧化还原电势,结合能斯特方程估算了反应平衡状态下电子转移反应的热力学驱动力 $\Delta G°$。计算结果发现 heme-RF/rGO 体系的电子转移速率比 heme-rGO 体系的电子转移速率高 10^{23} 倍(表 8.5),说明 RF 修饰在电极表面后可极大地促进 EET。该结果同样在分子水平证明了 RF 可作为分子桥梁连接 OM c-Cyts 和 rGO,促进电子传递[55]。尽管实际实验条件下 BES 的能量转化性能受不同因素影响,但 RF/rGO 阳极的促进效果依然显著(图 8.6)。

表 8.5　heme 与 RF/rGO 或者 rGO 之间的电子转移速率常数

	heme-RF/rGO	heme-rGO
λ_{in}(eV)[a]	-3.81	-3.91
r_D(Å)[a]	8.44	8.44
r_A(Å)[a]	10.94	10.94
R_{DA}(Å)[a]	10.30	11.54
λ_{out}(eV)[a]	4.29	9.95
$\lambda = \lambda_{in} + \lambda_{out}$(eV)[a]	0.48	6.04
V_{DA}(eV)[a]	0.12	0.12
k_{et}(s^{-1})	1.20×10^{11}	1.62×10^{13}

注:a. 各参数意义见 8.2.6 小节。

为更好地理解 RF 和 rGO 的协同作用对 EET 的促进作用,采用了分子动力学模拟解析 OM c-Cyts 在 RF/rGO 表面的最佳构型。在 EET 过程中,结构

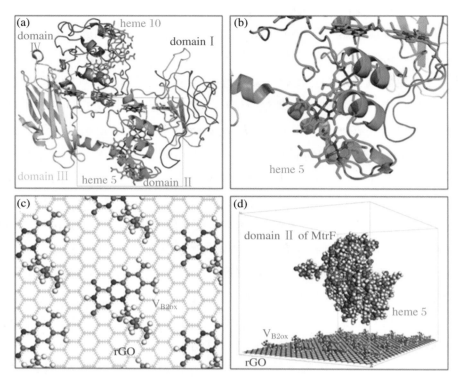

图 8.11 MtrF 和 RF/rGO 的结构示意图:MtrF 的晶体结构(a)(3PMQ),包含 4 个结构域和 10 个 heme;结构域 Ⅱ 中 heme 1～5 的放大结构(b);V_{B2ox} 与 rGO 共轭结构的俯视图(c);结构域 Ⅱ 以及 heme 5 朝向 RF/rGO 表面的分子构型示意图(d),模拟盒子的几何参数分别为 $a=b=59.04\text{Å}$, $c=64.59\text{Å}$, $\alpha=\beta=90°$, $\gamma=60°$

域 Ⅱ(氨基酸序号 187～318)的 heme 5 是电子传递链的终端,直接将电子传递给固体底物或者电子媒介等电子受体(图 8.11(b))。结构域 Ⅱ/RF/rGO 体系在水中的平衡状态[图 8.11(c),(d)]表明,尽管 rGO 表面发生褶皱,但依然与 RF 分子共轭[图 8.11(c)]。不仅如此,RF 分子的核糖醇基暴露在水溶液中,可与水分子或者结构域 Ⅱ 表面的氨基酸残基形成氢键,这缩短了 heme 5 中心 Fe 原子和底物的距离,有利于电子传递。与此同时,核糖醇基还可以增强电极表面的粗糙度和亲水性。因此,该电子媒介可在分子尺度上调节电极表面的电化学性质和物理性质,从而更好地服务于 EET。这些分子动力学模拟结果意味着天然具有氧化还原活性的分子或者经定向人工修饰后的分子都可以用于制备电极,提高 BES 性能。

随着廉价电极材料和 MFC 新构型的开发以及在分子水平和生物膜水平对 EAB 行为的深入理解,MFC 在未来的性能是值得期待的。相比于化学燃料电池,MFC 优势之一是运行条件温和,可显著降低能量消耗。在未来可通过优化反应器构型和水力学参数来进一步降低能量消耗[56]。MFC的另一个突出优点是可便捷地实现实时监控和控制[57]。虽然现阶段MFC产电很低，但这种生物电依然可原位应用于MEC、微生物脱盐或者其他低功率设备[58]。此外，MFC和厌氧消化、光电催化等水处理技术耦合，可极大提高污染物转化效率。在未来可通过扩大反应器，调控微生物代谢等方式促进MFC产电[59]。随着微生物学、工程学以及化学的发展，以MFC为核心的技术在能量转化、污染物处理等方面有很大的应用前景。

参考文献

[1] Li W W, Yu H Q, He Z. Towards sustainable wastewater treatment by using microbial fuel cells-centered technologies[J]. Environmental Science & Technology, 2014, 7（3）：911-924.

[2] Schröder U, Harnisch F. Life electric：nature as a blueprint for the development of microbial electrochemical technologies[J]. Joule, 2017, 1（2）：244-252.

[3] Badwal S P, Giddey S S, Munnings C, et al. Emerging electrochemical energy conversion and storage technologies[J]. Frontiers Chemistry, 2014, 2：79.

[4] Reguera G, McCarthy K D, Mehta T, et al. Extracellular electron transfer via microbial nanowires[J]. Nature, 2005, 435（7045）：1098.

[5] Zhao S, Li Y, Yin H, et al. Three-dimensional graphene/Pt nanoparticle composites as freestanding anode for enhancing performance of microbial fuel cells[J]. Science Advances, 2015, 1（10）：e1500372.

[6] Fredrickson J K, Romine M F, Beliaev A S, et al. Towards environmental systems biology of *Shewanella*[J]. Nature Reviews Microbiology, 2008, 6（8）：592-603.

[7] Gorby Y A, Yanina S, McLean J S, et al. Electrically conductive bacterial nanowires produced by *Shewanella oneidensis* strain MR-1 and other microorganisms[J]. Proceedings of the National Academy of Sciences of the United States of America, 2006, 103（30）：11358-11363.

[8] Newman D K, Kolter R. A role for excreted quinones in extracellular electron transfer[J]. Nature, 2000, 405（6782）：94.

[9] Marsili E, Baron D B, Shikhare I D, et al. *Shewanella* secretes flavins that mediate extracellular electron transfer[J]. Proceedings of the National Academy of Sciences of the United States of America, 2008, 105 (10): 3968-3973.

[10] Okamoto A, Saito K, Inoue K, et al. Uptake of self-secreted flavins as bound cofactors for extracellular electron transfer in *Geobacter species*[J]. Energy & Environmental Science, 2014, 7 (4): 1357-1361.

[11] Velasquez-Orta S B, Head I M, Curtis T P, et al. The effect of flavin electron shuttles in microbial fuel cells current production[J]. Applied Microbiology Biotechnology, 2010, 85 (5): 1373-1381.

[12] Hong J, Lee M, Lee B, et al. Biologically inspired pteridine redox centres for rechargeable batteries[J]. Nature Communications, 2014, 5: 5335.

[13] Lapinsonnière L, Picot M, Barrière F. Enzymatic versus microbial bio-catalyzed electrodes in bio-electrochemical systems[J]. ChemSusChem, 2012, 5(6): 995-1005.

[14] Okamoto A, Hashimoto K, Nealson K H, et al. Rate enhancement of bacterial extracellular electron transport involves bound flavin semiquinones[J]. Proceedings of the National Academy of Sciences of the United States of America, 2013, 110 (19): 7856-7861.

[15] Chen W, Chen J J, Lu R, et al. Redox reaction characteristics of riboflavin: A fluorescence spectroelectrochemical analysis and density functional theory calculation[J]. Bioelectrochemistry, 2014, 98: 103-108.

[16] Chen J J, Chen W, He H, et al. Manipulation of microbial extracellular electron transfer by changing molecular structure of phenazine-type redox mediators[J]. Environmental Science & Technology, 2012, 47 (2): 1033-1039.

[17] Wang Q Q, Wu X Y, Yu Y Y, et al. Facile in-situ fabrication of graphene/riboflavin electrode for microbial fuel cells[J]. Electrochimica. Acta, 2017, 232: 439-444.

[18] Huang Y X, Liu X W, Xie J F, et al. Graphene oxide nanoribbons greatly enhance extracellular electron transfer in bio-electrochemical systems[J]. Chemical Communications, 2011, 47 (20): 5795-5797.

[19] Chen D, Feng H, Li J. Graphene oxide: preparation, functionalization, and electrochemical applications[J]. Chemical Reviews, 2012, 112 (11):

6027-6053.

[20] Wang R, Yan M, Li H, et al. FeS$_2$ nanoparticles decorated graphene as microbial-fuel-cell anode achieving high power density[J]. Advanced Materials, 2018, 30 (22): 1800618.

[21] Hummers Jr W S, Offeman R E. Preparation of graphitic oxide[J]. Journal of the American Chemical Society, 1958, 80 (6): 1339-1339.

[22] Gale J D, Rohl A L. The general utility lattice program (GULP)[J]. Molecular Simulation, 2003, 29 (5): 291-341.

[23] Delley B. From molecules to solids with the DMol^3 approach[J]. The Journal of Chemical Physical, 2000, 113 (18): 7756-7764.

[24] Perdew J P, Chevary J A, Vosko S H, et al. Atoms, molecules, solids, and surfaces: applications of the generalized gradient approximation for exchange and correlation[J]. Physical Reviews B: Condens. Matter mater. Phys., 1992, 46 (11): 6671.

[25] Perdew J P, Burke K, Ernzerhof M. Generalized gradient approximation made simple[J]. Physical Reviews lett., 1996, 77 (18): 3865.

[26] Marcus R A, Sutin N. Electron transfers in chemistry and biology[J]. Biochimica et Biophysica Acta, 1985, 811 (3): 265-322.

[27] Marcus R A. On the theory of oxidation-reduction reactions involving electron transfer[J]. The Journal of Chemical Physical, 1956, 24 (5): 966-978.

[28] Nelsen S F, Blackstock S C, Kim Y. Estimation of inner shell Marcus terms for amino nitrogen compounds by molecular orbital calculations[J]. Journal of the American Chemical Society, 1987, 109 (3): 677-682.

[29] Marcus R A. Chemical and electrochemical electron-transfer theory[J]. Annual Review of Physical Chemistry, 1964, 15 (1): 155-196.

[30] Blomgren F, Larsson S, Nelsen S F. Electron transfer in bis (hydrazines), a critical test for application of the Marcus model[J]. Journal Computational Chemisty, 2001, 22 (6): 655-664.

[31] Essmann U, Perera L, Berkowitz M L, et al. A smooth particle mesh Ewald method[J]. The Journal of Chemical Physical, 1995, 103 (19): 8577-8593.

[32] Rappé A K, Casewit C J, Colwell K, et al. A full periodic table force field for molecular mechanics and molecular dynamics simulations[J]. Journal of the American Chemical Society, 1992, 114 (25): 10024-10035.

[33] Andersen H C. Molecular dynamics simulations at constant pressure and/or

temperature[J]. The Journal of Chemical Physical, 1980, 72 (4): 2384-2393.

[34] Wu J F, Xu M Q, Zhao G C. Graphene-based modified electrode for the direct electron transfer of cytochrome c and biosensing[J]. Electrochemistry Communications, 2010, 12 (1): 175-177.

[35] Wang J. Analytical electrochemistry [M]. Hoboken: John Wiley & Sons, 2006.

[36] He Z, Mansfeld F. Exploring the use of electrochemical impedance spectroscopy (EIS) in microbial fuel cell studies[J]. Energy & Environmental Science, 2009, 2 (2): 215-219.

[37] Dominguez-Benetton X, Sevda S, Vanbroekhoven K, et al. The accurate use of impedance analysis for the study of microbial electrochemical systems [J]. Chemical Society Reviews, 2012, 41 (21): 7228-7246.

[38] Call D F, Merrill M D, Logan B E. High surface area stainless steel brushes as cathodes in microbial electrolysis cells[J]. Environmental Science & Technology, 2009, 43 (6): 2179-2183.

[39] Dhar B R, Gao Y, Yeo H, et al. Separation of competitive microorganisms using anaerobic membrane bioreactors as pretreatment to microbial electrochemical cells[J]. Bioresource Technology, 2013, 148: 208-214.

[40] Sleutels T H, Ter Heijne A, Buisman C J, et al. Steady-state performance and chemical efficiency of microbial electrolysis cells[J]. Inter national Journal Hydrogen Energy, 2013, 38 (18): 7201-7208.

[41] Rozendal R A, Hamelers H V, Molenkamp R J, et al. Performance of single chamber biocatalyzed electrolysis with different types of ion exchange membranes[J]. Water Research, 2007, 41 (9): 1984-1994.

[42] Liu H, Grot S, Logan B E. Electrochemically assisted microbial production of hydrogen from acetate[J]. Environmental Science & Technology, 2005, 39 (11): 4317-4320.

[43] Yilmazel Y D, Zhu X, Kim K Y, et al. Electrical current generation in microbial electrolysis cells by hyperthermophilic archaea *Ferroglobus placidus* and *Geoglobus ahangari*[J]. Bioelectrochemistry, 2018, 119: 142-149.

[44] Zhang J, Zhang Y, Liu B, et al. A direct approach for enhancing the performance of a microbial electrolysis cell (MEC) combined anaerobic reactor by dosing ferric iron: enrichment and isolation of Fe (Ⅲ) reducing bacteria

[J]. Chemical Engineering Journal, 2014, 248: 223-229.

[45] Yasri N G, Nakhla G. Impact of interfacial charge transfer on the start-up of bioelectrochemical systems[J]. Journal of Environmental Chemical Engineering, 2017, 5 (4): 3640-3648.

[46] Kiely P D, Cusick R, Call D F, et al. Anode microbial communities produced by changing from microbial fuel cell to microbial electrolysis cell operation using two different wastewaters[J]. Bioresource Technology, 2011, 102 (1): 388-394.

[47] Okamoto A, Kalathil S, Deng X, et al. Cell-secreted flavins bound to membrane cytochromes dictate electron transfer reactions to surfaces with diverse charge and pH[J]. Scientific Reports, 2014, 4: 5628.

[48] Li R, Tiedje J M, Chiu C, et al. Soluble electron shuttles can mediate energy taxis toward insoluble electron acceptors[J]. Environmental Science & Technology, 2012, 46 (5): 2813-2820.

[49] Baron D, LaBelle E, Coursolle D, et al. Electrochemical measurement of electron transfer kinetics by *Shewanella oneidensis* MR-1[J]. Journal of Biological Chemistry, 2009, 284 (42): 28865-28873.

[50] Tan S L, Webster R D. Electrochemically induced chemically reversible proton-coupled electron transfer reactions of riboflavin (vitamin B_2)[J]. Journal of the American Chemical Society, 2012, 134 (13): 5954-5964.

[51] Kerisit S, Rosso K M, Dupuis M, et al. Molecular computational investigation of electron-transfer kinetics across cytochrome-iron oxide interfaces[J]. The Journal of Physical Chemistry C, 2007, 111 (30): 11363-11375.

[52] Liu Y, Wang Z, Liu J, et al. A trans-outer membrane porin-cytochrome protein complex for extracellular electron transfer by *Geobacter sulfurreducens* PCA[J]. Environmental Microbiology Reports, 2014, 6 (6): 776-785.

[53] Ding Y H R, Hixson K K, Giometti C S, et al. The proteome of dissimilatory metal-reducing microorganism *Geobacter sulfurreducens* under various growth conditions[J]. Biochimia et Biophysica Acta, 2006, 1764 (7): 1198-1206.

[54] Clarke T A, Edwards M J, Gates A J, et al. Structure of a bacterial cell Surface decaheme electron conduit[J]. Proceedings of the National Academy of Sciences of the United States of America, 2011, 108 (23): 9384-9389.

[55] Su T A, Neupane M, Steigerwald M L, et al. Chemical principles of single-

molecule electronics[J]. Nature Reviews Materials, 2016, 1 (3): 16002.

[56] Shannon M A, Bohn P W, Elimelech M, et al. Science and technology for water purification in the coming decades[C]//Nanoscience and technology: a collection of reviews from nature Journals, Republic of Singapore: World Scientific, 2010: 337-346.

[57] Peixoto L, Min B, Martins G, et al. In situ microbial fuel cell-based biosensor for organic carbon[J]. Bioelectrochemistry, 2011, 81 (2): 99-103.

[58] Sun M, Zhai L F, Li W W, et al. Harvest and utilization of chemical energy in wastes by microbial fuel cells[J]. Chemical Society Reviews, 2016, 45 (10): 2847-2870.

[59] Yuan S J, Sheng G P, Li W W, et al. Degradation of organic pollutants in a photoelectrocatalytic system enhanced by a microbial fuel cell[J]. Environmental Science & Technology, 2010, 44 (14): 5575-5580.

第 9 章

金属硫化物 FeNiS$_2$ 电催化氧化过程的界面电子转移机制及电催化性能

9.1 电催化转化模式反应与金属硫化物的电催化优势

在化学转化过程中，催化剂界面的电子传递可能引起底物的氧化或还原反应，但在实际体系中有机污染物种类繁多，反应过程复杂，直接以有机污染物为底物难以清晰地阐明催化剂界面的电子转移机制并以此优化催化剂的界面结构。因此，以产氧反应(OER)与氧还原反应(ORR)等相对简单的过程作为催化剂界面电子传递的模式反应进行分析将有利于清晰地获得催化剂界面结构变化的影响规律。而且，OER催化反应作为水电解和金属空气电池等装置的重要反应，也值得对其进行详细解析，其反应速率决定了装置的能量转化效率[1]，因此，需要高活性的催化剂来克服过高的能垒，促使反应快速发生。迄今为止，基于贵金属Ru和Ir的催化剂依然被认为是OER反应最优异的催化剂[2]，但是其低储量和高成本极大地阻碍了它们的广泛应用。因此，开发具有高催化活性，且具有良好稳定性和选择性的不含贵金属的催化剂是一项具有挑战性和前瞻性的工作。

在众多非贵金属OER电催化剂中，镍金属氧化物被认为是碱性条件下OER催化活性非常高的催化剂[3-5]，但是很多研究表明镍氧化物在催化过程中表现出不稳定性，在反应时较高的电位下其表面被氧化，生成真正的催化活性中心，经过OER长时间循环之后其结构和物相都发生了剧烈转变，导致材料的稳定性变弱，并且氧化物导电能力较差，不利于电子在材料体相的传递。与氧化物相比，镍硫化物的金属硫共价键具有更好的电子传导能力，也更耐电解质腐蚀，从而使催化剂具有良好的导电性以及稳定性。

另外，研究发现将Fe掺入氢氧化镍/氧化物(LDH)中可以大大提高其OER催化活性[6,7]。通过各类原位表征和分析发现Fe^{3+}可对周围的Ni^{3+}位点起到部分电荷转移激活作用，从而改变其局部氧化态，使Ni^{3+}位点活动增加[8-11]。此外，优化催化剂的几何结构也是提升其催化性能的理想途径，密度泛函理论计算和实验结果均表明，与块体材料相比，二维过渡金属硫化物可以暴露出更多的活性位点，表现出更强的导电性，从而提升其电化学催化性能[12-14]。这些结果表明，掺入Fe的二维镍基金属硫化物纳米材料有望成为一种良好的电化学催化剂。

然而，目前的研究中关于二维Fe-Ni硫化物材料的信息很少，主要是由于缺乏获得高纯度 Fe-Ni 硫化物纳米片的有效途径。因此,本章提出了一种基于油相反应的合成路线,利用一锅法合成出二维 $FeNiS_2$ 纳米片和相应的二元硫化物 FeS 和 Ni_9S_8。本章对这些产物的物相、组成和结构进行了详细表征,并研究了三元 $FeNiS_2$ 纳米片的形成过程。最后对其性能进行测试分析,以了解制备的三元 $FeNiS_2$ 在碱性条件下的 OER 催化活性和稳定性,并为合成或者优化得到性能优异的电催化剂提供新思路和新方法。

9.2 金属硫化物的制备、表征及界面电子转移性能测试方法

9.2.1 $FeNiS_2$ 纳米片、FeS 纳米片和 Ni_9S_8 纳米棒的合成

$FeNiS_2$ 二维超薄纳米片的制备:称取 0.0375 g(0.1 mmol) Fe(acac)$_3$ 与 0.0257 g(0.1 mmol) Ni(acac)$_2$ 置于 100 mL 三颈烧瓶中,将其放在加热套上并固定在铁架台上,随后加入 1.0 mL 油胺、5.0 mL 十八烯和 0.2 mL 辛硫醇,整个体系通入氮气 10 min 以排除装置内的氮气,并加入磁子搅拌,转速为 450 r·min^{-1},在氮气保护条件下,该体系在 20 min 内自室温升至 120 ℃,在 120 ℃保留 60 min 以除去体系中低沸点溶剂杂质,随后以 10 ℃·min^{-1} 的升温速率升温至 220 ℃,并在该温度下反应 60 min。反应完后自然冷却至室温,得到黑色混合物,加入 10 mL 乙醇后经 8000 r·min^{-1} 离心除去混合液得到黑色固体,然后用乙醇和正己烷的混合溶剂(乙醇与正己烷体积比为 4∶1)洗涤离心 3~4 次,最后在 60 ℃的真空干燥箱中干燥,得到最终的产物。

FeS 纳米片的制备:FeS 纳米片的制备方法与上述类似,区别在于,仅称取 0.0650 g(0.2 mmol) Fe(acac)$_3$ 结晶置于 100 mL 三颈烧瓶中,且加入的溶剂为

3.0 mL 油胺和 3.0 mL 十八烯；降低最终反应温度至 200 ℃，并将反应时间缩短为 30 min，其他条件不变。

Ni_9S_8 纳米棒的制备：Ni_9S_8 纳米棒的制备方法也与上述类似，区别在于，仅称取 0.0516 g Ni(acac)$_2$ 结晶置于 100 mL 三颈烧瓶中，加入的溶剂为 6.0 mL 油胺，反应时间为 30 min。

9.2.2
材料结构和形貌表征

XRD 测试采用 Philips X′Pert PRO SUPER 衍射仪，阳极靶为铜靶，衍射仪装配有石墨单色化的 Cu-K 放射源（$\lambda = 1.541874$ Å，Rigaku TTR-Ⅲ，PHLIPS Co.，the Netherlands），用来分析样品的纯度、物相和晶格参数等信息。样品的形貌分析由场发射扫描电子显微镜（SEM，Supra 40，Zeiss Co.，Germany）和透射电子显微镜（TEM，H-7650，Hitachi Co.，Japan，加速电压 100 kV）测试得到。高分辨透射电镜（HRTEM，JEM. ARM 200F，工作电压 200 kV）则用以分析材料的形貌、晶格结构和元素组成等信息，得到包括高分辨透射显微（HRTEM）图、高角度环型暗场扫描透射（HAADF-STEM）图和对应的元素分布（EDS）图等结果。样品的化学组成和元素价态分析由 X-射线光电子能谱仪（XPS，Thermo Fisher ESCALAB250 型）测试得到。拉曼谱图则由共聚焦激光显微拉曼光谱仪（LABRAM-HR，Jobin-Yvon Co.，France）测试得到，设置激发波长为 514.5 nm。材料的比表面积分析采用 BET 分析方法，利用 Builder 4200 instrument（Tristar Ⅱ 3020M，Microme-ritics Co.，USA）测试得到结果。

9.2.3
材料界面电子转移及电化学性能测试

所有的电化学测试均在旋转圆盘装置（Pine Research Instrumentation Inc.，USA）上进行，参数设定和数据采集使用 CHI 760e 电化学工作站（Chenhua Instrument Co.，China）。测试分析采用三电极系统，其中工作电极（Working electrode）为美国 Pine 公司原装玻碳（GC）电极，电极有效直径为

5.0 mm,几何面积为 0.196 cm²,参比电极为 Ag/AgCl 电极(内装饱和 KCl 溶液),对电极为铂丝电极。

对材料进行电化学测试:称取 2.0 mg 样品,再加入 2.0 mg 导电炭黑(cabot carbon vxc-72),一起分散至 1.0 mL 异丙醇和水的混合溶剂中(体积比为 1∶3),随后加入 10 μL Nafion 膜溶液,超声 60 min 使其混合分散均匀形成墨水状溶液。选取商业 RuO_2(Sigma Aldrich Co.,USA)作为性能对比材料,按照相同方法制备均匀分散液,其详细过程不再赘述。

用移液管移取 10 μL 上述墨水状催化剂溶液滴到玻碳电极表面,待其自然风干成膜,催化剂在电极上负载量为 0.10 mg/cm²。

在测试过程中采用 Ag/AgCl 电极作为参比电极,通过以下公式将得到的电位转换为标准氢电极电位(RHE):

$$E_{\text{vs. RHE}} = E_{\text{vs. Ag/AgCl}} + E^{\ominus}_{\text{Ag/AgCl}} + 0.059 \times \text{pH} \tag{9.1}$$

OER 催化性能测试在 O_2 饱和的 0.1 mol·L⁻¹ KOH 溶液中进行。LSV 测试电压范围为 1.2~1.8 V,扫速为 5 mV/s,旋转圆盘转速为 1600 r·min⁻¹。CV 测试的电压范围为 0.7~1.0 V,并设置不同扫速进行测试。电化学阻抗谱(EIS)的测试在 1.6 V 电压下进行,频率范围为 0.1 Hz~100 kHz,外加电压振幅为 10 mV。稳定性测试采用计时电流法(i-t),测试电压为 1.7 V。

9.3 金属硫化物电催化界面电子转移性能与机制

9.3.1 FeNiS₂ 二维纳米片的催化界面的生长过程

图 9.1(a)为油相法制备 $FeNiS_2$ 二维纳米片的合成路线示意图,图 9.1(b)

为 $FeNiS_2$、FeS 和 Ni_9S_8 的 XRD 图谱。对 XRD 结果进行分析,$FeNiS_2$ 的主要衍射峰值为 $2\theta = 29.5°、30.8°、35.8°、46.9°$ 和 $51.5°$,同标准的 $FeNiS_2$(JCPDS Card,No. 75-0611)卡片数据基本一致,分别对应于 $FeNiS_2$ 面心立方相的(311)、(222)、(400)、(511)和(440)晶面,谱图中没有对应二元硫化物杂峰出现,因此,合成的产物纯度很高。同样地,分析二元产物 FeS 和 Ni_9S_8 的 XRD 谱图,根据对应的峰位可推出 FeS 为六方晶相,与标准卡片(JCPDS Card,No. 75-0602)对应,而 Ni_9S_8 是底心正交晶相,对应标准卡片为(JCPDS Card,No. 78-1886)。图 9.1(c)是 $FeNiS_2$、FeS 和 Ni_9S_8 的 Raman 谱图,分析结果发现 $FeNiS_2$ 的峰位置与两种二元产物的峰位置无重合的部分,两种二元产物的峰位置与文献中所报道的拉曼结果有较好的对应[15-17],表明通过有效的合成方法合成了纯度非常高的三元产物 $FeNiS_2$ 与其对应的二元产物 FeS 和 Ni_9S_8。

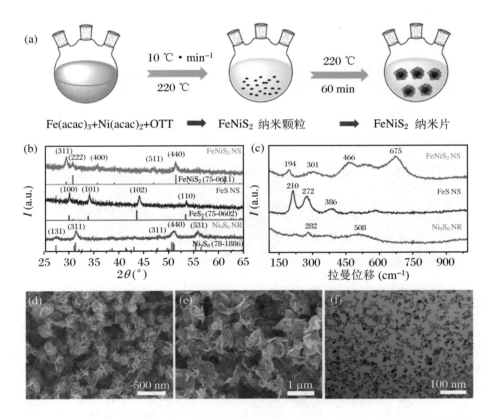

图 9.1　$FeNiS_2$ 二维纳米片的合成路线示意图(a);$FeNiS_2$、FeS 和 Ni_9S_8 的 XRD 图谱(b);$FeNiS_2$、FeS 和 Ni_9S_8 的拉曼谱图(c);$FeNiS_2$ 纳米片、FeS 纳米片的 TEM 图像和 Ni_9S_8 纳米棒的 SEM 图像(d)~(f)

图 9.1(d)、(e)和(f)分别对应所合成的 $FeNiS_2$ 的 SEM 图、FeS 的 SEM 图和 Ni_9S_8 的 TEM 图。对产物形貌特征进行分析,图 9.1(d)表明三元 $FeNiS_2$ 主要为超

薄纳米片状结构,可见其边缘有卷曲的部分,并且多个片状结构可以组合成花球,每个花球的直径在 300 nm 左右。如图 9.1(e)所示,与 FeNiS$_2$ 相似的形貌结构在 FeS 的 SEM 图中也可以看到,但是 FeS 纳米片团成的花球结构较为松散,且直径较大,已达到微米级。图 9.1(f)中 Ni$_9$S$_8$ 的 TEM 图展示的形貌与 FeNiS$_2$ 和 FeS 都不相同,所得到的产物为纳米棒状,而且尺寸非常小,长度约为 10 nm。

图 9.2 是为了对目标产物 FeNiS$_2$ 的形貌特征做进一步探究所做的相关表征。图 9.2(a)～(c)为常规的低倍率和高倍率 TEM 图,可以清晰地看到纳米片组成的花球以及花球边缘薄片状的 FeNiS$_2$ 二维纳米片。图 9.2(d)为 FeNiS$_2$ 二维纳米片的高分辨透射图,由图可以看出纳米片厚度非常薄。图 9.2(e)～(f)为 FeNiS$_2$ 二维纳米片的 HRTEM 图像,选择晶格条纹较为清晰的区域进行放大分析,呈现于右上角小图中,由此可以确定材料暴露的晶面,图 9.2(e)测量得到的晶面间距为 0.248 nm,对应于立方相 FeNiS$_2$ 的(400)晶面,同样地,在图 9.2(f)中测量得到的晶面间距为 0.253 nm,在测量的误差范围内,该晶面也可以归为 FeNiS$_2$ 的(400)晶面。因此,结合 XRD 的分析结果可以得出结论,同轴晶面(400)和(440)均为优势暴露晶面。图 9.2(g)为高角度环型暗场扫描透射图(HAADF-STEM)和与之对应的元素分布图(EDS mapping),由图给出的信息可以得到:Fe,Ni 和 S 元素在 FeNiS$_2$ 超薄纳米片中的分布是均匀一致的。

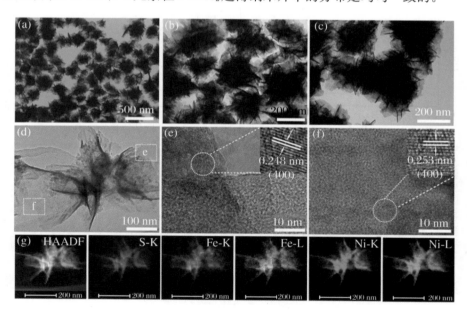

图 9.2　FeNiS$_2$ 二维纳米片的低倍率和高倍率透射图(a)～(d);(e)～(f)为(d)图所选区域的高分辨透射图;FeNiS$_2$ 的 HAADF 图和对应的 S 元素(红色)、Fe 元素(橙色 K 边和黄色 L 边)和 Ni 元素(绿色 K 边和靛青 L 边)分布图(g)

为了对比分析三元 FeNiS$_2$ 超薄纳米片与二元 FeS 纳米片和 Ni$_9$S$_8$ 纳米棒的表面化学组成和电子结构差异,对这三种产物做了光电子能谱(XPS)表征和分析。图 9.3(a)为 FeNiS$_2$、FeS 和 Ni$_9$S$_8$ 的 XPS 总谱图,对 FeNiS$_2$ 的谱线进行分析,谱中出现 Fe、Ni、S、C 和 O 元素的峰。对于二元化合物,FeS 谱线中未出现 Ni 峰信号,Ni$_9$S$_8$ 谱线中也未出现 Fe 峰信号。对不同元素单独进行分析,图 9.3(b)中 FeNiS$_2$ 纳米片和 FeS 纳米片中 Fe 元素的 2p 信号峰位和峰强度是相同的,两个信号峰结合能为 710.8 eV 和 724.5 eV,分别与 Fe 原子的 $2p^{3/2}$ 和 $2p^{1/2}$ 轨道能量对应[18]。图 9.3(c)中 FeNiS$_2$ 纳米片和 Ni$_9$S$_8$ 纳米棒的 Ni 元素 2p 信号峰位相同,信号峰位于结合能 853.1 eV 和 870.1 eV 处,正好对应 Ni 原子的 $2p^{3/2}$ 和 $2p^{1/2}$ 轨道能量[19],而位于 855.6 eV 和 873.2 eV 处的峰则是 Ni(Ⅱ)—O 峰,表明样品表面有镍的氧化物成分[19],这是因为 Ni 元素本身性质较为活泼,暴露在材料表面的 Ni 原子容易与空气中的氧气反应生成少量氧化镍。对三种化合物中的 S 的 2p 谱图进行分析,S 原子 $2p^{3/2}$ 轨道峰位结合能约为 161.3 eV,$2p^{1/2}$ 轨道峰位结合能约为 162.5 eV,由此可推断出材料中硫元素为二价硫离子态(S^{2-})。而在 166 eV 和 170 eV 之间,存在一个高结合能峰位,属于 S—O 峰,表明三种样品表面均有硫氧化物存在[19],但是样品在制备过程中通入氮气以隔绝空气,所以很可能是进行样品后处理时与空气中氧气接触发生反应导致。

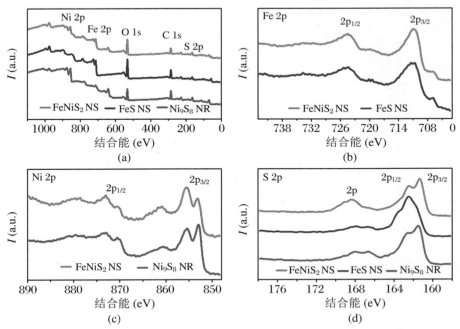

图 9.3 三元 FeNiS$_2$、二元 FeS 和 Ni$_9$S$_8$ 的 XPS 总谱图(a);FeNiS$_2$ 和 FeS 的 Fe 原子 2p 轨道对比谱图(b);FeNiS$_2$ 和 Ni$_9$S$_8$ 的 Ni 原子 2p 轨道对比谱图(c);FeNiS$_2$、FeS 和 Ni$_9$S$_8$ 的 S 原子 2p 轨道对比谱图(d)

对于催化剂材料,其比表面积和孔径分布也是衡量材料物化性能的一个重要因素,关系到它在反应中与溶液或者反应物进行接触的有效面积,因此,对制备得到的三种材料 $FeNiS_2$ 纳米片、FeS 纳米片和 Ni_9S_8 纳米棒都进行了比表面积分析。实验中采用 BET 比表面积测定法,利用 N_2 进行吸附和解吸。图 9.4 (a)~(c)分别为三元 $FeNiS_2$ 纳米片、二元 FeS 纳米片和 Ni_9S_8 纳米棒的吸附等温曲线和对应的孔径分布图。黑色点状曲线为吸附曲线,红色点状曲线为脱附曲线,图中插入的蓝色曲线为样品的孔径分布曲线。由吸附曲线得到的 $FeNiS_2$ 纳米片、FeS 纳米片和 Ni_9S_8 纳米棒的比表面积分别为 57.6 $m^2 \cdot g^{-1}$,134.0 $m^2 \cdot g^{-1}$ 和 30.7 $m^2 \cdot g^{-1}$。对比块状材料可以看出,所合成的三种材料都具有较大的比表面积。这意味着它们的表面都有很多的活性中心位点,并且有很大的吸附活性,对催化剂的催化反应非常有利。

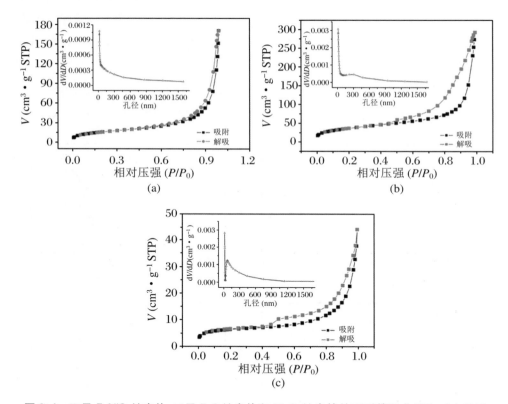

图 9.4 三元 $FeNiS_2$ 纳米片、二元 FeS 纳米片和 Ni_9S_8 纳米棒的吸附等温曲线和对应的孔径分布图(a)~(c)

在材料制备的过程中,为了探究 $FeNiS_2$ 纳米片的生长过程,控制了反应时间,在反应发生特定时间后就停止反应并取出产物,随后通过 TEM 和 XRD 对产物进行形貌和物相分析,以此来明确材料在合成过程中的演化过程。如图

9.5(a)~(d)所示,反应 1 min 后,反应体系中已经出现结晶,但多数为团聚的颗粒状且尺寸较小,反应 5 min 后,颗粒边缘开始生长出二维片状结构,当反应时间延长至 30 min,大量的二维片状结构已经形成,仅残余小部分颗粒,最后,当反应达到 60 min 时,几乎所有的纳米颗粒表面都生长出界面清晰的超薄纳米片,即为得到的最终产物。图 9.6 的 XRD 结果也给出了生长过程的相关信息,在反应 1 min 后,从产物呈现的小峰的峰位可以判断开始形成的纳米颗粒是三元 $FeNiS_2$,随着反应时间的增长,峰的强度不断增加,峰位不移动,表明整个反应不存在相转移过程,因此,三元 $FeNiS_2$ 超薄纳米片的形成过程是通过自下而上的途径。

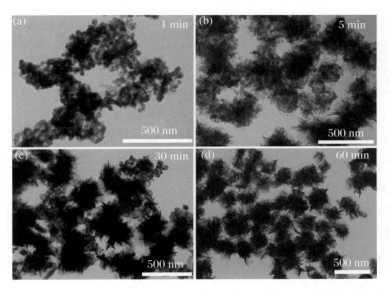

图 9.5 合成 $FeNiS_2$ 纳米片的反应发生 1 min、5 min、30 min 和 60 min 得到的产物 TEM 图(a)~(d)

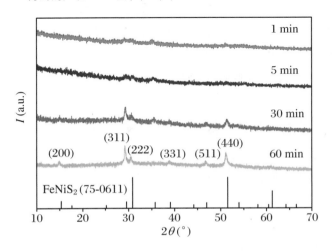

图 9.6 合成 $FeNiS_2$ 纳米片的反应发生 1 min、5 min、30 min 和 60 min 的产物 XRD 谱图

9.3.2
金属硫化物界面催化及电子转移性能分析

经油相合成的三元 $FeNiS_2$ 纳米片、二元 FeS 纳米片和 Ni_9S_8 纳米棒在 400 ℃ 条件下 N_2 气氛中退火两个小时,随后在 O_2 饱和的 $0.1\ mol \cdot L^{-1}$ KOH 溶液中进行产氧反应测试。图 9.7(a)为 $FeNiS_2$ 纳米片、FeS 纳米片、Ni_9S_8 纳米棒、商业 RuO_2 和玻碳电极(GC)的线性扫描伏安(LSV)曲线,电流密度经过内阻校正和电极有效面积归一化。根据 LSV 曲线,得到纯玻碳电极几乎无 OER 催化活性,对样品进行分析发现,三元 $FeNiS_2$ 纳米片具有远低于其对应二元化合物 FeS 纳米片和 Ni_9S_8 纳米棒的起始电位,并且也要低于商业 RuO_2 的起始电位。图 9.7(b)为(a)的 LSV 曲线中不同材料电流密度达到 $10\ mA \cdot cm^{-2}$ 时所需电势,$FeNiS_2$ 纳米片、RuO_2、二元 FeS 纳米片和 Ni_9S_8 纳米棒所需电势分别为 1.54 eV、1.59 eV、1.76 eV 和 1.64 V。$FeNiS_2$ 纳米片所需电势比 RuO_2 所需电势要低,同样也远低于 FeS 纳米片和 Ni_9S_8。

图 9.7(c)为不同材料的塔菲尔斜率图,$FeNiS_2$ 纳米片、RuO_2、FeS 纳米片和 Ni_9S_8 纳米棒的塔菲尔斜率分别为 $46\ eV \cdot dec^{-1}$、$81\ eV \cdot dec^{-1}$、$51\ eV \cdot dec^{-1}$ 和 $58\ eV \cdot dec^{-1}$。$FeNiS_2$ 纳米片具有最低的塔菲尔斜率值,因此,当进行 OER 反应时,在电势增量相同的情况下,利用 $FeNiS_2$ 纳米片作为催化剂能获得更大的电流密度。将以上结果与文献所报道的材料进行对比[20-29],如表 9.1 所示,对比其他以 Ni、Fe 或者 Co 等非贵金属为基础的催化剂材料,所合成的 $FeNiS_2$ 纳米片在 $10\ mA \cdot cm^{-2}$ 的过电势值和塔菲尔斜率值上仍具有一定的优势。

图 9.7(d)为不同扫速下 LSV 曲线在电势为 0.85 V 时对应的氧化电流与还原电流密度差值线性回归曲线,曲线斜率即为双层电容值(C_{dl})的两倍,双层电容与材料的电化学活性面积(ECSA)成正比例关系[30],图中三元 $FeNiS_2$ 纳米片对应斜率明显高于二元 FeS 纳米片和 Ni_9S_8 纳米棒对应斜率,表明三元 $FeNiS_2$ 纳米片具有最大的电化学活性表面积。为了进一步证明 $FeNiS_2$ 纳米片具有更优异的 OER 催化活性,通过交流阻抗技术得到材料的电荷传递阻抗,如图 9.7(e)即为电化学阻抗谱(EIS),展示出三种不同材料在 1.6 V 电势条件下的阻抗,由图可得知三元 $FeNiS_2$ 纳米片的阻抗最小,因此在反应中具有更好的电子传导性,这也是 $FeNiS_2$ 纳米片催化性能如此突出的原因之一。此外,稳定性也是衡量催化剂性能好坏的关键因素,图 9.7(f)为 $FeNiS_2$、FeS 和 Ni_9S_8 三种

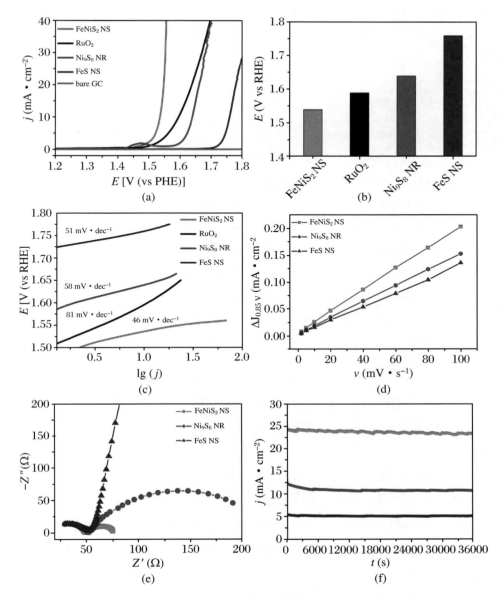

图 9.7 FeNiS$_2$ 纳米片、FeS 纳米片、Ni$_9$S$_8$ 纳米棒、商业 RuO$_2$ 和玻碳电极 (GC) 的线性扫描伏安 (LSV) 曲线 (a); 各催化材料电流密度达到 10 mA·cm^{-2} 时所需电势 (b); 各催化材料的塔菲尔斜率图 (c); 不同扫速下电势为 0.85 V 时对应的氧化电流与还原电流密度差值线性回归曲线图 (d); 1.6 V 电势条件下 FeNiS$_2$ 纳米片、FeS 纳米片和 Ni$_9$S$_8$ 纳米棒的电化学阻抗图 (e); 1.7 V 电势条件下三种材料的时间电流曲线,红色代表 FeNiS$_2$ 纳米片 (最上),绿色代表 Ni$_9$S$_8$ 纳米棒 (中间),蓝色代表 FeS 纳米片 (最下)。以上测试均在 0.1 mol·L^{-1} KOH 溶液中进行 (f)

样品在电势为 1.7 V 条件下进行反应的时间-电流曲线,恒压下反应时间为 10 h,其中三元 FeNiS$_2$ 纳米片对应的电流密度最大,几乎是 Ni$_9$S$_8$ 纳米棒电流密度的两倍和 FeS 纳米片电流密度的五倍,而且三种材料在 10 h 内均并无明显衰减,表明三种材料都能在长时间的电化学反应过程中保持稳定。

表 9.1 文献报道的部分硫族化合物和 $FeNiS_2$ 纳米片的 OER 催化性能比较

OER 催化剂	过电势 (mV)	塔菲尔斜率 ($mV \cdot dec^{-1}$)	参考文献
Ni_3S_2 nanorods/Ni foam	187 ($10\ mA \cdot cm^{-2}$)	159.3	[3]
Ni_3S_2/NF	260 ($10\ mA \cdot cm^{-2}$)	—	[4]
$NiCo_2S_4$ NW/NF	260 ($10\ mA \cdot cm^{-2}$)	40.1	[22]
NiS/Ni foam	335 ($50\ mA \cdot cm^{-2}$)	89	[23]
$NiCo_2S_4$ NA/CC	340 ($100\ mA \cdot cm^{-2}$)	89	[24]
NiS@SLS	297 ($10\ mA \cdot cm^{-2}$)	47	[25]
NF-Ni_3Se_2/Ni	353 ($100\ mA \cdot cm^{-2}$)	—	[26]
NiS@N,S-C/GC	417 ($10\ mA \cdot cm^{-2}$)	48	[27]
$NiCo_2S_4$ NPs@graphene/GC	470 ($10\ mA \cdot cm^{-2}$)	—	[28]
CoS/N,S-C/GC	470 ($10\ mA \cdot cm^{-2}$)	—	[29]
$FeNiS_2$ NSs/GC	310 ($10\ mA \cdot cm^{-2}$)	46	本书

对稳定性测试之后的材料再次进行物相和形貌表征,如图 9.8 所示。根据图 9.8(c),得到三元 $FeNiS_2$ 纳米片在稳定性测试后会变成无定形纳米片,这是因为材料表面的 Ni 在施加电压为 1.45 V 左右时被部分氧化为 NiOOH,从 Ni(Ⅱ)转变为 Ni(Ⅲ),作为催化 OER 反应实际的催化活性位点[31]。同时,化合物中的 Fe 对 Ni 具有部分电荷转移激活作用,可促进富氧中间产物的结合以增强 OER 催化活性,图 9.8(a)和(b)也给出了形貌上的相关证据,在反应前 $FeNiS_2$ 纳米片的片状结构非常清晰,但是在反应后,边缘的片状结构在电镜下与基底的界限变得模糊,也说明材料在反应后表面发生无定性化转变。虽然该变化不可逆转,但是在 10 h 的反应时间内 $FeNiS_2$ 纳米片的性能仍然没有太多衰减,表明材料发生的缓慢氧化过程在较长时间内不会导致材料结构塌陷,且随着氧化过程的进行,材料可以暴露更多的活性位点。因此,$FeNiS_2$ 自氧化过程一方面促使材料增强催化性能,另一方面也会导致材料结构发生变化,但是在较长时间内两者处于较为平衡的状态,因而,材料可以展现出优异的催化性能和良好的稳定性。

因此,现在发展了一种二维三元 $FeNiS_2$ 超薄纳米片的油相合成新方法,所制得的 $FeNiS_2$ 纳米片在碱性溶液中具有优异的 OER 催化活性,性能超过商业 RuO_2。过渡金属 Fe、Ni 与 S 结合形成三元的硫化物 $FeNiS_2$,通过组分调控改

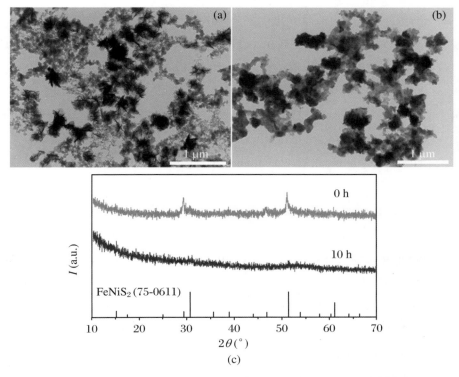

图 9.8　稳定性测试前 FeNiS$_2$ 纳米片的 TEM 图(a)；稳定性测试 10 h 后 FeNiS$_2$ 纳米片的 TEM 图(b)；FeNiS$_2$ 纳米片在稳定性测试前后的 XRD 对比图(c)

变了材料的电子结构，使材料固有的催化活性得到强化提升，并且二维的薄片结构易于与溶液中的反应物接触，暴露出更多活性位点，并且提高了材料的电导率，因而，三元 FeNiS$_2$ 超薄纳米片展现出了优越的 OER 催化性能，同时兼具高选择性和高稳定性。FeNiS$_2$ 超薄纳米片的 OER 性能也远超与其对应的二元化合物 FeS 纳米片和 Ni$_9$S$_8$ 纳米棒。因此，目前研发了一种优化金属硫化物 OER 性能的新策略。此外，三元 FeNiS$_2$ 超薄纳米片可以成为替代贵金属的高效廉价催化剂候选材料，应用于能源储存和转化领域。

参考文献

[1] Jiao Y，Zheng Y，Jaroniec M，et al. Design of electrocatalysts for oxygen- and hydrogen-involving energy conversion reactions [J]. Chemical Society Reviews，2015，44(8)：2060-2086.

[2] Liu H L，Nosheen F，Wang X. Noble metal alloy complex nanostructures：

controllable synthesis and their electrochemical property [J]. Chemical Society Reviews, 2015, 44 (10): 3056-3078.

[3] Wang X P, Wu H J, Xi S B, et al. Strain stabilized nickel hydroxide nanoribbons for efficient water splitting [J]. Energy & Environmental Science, 2020, 13(1): 229-237.

[4] Wang Q, Huang X, Zhao Z L, et al. Ultrahigh-loading of Ir single atoms on NiO matrix to dramatically enhance oxygen evolution reaction [J]. Journal of the American Chemical Society, 2020, 142(16): 7425-7433.

[5] Lee S, Bai L C, Hu X L. Deciphering iron-dependent activity in oxygen evolution catalyzed by Nickel-iron layered double hydroxide [J]. Angewandte Chemie-International Edition, 2020, 59(21): 8072-8077.

[6] Long X, Li J, Xiao S, et al. A strongly coupled graphene and FeNi double hydroxide hybrid as an excellent electrocatalyst for the oxygen evolution reaction [J]. Angewandte Chemie International Edition, 2014, 53(29): 7584-7588.

[7] Liang H, Meng F, Cabán-Acevedo M, et al. Hydrothermal continuous flow synthesis and exfoliation of NiCo layered double hydroxide nanosheets for enhanced oxygen evolution catalysis [J]. Nano letters, 2015, 15(2): 1421-1427.

[8] Trotochaud L, Young S L, Ranney J K, et al. Nickel-iron oxyhydroxide oxygen-evolution electrocatalysts: the role of intentional and incidental iron incorporation [J]. Journal of the American Chemical Society, 2014, 136(18): 6744-6753.

[9] Hunter B M, Blakemore J D, Deimund M, et al. Highly active mixed-metal nanosheet water oxidation catalysts made by pulsed-laser ablation in liquids [J]. Journal of the American Chemical Society, 2014, 136 (38): 13118-13121.

[10] Smith A M, Trotochaud L, Burke M S, et al. Contributions to activity enhancement via Fe incorporation in Ni-(oxy) hydroxide/borate catalysts for near-neutral pH oxygen evolution [J]. Chemical Communications, 2015, 51(25): 5261-5263.

[11] Long X, Li G, Wang Z, et al. Metallic iron-nickel sulfide ultrathin nanosheets as a highly active electrocatalyst for hydrogen evolution reaction in acidic media [J]. Journal of the American Chemical Society, 2015,

137(37): 11900-11903.

[12] Liu X, Liu W, Ko M, et al. Metal (Ni, Co)-Metal oxides/graphene nanocomposites as multifunctional electrocatalysts [J]. Advanced Functional Materials, 2015, 25 (36): 5799-5808.

[13] Wu C, Lu X, Peng L, et al. Two-dimensional vanadyl phosphate ultrathin nanosheets for high energy density and flexible pseudocapacitors [J]. Nature Communications, 2013, 4: 2431.

[14] Xie J, Zhang J, Li S, et al. Controllable disorder engineering in oxygen-incorporated MoS2 ultrathin nanosheets for efficient hydrogen evolution [J]. Journal of the American Chemical Society, 2013, 135(47): 17881-17888.

[15] Li L, Cabán-Acevedo M, Girard S N, et al. High-purity iron pyrite (FeS_2) nanowires as high-capacity nanostructured cathodes for lithium-ion batteries [J]. Nanoscale, 2014, 6(4): 2112-2118.

[16] Guillaume F, Huang S, Harris K D, et al. Optical phonons in millerite (NiS) from single-crystal polarized Raman spectroscopy [J]. Journal of Raman Spectroscopy: An International Journal for Original Work in all Aspects of Raman Spectroscopy, Including Higher Order Processes, and also Brillouin and Rayleigh Scattering, 2008, 39(10): 1419-1422.

[17] Yang Y, Yao H Q, Yu Z H, et al. Hierarchical Nanoassembly of MoS_2/Co_9S_8/Ni_3S_2/Ni as a Highly Efficient Electrocatalyst for Overall Water Splitting in a Wide pH Range [J]. Journal of the American Chemical Society, 2019, 141(26): 10417-10430.

[18] Kitajou A, Yamaguchi J, Hara S, et al. Discharge/charge reaction mechanism of a pyrite-type FeS_2 cathode for sodium secondary batteries [J]. Journal of Power Sources, 2014, 247: 391-395.

[19] Lai W, Chen Z, Zhu J, et al. A NiMoS flower-like structure with self-assembled nanosheets as high-performance hydrodesulfurization catalysts [J]. Nanoscale, 2016, 8(6): 3823-3833.

[20] Zhou W, Wu X J, Cao X, et al. Ni_3S_2 nanorods/Ni foam composite electrode with low overpotential for electrocatalytic oxygen evolution [J]. Energy & Environmental Science, 2013, 6(10): 2921-2924.

[21] Feng L L, Yu G, Wu Y, et al. High-index faceted Ni_3S_2 nanosheet arrays as highly active and ultrastable electrocatalysts for water splitting [J]. Journal of the American Chemical Society, 2015, 137(44): 14023-14026.

[22] Sivanantham A, Ganesan P, Shanmugam S. Hierarchical NiCo$_2$S$_4$ nanowire arrays supported on Ni foam: an efficient and durable bifunctional electrocatalyst for oxygen and hydrogen evolution reactions [J]. Advanced Functional Materials, 2016, 26(26): 4661-4672.

[23] Zhu W, Yue X, Zhang W, et al. Nickel sulfide microsphere film on Ni foam as an efficient bifunctional electrocatalyst for overall water splitting [J]. Chemical Communications, 2016, 52(7): 1486-1489.

[24] Liu D, Lu Q, Luo Y, et al. NiCo$_2$S$_4$ nanowires array as an efficient bifunctional electrocatalyst for full water splitting with superior activity [J]. Nanoscale, 2015, 7 (37): 15122-15126.

[25] Chen J S, Ren J, Shalom M, et al. Stainless steel mesh-supported NiS nanosheet array as highly efficient catalyst for oxygen evolution reaction [J]. ACS Applied Materials & Interfaces, 2016, 8(8): 5509-5516.

[26] Xu R, Wu R, Shi Y, et al. Ni$_3$Se$_2$ nanoforest/Ni foam as a hydrophilic, metallic, and self-supported bifunctional electrocatalyst for both H$_2$ and O$_2$ generations [J]. Nano Energy, 2016, 24: 103-110.

[27] Yang L, Gao M, Dai B, et al. An efficient NiS@ N/SC hybrid oxygen evolution electrocatalyst derived from metal-organic framework [J]. Electrochimica Acta, 2016, 191: 813-820.

[28] Liu Q, Jin J, Zhang J. NiCo2S4@ graphene as a bifunctional electrocatalyst for oxygen reduction and evolution reactions [J]. ACS Applied Materials & Interfaces, 2013, 5(11): 5002-5008.

[29] Chen B, Li R, Ma G, et al. Cobalt sulfide/N, S codoped porous carbon core-shell nanocomposites as superior bifunctional electrocatalysts for oxygen reduction and evolution reactions [J]. Nanoscale, 2015, 7 (48): 20674-20684.

[30] Fan X, Peng Z, Ye R, et al. M$_3$C (M: Fe, Co, Ni) nanocrystals encased in graphene nanoribbons: An active and stable bifunctional electrocatalyst for oxygen reduction and hydrogen evolution reactions [J]. ACS Nano, 2015, 9(7): 7407-7418.

[31] Juodkazis K, Juodkazytė J, Vilkauskaitė R, et al. Nickel surface anodic oxidation and electrocatalysis of oxygen evolution [J]. Journal of Solid State Electrochemistry, 2008, 12(11): 1469-1479.

第 10 章

金属硫化物/石墨烯纳米复合材料界面电子转移的自氧化调控

10.1
金属硫化物与石墨烯的复合与调控作用

在第 9 章中已经提及 OER 过程可以作为污染物转化催化剂界面优化的模式反应，同时对于许多能源转换技术而言，例如电解水和金属空气电池[1-2]，OER 都是其中重要半电池反应，因此，界面 OER 过程的机制解析与性能优化将在环境与能源两个领域发挥重要作用。一般情况下，OER 四电子-质子耦合过程使得反应有较高的过电位，因此，析氧电位远高于理论水分解电压（1.23 V）。近几十年来，科研工作者们一直致力于开发廉价且高效的催化剂来降低 OER 的过电位，提高其反应速率。早期研究表明，氢氧化镍电极中的 Fe 杂质能够使 Ni 的氧化还原峰往高电位偏移，同时 $Ni(OH)_2$ 的 OER 过电位降低[3-5]，可极大地提升氢氧化镍的催化活性。但 Fe 的引入量对于性能的影响是不同的，适量的 Fe 可以促使 Ni 元素向高价态转化，产生活性更高的催化中心 Ni(Ⅳ)，但是当 Fe 引入量过多时，富含 Fe 的相容易与富含 Ni 的相分离[6]，使材料晶格和物相发生改变。

随着近年来对金属空气电池和 OER 的关注度逐渐增加，出现了很多关于 FeNi 氧化物、氢氧化物和硫化物电催化剂的研究[7-12]。研究者们设计不同 Fe 含量的化合物，并通过调控其物相结构和暴露晶面以寻找最优的非贵金属 OER 催化剂，但是低导电性和不稳定性依然是这些催化剂所面临的关键问题。为了解决该问题，许多研究者对 FeNi 氧化物或者氢氧化物进行尺寸调整[13]，表面原子掺杂[14]和与导电性基底复合[15]等方法进行改进。

研究表明，铁镍硫化物纳米片（$FeNiS_2$ NS）催化剂因其独特的二维硫化物结构和 Fe 与 Ni 的合金协同作用[16]，材料的整体导电性得到增强，在碱性条件下表现出优异的 OER 催化性能。虽然超薄二维硫化物的结构有利于材料中的电子传递，但其在 OER 催化过程中存在的自动氧化过程仍然无法避免[17-18]。前期实验结果表明，在未达到产氧过电位时，LSV 曲线在稍低的电位处出现了一个非常明显的氧化峰，这与 OER 催化的实际活性位点有关[19]。例如，有研究者利用原位拉曼光谱表征分析电化学沉积的 Ni—Fe 薄膜，发现在催化反应过程中 Fe 含量较高的样品出现 FeOOH 和 Fe_2O_3 的量增加[19-20]。此外，利用高能量分辨率荧光检测的 X 射线吸收光谱（XAS）分析 $Ni_{1-x}Fe_xOOH$ 催化剂，发现 Fe^{3+} 占据其中八面体位点，且 Fe—O 键距离非常短，这是与周围 NiO_6 八面体进行边缘共享所致[6]。另外 XAS 还评估了 OER 过程中 Ni—Fe 催化剂的 Ni 和 Fe 的 K 边的氧化态和局部原子结构[21]。数据表明，高达 75% 的 Ni 活性中心价态从 +2 增加到 +3，而高达 25% 的 NiOOH 催化剂中 Ni 达到 +4 价，高价态 Ni 在 OER 催化下被快速消耗；而不管催化剂中 Ni 的价态如何变化，Fe 中心始终处于 +3 价。这些结果表明 FeNi 基催化剂的自动氧化在催化过程中会不可避免地发生。该过程会引起催化剂结构、化合价和组成的变化，导致真实活性位点的转变和 OER 性能的提升。然而，这种自氧化现象同时也有一些负面后果，过度氧化会导致材料活性成分溶出、稳定性下降、导电性变差[21]。因此，首先需要从微观尺度对 $FeNiS_2$ 在反应中的自氧化过程进行具体分析，随后通过有效手段在利用自氧化过程所带来的 OER 性能提升的同时又避免其氧化过度带来的负面效应。

基于上述考虑，首先利用非原位 HRTEM 和 EDX 表征 $FeNiS_2$ 纳米片在反应前后的结构和组成变化，通过各项表征结果评估 $FeNiS_2$ 纳米片在 OER 过程中自动氧化引起的问题，例如结构的非晶化、氧化物的形成和活性组分的溶解。随后通过可控合成，利用油相一锅法和后续 NH_3 气氛下的退火过程实现 $FeNiS_2$ 纳米片与还原氧化石墨烯（rGO）的原位复合（表示为 $FeNiS_2$ NS/rGO）。利用电化学测试评估了 $FeNiS_2$ NS/rGO 的 OER 催化活性和稳定性，并阐明了其超高活性和优异稳定性的机制，通过这种方式，不仅可以制备具有 OER 超高活性的新型复合物电催化剂 $FeNiS_2$ NS/rGO，还有助于研发解决 Ni 基催化剂自动氧化问题和制定优化其 OER 性能的有效策略。

10.2
金属硫化物与石墨烯复合材料的制备、表征与性能测试方法

10.2.1
石墨烯、$FeNiS_2$纳米片和$FeNiS_2/rGO$的制备

石墨烯的制备:采用改进的Hummers方法[22-23],步骤如下:0.3 g石墨粉(粒径<30 μm)、0.5 g $K_2S_2O_8$(过二硫酸钾)和0.5 g P_2O_5混合均匀,随后缓慢向其中滴加15 mL浓硫酸并搅拌均匀,随后将混合体系转移至水浴或油浴中,在85 ℃条件下搅拌反应4.5 h,待反应完成后将混合体系缓慢倒入100 mL去离子水中,待冷却至室温后进行抽滤得到黑色固体中间体。中间体干燥除水后,在冰浴和搅拌条件下先缓慢滴加30 mL浓硫酸,随后非常缓慢地向其中加入3 g $KMnO_4$,之后再缓慢向其中加入0.5 g $NaNO_3$,随后在35 ℃水浴中搅拌反应2 h,反应完成后,将混合液缓慢倒入200 mL去离子水中,待冷却后向其中缓慢滴加H_2O_2溶液(10%),溶液由墨绿转变为土褐色最终到金黄色,且滴加H_2O_2溶液再无气泡产生,再将金黄色溶液超声24 h,得到单层或多层石墨烯溶液,随后离心洗涤至石墨烯溶液酸碱度为中性以除去多余的酸和金属离子,最后将石墨烯溶液冷冻干燥得到除水的石墨烯。

$FeNiS_2$纳米片的制备:$FeNiS_2$超薄纳米片的制备在第9章的实验部分做过详细介绍。

$FeNiS_2$纳米片/rGO的制备:$FeNiS_2$超薄纳米片/rGO的制备方法与$FeNiS_2$超薄纳米片的制备方法类似,只是在前驱中多加入了GO,并且在得到产物后还需要将GO还原为rGO。具体制备方法如下:称取0.0375 g(0.1 mmol) $Fe(acac)_3$与0.0257 g(0.1 mmol)$Ni(acac)_2$置于100 mL三颈烧瓶中,将其放在加热套上并固定在铁架台上,随后加入6.0 mL油胺和0.2 mL辛硫醇,整个体系通入氮气10 min以排除装置内的空气,加入磁子进行搅拌,转速为450 r·min^{-1},在氮气保护条件下,该体系在20 min内室温升至120 ℃,在120 ℃保留

60 min 以除去整个体系中的低沸点溶剂杂质,随后以 10 ℃·min^{-1} 的升温速率升温至 220 ℃,并在该反应温度下反应 60 min。反应完后自然冷却至室温,得到黑色混合物,加入 10 mL 乙醇后经 8000 r·min^{-1} 离心除去混合液得到黑色固体,然后用乙醇和正己烷的混合溶剂(乙醇与正己烷比例为 4∶1)洗涤离心 3~4 次,最后在 60 ℃ 的真空干燥箱中干燥,得到产物。干燥后的 FeNiS$_2$ 纳米片/GO 研磨之后放入管式炉,通入 NH$_3$/Ar 混合气(氨气体积分数为 10%),以 5 ℃·min^{-1} 的速率升温至 400 ℃,在该温度下保持 2 h,随后自然降温冷却。该步骤可以去除产物粉末表面的溶剂和表面活性剂,将其转化为亲水性产物,并且将 GO 还原为 rGO,从而提高其导电性。

10.2.2
材料结构和形貌表征

样品的纯度、物相、晶格参数、形貌分析、元素组成等信息在第 9 章的实验部分做过详细介绍。热重分析采用 TGA 模式,在空气中从室温升温至 800 ℃ 进行测试,升温速率为 10 ℃·min^{-1}。

10.2.3
材料电化学性能测试

电极制备、电化学测试步骤在第 9 章的实验部分做过详细介绍,不再赘述。其中,含有石墨烯的复合催化材料配比为:4 mg FeNiS$_2$ 纳米片,FeNiS2 纳米片与 rGO 的混合物(物理混合,质量比为 3∶1)、FeNiS$_2$ 纳米片/rGO 复合材料和商业 RuO$_2$,分别分散至 1.0 mL 异丙醇和水的混合溶剂中(体积比为 1∶3),随后加入 10 μL Nafion 膜溶液,超声 60 min 使其混合分散均匀形成墨水状溶液。

10.3
金属硫化物复合石墨烯对界面电子传递的调控作用

10.3.1
FeNiS$_2$纳米片自氧化过程的弊端

为了探究 FeNiS$_2$ 纳米片在催化 OER 过程中发生的自氧化现象,对 FeNiS$_2$ 纳米片在反应前后的结构和组成进行了表征。图 10.1(a)~(b)为新制的 FeNiS$_2$ 纳米片的 TEM 图,可以看到纳米片组成的花球以及边缘薄片状的二维纳米片。在 1.0 mol·L^{-1} KOH 溶液中进行 5000 圈循环伏安测试,得到如图 10.1(c)~(d)所示的测试之后的 FeNiS$_2$ 纳米片 TEM 图,从图像可知,FeNiS$_2$ 纳米片的基本结构仍然保持,但边缘的薄片结构出现模糊,说明二维片表面结构和成分发生了变化,而且循环测试之后的纳米片相对于测试之前尺寸有所减小。除此之外,还对单个的花球做了表征,图 10.2(a)为 FeNiS$_2$ 纳米片 5000 圈 CV 测试之后的 TEM 图,与图 10.1(d)相似,花球边缘的纳米片结构模糊,从更微观的结构分析,图 10.2(b)为循环之后 FeNiS$_2$ 纳米片的 HRTEM 图,从中无法得到清晰的晶格条纹,表明在稳定性循环测试之后 FeNiS$_2$ 纳米片由结晶态转变为无定形态,傅里叶变换后得到的 SAED 图也验证了这一点,得到的衍射环为晕环,说明结晶性非常差或者是无定形态。反应后 FeNiS$_2$ 纳米片产物的组成和成分也通过 EDX 进行了分析,如图 10.2(c)所示,Fe、Ni、S、O 元素原子比例为 22.17∶11.80∶1.42∶64.59,除了吸附的氧元素和环境中存在的氧,有大量的氧元素出现在循环测试后的产物中,同时 Ni 和 S 的含量相对于 OER 循环测试之前有所减少,特别地,S 原子含量大幅度下降,说明在反应过程中有金属氧化物生成以及少量 Ni 原子和大量 S 原子溶出。图 10.2(d)是 FeNiS$_2$ 纳米片在循环之后的元素分布图,由图可知 Fe、Ni、S 和 O 四种元素在材料中是均匀分布的。

▷ 第10章

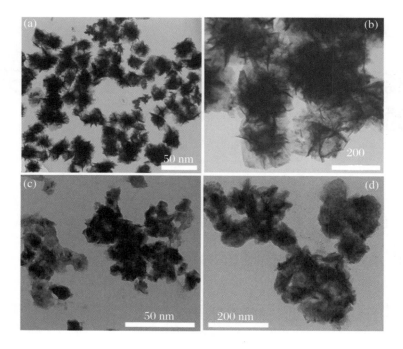

图 10.1　新制的 FeNiS$_2$ 纳米片的低分辨率和高分辨 TEM 图(a)～(b);
FeNiS$_2$ 纳米片在 5000 圈循环伏安测试后的 TEM 图(c)～(d)

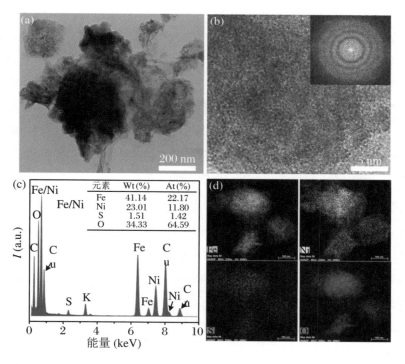

图 10.2　FeNiS$_2$ 纳米片在经历 5000 圈循环伏安测试之后:TEM 图(a);
HRTEM 图和对应的傅里叶变换(FFT)图(b);EDX 元素分布(c);对
应的元素分布像(d),包括 Fe(靛蓝),Ni(黄色),S(红色)和 O(紫色)

以上的测试结果都说明,新制的 FeNiS$_2$ 纳米片在 OER 稳定性测试之后会由结晶态转变为无定形态,这样的转变有利于提升材料的 OER 催化活性,因为表面晶格无序化的结构会包含更多的缺陷和空位,而这些缺陷和空位可以作为催化 OER 反应的活性位点[11,24-25]。但是,FeNiS$_2$ 纳米片在催化反应过程中的自氧化会生成大量的金属氧化物,会因此降低材料的电导率。除此之外,氧元素进入晶格形成氧化物和 Ni、S 元素的溶出会对 FeNiS$_2$ 纳米片的结构产生一定破坏,从而导致催化剂的稳定性变差,无法在反应体系中持续稳定进行催化。因此,需要寻找合适的方法,一方面,保留 FeNiS$_2$ 纳米片在发生无定形转化之后可以提高催化活性的优势,另一方面,减缓自氧化的速度可以避免材料电导率的下降和结构的破坏。

10.3.2
FeNiS$_2$ 纳米片和还原氧化石墨烯(rGO)的原位结合

还原氧化石墨烯因为具备特殊的二维结构和理化性质,例如比表面积大、导电性能优异、机械性能好和化学稳定性高等优点,故被公认为是一种理想的电催化剂载体,可提高催化剂的导电性能和稳定性[26-27]。因此,将 FeNiS$_2$ 纳米片与还原氧化石墨烯进行原位复合,可以成为一种解决上面所述问题的有效方法,具体思路可参考图 10.3,在有 rGO 作为基底的情况下,可以增加材料的导电性,并同时减慢材料的氧化速率,有利于保持材料的结构稳定。

但是要实现 rGO 的原位复合并不是一项容易的工作,尤其还需要同时可控地保留 FeNiS$_2$ 纳米片原本的优势结构和形貌。因此,开发了一种可以一步实现 FeNiS$_2$ 纳米片与 rGO 进行原位复合的油相合成方法,图 10.4 所展示的是合成步骤图,首先通过改进的 Hummers 方法合成氧化石墨烯(GO),处理之后得到冷冻干燥的 GO 气凝胶,随后将金属前驱、硫源和 GO 气凝胶一起加入到溶剂油胺(OAM)中,因为油胺在长碳链端有一个氨基,易于与 GO 表面的含氧官能团结合,因此,GO 可以在其中均匀分散并铺展,为催化剂的后续成核生长提供大量的位点。在反应之后,FeNiS$_2$ 纳米片原位负载在 GO 上,形成三维的复合结构,随后在 NH$_3$/Ar 气氛中进行退火处理,利用 NH$_3$ 的还原性将 GO 还原为 rGO。

对油相一次性合成得到的 FeNiS$_2$ 纳米片/GO 复合材料进行了详细的形貌表征。图 10.5(a)~(b)为 FeNiS$_2$ 纳米片/GO 的低放大倍数和高放大倍数的 SEM 图,花球状的 FeNiS$_2$ 纳米片均匀铺展在氧化石墨烯上,两者得到了有效结

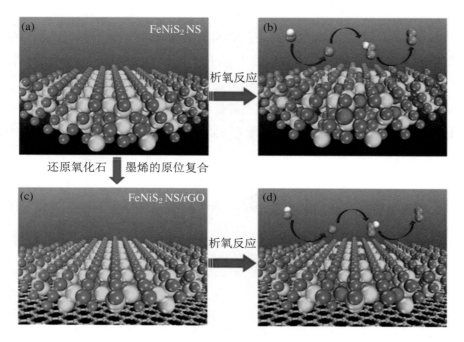

图 10.3　实现 FeNiS$_2$ 纳米片与 rGO 复合的策略示意图

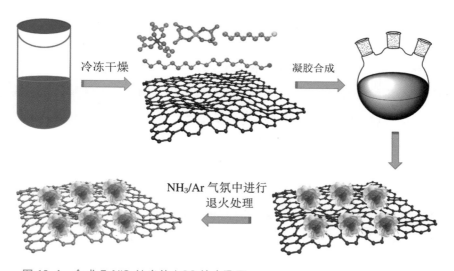

图 10.4　合成 FeNiS$_2$ 纳米片/rGO 的步骤图

合且无明显团聚。同时,通过图 10.5(c)(TEM 图)可以很清晰地看到,直径为 200 nm 左右的花球生长在很薄的氧化石墨烯片层上,意味着利用两者构建了较好的三维纳米复合结构。为了从更小尺度上对形貌结构有进一步的了解,如图 10.5(d)所示,从左下角的 TEM 图中挑选合适的区域进行 HRTEM 表征,结合两图分析,氧化石墨烯基底为无定形结构,与之结合的 FeNiS$_2$ 纳米片有很清晰的晶格条纹,晶格间距为 0.247 nm,正好对应于立方 FeNiS$_2$ 的(400)晶面,右侧深色区域 FeNiS$_2$ 纳米片的侧面结构,测量得到其厚度约为 5 nm。图 10.5(e)为退火处理之后通过 EDX 测试得到的 FeNiS$_2$ 纳米片/rGO 中各元素的含量情况,Fe、Ni、S、C 元素原子比例为 21.64∶22.12∶37.51∶18.73。其中 Fe、Ni 和 S 元素原子的比例接近 FeNiS$_2$ 的化学计量比,另外,复合材料中 O 元素的峰非常弱,说明合成的复合物中氧化物或含氧基团非常少。通过 STEM 测试的 HAADF 模式得到元素分布像,如图 10.5(f)所示,Fe、Ni 和 S 均匀分布在花球状的 FeNiS$_2$ 纳米片区域,C 元素则均匀分布在片状的 rGO 基底上。

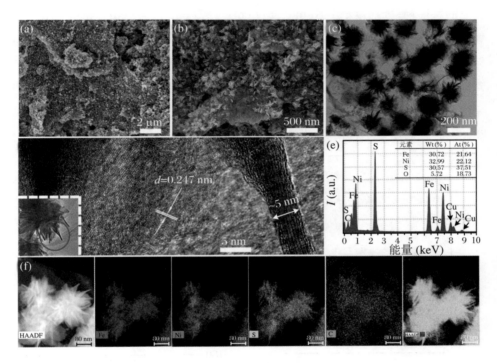

图 10.5 FeNiS$_2$ 纳米片/rGO 的结构表征:低放大倍数和高放大倍数的 SEM 图(a)～(b); TEM 图(c);HRTEM 图和高倍率 TEM 图(左下角插入)(d);EDX 元素分布(e); HAADF 图像和对应的元素分布像(f),包括 Fe(紫色)、Ni(绿色)、S(黄色)和 C (靛蓝)以及所有元素的总分布像

还采用了其他表征手段进一步探究 FeNiS$_2$ 纳米片/rGO 的结构信息。图 10.6(a)为 FeNiS$_2$ 纳米片/rGO 和单独的 FeNiS$_2$ 纳米片的 XRD 对比图,两者的 XRD 峰基本一致,可以归属为立方 FeNiS$_2$ 的衍射峰(JCPDS Card No. 75-0611),唯一的区别在于,FeNiS$_2$ 纳米片/rGO 在 $2\theta = 22°$ 左右存在一个较小的衍射峰,经过调研确认该峰为 rGO 的衍射峰,两者 XRD 谱图中均不存在其他杂质峰,说明成功实现 FeNiS$_2$ 纳米片和 rGO 的原位复合而不产生其他杂质,例如金属单质、单金属硫化物或者其他晶型的铁硫化合物。另外,利用拉曼光谱验证

还原氧化石墨烯的存在并检验其石墨化程度,如图10.6(b)所示,FeNiS$_2$纳米片/rGO 的拉曼谱中存在两个主要的拉曼峰,峰位大致为 1388 cm^{-1} 和 1582 cm^{-1},分别对应着石墨烯的 D 峰和 G 峰,D 峰比 G 峰弱,因此,I_D/I_G 值较小,说明经 NH$_3$ 还原的 rGO 结晶度较高(通过 G 峰和 D 峰的强度之比 $R = I_D/I_G$,可以定性地表征石墨烯的晶化情况,R 值越小,表明结晶越完善)。除此之外,在 100~1000 cm^{-1} 波段范围内,FeNiS$_2$ 纳米片/rGO 在 480 cm^{-1} 和 680 cm^{-1} 处存在两个不明显的小峰。对于单独的 FeNiS$_2$ 纳米片,在低波段范围内存在 5 个明显的特征峰,分别位于 188 cm^{-1}、317 cm^{-1}、477 cm^{-1}、548 cm^{-1} 和 681 cm^{-1} 位置处,其中两个峰与 FeNiS$_2$ 纳米片/rGO 的峰位置可以对应,由此也可以说明,与还原氧化石墨烯复合的产物为 FeNiS$_2$。为了对复合物中还原氧化石墨烯的含量做定量检测并同时评估复合物的热稳定性,利用热重分析的 TGA 模式进行检测,在 25~400 ℃ 之间,质量损失仅为 6.35%,加热至 800 ℃ 后,质量损失为 23.89%,之后质量不再损失,由此可判断氧化石墨烯的含量大致为 24%。复合

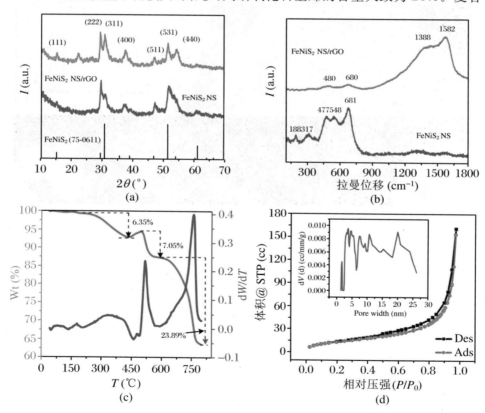

图 10.6　FeNiS$_2$ 纳米片/rGO 和单独 FeNiS$_2$ 纳米片的 XRD 对比图(a);FeNiS$_2$ 纳米片/rGO 和单独 FeNiS$_2$ 纳米片的拉曼对比图(b);FeNiS$_2$ 纳米片/rGO 的 TGA 曲线(c);FeNiS$_2$ 纳米片的 N$_2$ 吸脱附等温曲线和对应的孔径分布图(d)

物比表面积也是衡量材料理化性能的重要因素，采用 BET 分析方法，测试得到图 10.6(d) 的 N_2 吸脱附等温曲线和对应的孔径分布图，$FeNiS_2$ 纳米片/rGO 的比表面积为 81.9 $m^2 \cdot g^{-1}$，比单独的 $FeNiS_2$ 纳米片比表面积更大，而这可以归功于比表面积大的还原氧化石墨烯的紧密复合结构。

10.3.3
$FeNiS_2$ 纳米片/rGO 复合材料的 OER 催化性能

为了证明 $FeNiS_2$ 纳米片与 rGO 的原位复合结构在 OER 催化上更有优势，对 $FeNiS_2$ 纳米片/rGO 复合物、$FeNiS_2$ 纳米片与 rGO 的混合物、单独的 $FeNiS_2$ 纳米片和商业化的 RuO_2 进行了 OER 测试，图 10.7(a) 是上述材料在 O_2 饱和的 0.1 $mol \cdot L^{-1}$ KOH 溶液中经过内阻校正的 LSV 曲线。对于 $FeNiS_2$ 纳米片/rGO 复合物和 $FeNiS_2$ 纳米片与 rGO 的混合物，尤其是 $FeNiS_2$ 纳米片/rGO 复合物，在发生 OER 反应之前，于 1.4~1.5 V 电位之间会出现明显的氧化峰，而单独的 $FeNiS_2$ 纳米片则几乎没有，这一现象与文献中报道的在碳基底存在的情况下，氧化过程电流更大是一致的[21]，因为 rGO 的存在显著改善了材料的导电性能，另一点明显的现象是，有 rGO 存在的催化剂，不论是原位复合还是物理混合，都比单纯的 $FeNiS_2$ 纳米片具有更低的起始电位，甚至低于公认的 OER 优良的商业化催化剂 RuO_2 的起始电位，这也体现 rGO 基底在提升催化剂性能上有明显的优势。但是对比 $FeNiS_2$ 纳米片/rGO 复合物和 $FeNiS_2$ 纳米片与 rGO 的混合物，也可以发现它们在起始电势上存在较大的差异，$FeNiS_2$ 纳米片/rGO 复合物的起始电势为 1.40 V，$FeNiS_2$ 纳米片与 rGO 混合物的起始电势为 1.48 V，由此可以说明，原位复合的 $FeNiS_2$ 和 rGO 在催化性能上有更大的优势。同时，通过 LSV 曲线得到以上材料的塔菲尔斜率值，$FeNiS_2$ 纳米片/rGO 复合物、$FeNiS_2$ 纳米片与 rGO 的混合物、商业化的 RuO_2 和单独的 $FeNiS_2$ 纳米片的塔菲尔斜率值分别为 38 $mV \cdot dec^{-1}$、54 $mV \cdot dec^{-1}$、71 $mV \cdot dec^{-1}$ 和 93 $mV \cdot dec^{-1}$，$FeNiS_2$ 纳米片/rGO 复合物具有最小的塔菲尔斜率值，意味着从动力学上考虑，$FeNiS_2$ 纳米片/rGO 复合物依然具有最大的优势。值得注意的是，即使 rGO 的含量相同，$FeNiS_2$ 纳米片/rGO 复合物相对于 $FeNiS_2$ 纳米片与 rGO 的混合物，在塔菲尔斜率上进一步下降，具有非常明显的优势。造成差异的原因可能是原位复合的 $FeNiS_2$ 纳米片和 rGO 在原子尺度上形成良好结合的界面，使电子在两者之间传递的效率更高，且原位复合可以形成 $FeNiS_2$ 纳米片在 rGO 上均匀分散的三维结构,产生更多的反应活性位点,而 FeNiS$_2$ 纳米片与 rGO 的简单物理混合并不均匀,两者之间的结合也稍弱。

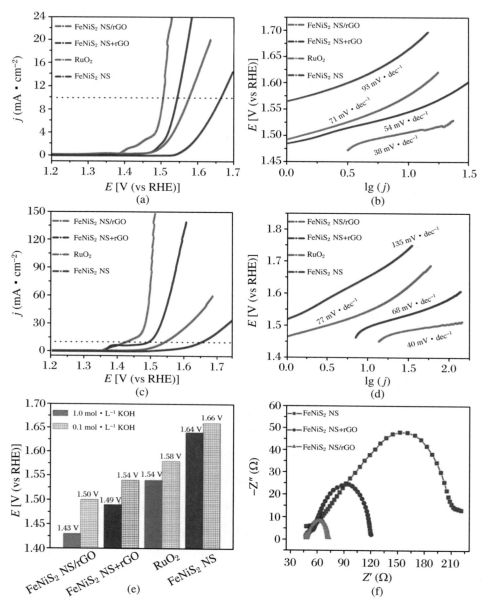

图 10.7　FeNiS$_2$ 纳米片/rGO 复合物、FeNiS$_2$ 纳米片与 rGO 的混合物、单独的 FeNiS$_2$ 纳米片和商业 RuO$_2$ 的 OER 性能测试:在 0.1 mol·L^{-1} KOH 溶液中的 LSV 曲线(a); 在 0.1 mol·L^{-1} KOH 溶液中的塔菲尔斜率(b);在 1 mol·L^{-1} KOH 溶液中的 LSV 曲线(c);在 1 mol·L^{-1} KOH 溶液中的塔菲尔斜率(d);在 0.1 mol·L^{-1} KOH 和 1 mol·L^{-1} KOH 溶液中达到 10 mA·cm^{-2} 电流密度时的过电势(e); 0.6 V(vs Ag/AgCl)电势下 FeNiS$_2$ 纳米片/GO 复合物、FeNiS$_2$ 纳米片与 GO 的混合物和单独 FeNiS$_2$ 纳米片的电化学阻抗图(d)

在此前提下,为了确定复合物中 rGO 的最佳比例,对油相反应加入 GO 的量进行调控,在固定金属前驱加入量为 0.2 mmol 的情况下,控制 GO 的加入量分别为 2.5 mg、5 mg 和 7.5 mg,随后得到相应的 rGO 含量不同的 $FeNiS_2$ 纳米片/rGO 复合物,图 10.8 为对以上 rGO 含量不同的样品在 $0.1\ mol \cdot L^{-1}$ KOH 溶液中进行的性能测试,如图 10.8(a)所示,GO 加入量为 2.5 mg 时得到的复合物催化剂性能要远弱于加入 5 mg GO 时的性能,但 GO 加入量为 7.5 mg 时,其性能与加入 5 mg GO 的催化剂性能一致,同时也对以上样品做了电化学阻抗谱测试,如图 10.8(b)所示,结果表明加入 5 mg 和 7.5 mg GO 的样品电化学阻抗值几乎相同,且小于加入 2.5 mg GO 样品的电化学阻抗,由此可以推断,在 GO 加入量太低时,基底对催化剂整体的性能提升有限,但是加入 GO 达到一定量之后,对于催化剂电导率和稳定性的提升达到一个上限,在以上实验的基础上,可以认为 5 mg 是一个较优的条件,因此在后续实验中都以 5 mg 作为最佳加入量。

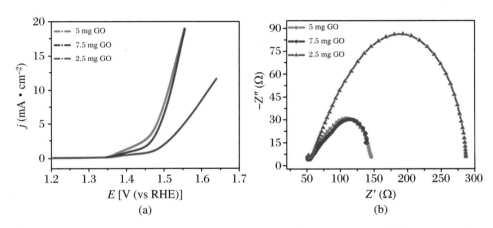

图 10.8　加入不同质量 GO 作为前驱得到 $FeNiS_2$ 纳米片/rGO 的 OER 性能测试:LSV 曲线,电解液为 $0.1\ mol \cdot L^{-1}$ KOH 溶液(a);电化学阻抗谱,测试电压为 0.5 V (vs Ag/AgCl)(b)

在 $0.1\ mol \cdot L^{-1}$ KOH 条件下进行性能测试之后,为了评估材料在碱性更强的环境中的耐受力和性能表现,在 $1\ mol \cdot L^{-1}$ KOH 溶液条件下也对材料进行了测试。图 10.7(c)为 $FeNiS_2$ 纳米片/rGO 复合物、$FeNiS_2$ 纳米片与 rGO 的混合物、商业 RuO_2 和单独的 $FeNiS_2$ 纳米片在 $1\ mol \cdot L^{-1}$ KOH 溶液中的 LSV 曲线,与 $0.1\ mol \cdot L^{-1}$ KOH 溶液中的测试结果趋势一致,$FeNiS_2$ 纳米片/rGO 复合物具有最小的起始电位,同样地,图 10.7(d)显示 $FeNiS_2$ 纳米片/rGO 复合物也具有最小的塔菲尔斜率,具体为 $40\ mV \cdot dec^{-1}$,远小于 $FeNiS_2$ 纳米片与 rGO 的混合物的 $68\ mV \cdot dec^{-1}$、商业催化剂 RuO_2 的 $77\ mV \cdot dec^{-1}$ 和纯 $FeNiS_2$ 纳米片的 $135\ mV \cdot dec^{-1}$。

综合以上结果,以 10 mA·cm^{-2} 电流密度为衡量标准,通过 LSV 图得到各催化剂在该电流密度下的过电势,由此得到如图 10.7(e)所示的结果,在 0.1 mol·L^{-1} KOH 中,FeNiS$_2$ 纳米片/rGO 复合物、FeNiS$_2$ 纳米片与 rGO 的混合物、商业 RuO$_2$ 和单独的 FeNiS$_2$ 纳米片达到 10 mA·cm^{-2} 电流密度的电势分别为 1.50 V、1.54 V、1.58 V 和 1.66 V,在 1 mol·L^{-1} KOH 中,达到 10 mA·cm^{-2} 电流密度的电势依次为 1.43 V、1.49 V、1.54 V 和 1.64 V。显然在两个不同浓度的 KOH 电解液中,FeNiS$_2$ 纳米片/rGO 复合物都具有最优的性能,表现出其良好的催化活性和耐电解液腐蚀性。事实上,在与文献中报道的材料进行比较时,FeNiS$_2$ 纳米片/rGO 复合物也具有非常显著的优势,在碱性条件下不论过电势还是塔菲尔斜率,FeNiS$_2$ 纳米片/rGO 复合物都优于大部分之前已报道的高效 OER 催化剂,表 10.1 即为 FeNiS$_2$ NS/rGO 与已报道的典型高级 OER 催化剂的电化学性能比较[7-10,13,15,17-18,28-46]。

表 10.1 FeNiS$_2$ NS/rGO 与已报道的典型高级 OER 催化剂的电化学性能比较

OER 催化剂	10 mA·cm^{-2} 条件下的过电势 (mV)	塔菲尔斜率 (mV·dec^{-1})	电解液	参考文献
FeNi-rGO LDH 复合体	206	39	1 mol·L^{-1} KOH	[7]
FeNi-rGO LDH	247	31	1 mol·L^{-1} KOH	[8]
无定形 Fe-Ni-O	286	48	1 mol·L^{-1} KOH	[9]
NiFe LDH/NF	240	—	1 mol·L^{-1} NaOH	[28]
NiFe-LDH NSs	302	39	1 mol·L^{-1} KOH	[10]
NiFeO$_x$/CFP	280	—	1 mol·L^{-1} KOH	[11]
Ni$_3$N NSs CC	256	41	1 mol·L^{-1} KOH	[13]
Ni-P 多孔细米片	300	64	1 mol·L^{-1} KOH	[15]
Ni/Ni$_x$M$_y$(M = P,S)	270 (30 mA·cm^{-2})	73.2	1 mol·L^{-1} KOH	[17]

续表

OER 催化剂	10 mA·cm^{-2} 条件下的过电势 (mV)	塔菲尔斜率 (mV·dec^{-1})	电解液	参考文献
$Ni_xCo_{3-x}S_4/Ni_3S_2/NF$	160	95	1 mol·L^{-1} KOH	[18]
Ni-Co PBA 纳米笼	380	50	0.1 mol·L^{-1} KOH	[29]
$NiFeOH@Ni_3S_2/NF$	165	93	1 mol·L^{-1} KOH	[30]
$CoNi(OH)_x$ 纳米管	280	77	1 mol·L^{-1} KOH	[31]
$A\text{-}CoS_{4.6}O_{0.6}$ PNCs	290	67	1 mol·L^{-1} KOH	[32]
$NiCo_2O_4$ 纳米管	290	53	1 mol·L^{-1} KOH	[33]
VOOH 纳米球	270	68	1 mol·L^{-1} KOH	[34]
超薄 CoFe LDHs-Ar NSs	266	37.85	1 mol·L^{-1} KOH	[35]
CoNi(20:1)-P-NS	209	43	1 mol·L^{-1} KOH	[36]
$CoSe_2$ NSs	320	44	0.1 mol·L^{-1} KOH	[37]
Co-Mn LDH NSs	334	41	1 mol·L^{-1} KOH	[38]
PBSCF-Ⅲ	358	52	0.1 mol·L^{-1} KOH	[39]
$Ti/RuO_2\text{-}Sb_2O_5\text{-}SnO_2$	~1.9 V(vs NHE) (500 mA·cm^{-2})	63	3 mol·L^{-1} H_2SO_4	[40]
$Ti/Cu_xCo_{3-x}O_4$	~0.75 V(vs Ag/AgCl) (10 mA·cm^{-2})	—	1 mol·L^{-1} NaOH	[41]

续表

OER 催化剂	10 mA·cm^{-2} 条件下的过电势（mV）	塔菲尔斜率（mV·dec^{-1}）	电解液	参考文献
Ti/IrO$_x$-Sb$_2$O$_5$-SnO$_2$	~1.9 V(vs Ag/AgCl)(1000 mA·cm^{-2})	—	3 mol·L^{-1} H$_2$SO$_4$	[42]
Fe$_{0.1}$NiS$_2$ NA/Ti	231 (100 mA·cm^{-2})	43	1 mol·L^{-1} KOH	[43]
NiFe-LDH@NiFe-Bi/CC	294 (50 mA·cm^{-2})	96	1 mol·L^{-1} KOH	[44]
CoCH/NF	332	126	1 mol·L^{-1} KHCO$_3$	[45]
FeNiS$_2$ NS/rGO	270 200	38 40	0.1 mol·L^{-1} KOH 1 mol·L^{-1} KOH	本书

为了更深入地了解FeNiS$_2$纳米片/rGO复合物活性增强背后的机制,同时研究了影响OER活性的两个重要因素——材料电导率和材料电化学活性面积。为此在0.1 mol·L^{-1} KOH电解液中测试得到FeNiS$_2$纳米片/rGO复合物、FeNiS$_2$纳米片与rGO的混合物和单独FeNiS$_2$纳米片的电化学阻抗谱,如图10.7(f)所示。Nyquist曲线显示,FeNiS$_2$纳米片/rGO复合物和FeNiS$_2$纳米片与rGO的混合物在电荷传递电阻上有非常明显的下降,可以归因于rGO原位复合或者混合之后导致的区域电子传导效率的提高,FeNiS$_2$纳米片/rGO复合物具有最小的电化学阻抗,由此也可以推断出,催化剂与rGO原位复合所形成的独特三维结构,有利于提高两者之间的电荷传递效率,从而增强催化活性。更进一步地,可以通过测量电化学双层电容的方法来评估材料的电化学活性面积。图10.9(d)是由图10.9(a)~(c)中结果得到的线性回归曲线,所取点为不同扫速下LSV曲线在电势为0.30 V时对应的氧化电流与还原电流的差值,曲线斜率即为双层电容值(C_{dl})的两倍,而双层电容与材料的电化学活性面积(ECSA)成正比例关系,因此,根据所显示的结果,FeNiS$_2$纳米片/rGO复合物具有最大的电化学活性面积。根据以上结果,可以得出如下结论:FeNiS$_2$纳米片/rGO复合物同时具有高电导率和更大的电化学反应活性面积,由此表现出优异的OER催化活性。

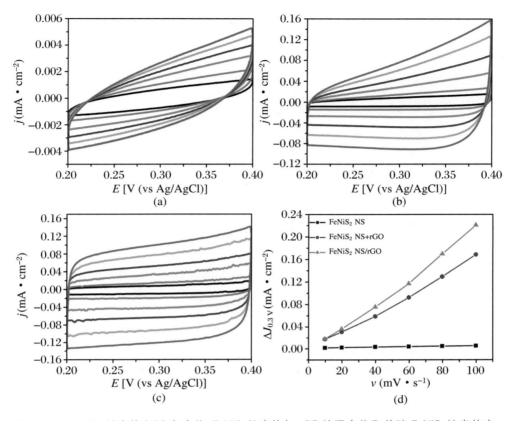

图 10.9 FeNiS$_2$ 纳米片/rGO 复合物、FeNiS$_2$ 纳米片与 rGO 的混合物和单独 FeNiS$_2$ 纳米片在 0.2～0.4 V(vs Ag/AgCl)电压范围内不同扫速(10 mV·s^{-1}、20 mV·s^{-1}、40 mV·s^{-1}、60 mV·s^{-1}、80 mV·s^{-1} 和 100 mV·s^{-1})下的 CV 曲线,电解液为 0.1 mol·L^{-1} KOH(a)～(c);不同扫速下电势为 0.3 V 时对应的氧化电流与还原电流密度差值线性回归曲线图(d)

10.3.4

FeNiS$_2$ 纳米片/rGO 复合材料的催化稳定性

10.3.1 节对 FeNiS$_2$ 纳米片在 OER 催化过程中发生的结构非晶化、氧化物的生成和有效成分的溶出做了详细的分析。一方面,FeNiS$_2$ 的自氧化导致的非晶化转变有利于提升材料的 OER 催化活性,另一方面,持续的自氧化过程会导致催化剂结构被破坏,使材料的稳定性下降。因此,为了验证 FeNiS$_2$ 纳米片与 rGO 原位复合之后是否在稳定性上有所提升,在 1.0 mol·L^{-1} KOH 溶液中,在 1.2～1.6 V(vs. RHE)电压下采用 100 mV s^{-1} 扫速进行 5000 圈 CV 循环,并对稳定性测试之后的催化剂进行表征。图 11.10(a) 为新制的 FeNiS$_2$ 纳米片/rGO 与循环伏安扫描 5000 圈之后 FeNiS$_2$ 纳米片/rGO 的 LSV 曲线对比图,两条曲线几乎无差别,表明 FeNiS$_2$ 纳米片/rGO 在长时间的循环过程中保持了良好的稳定性。

为了对循环测试之后材料的形貌结构、表面元素组成以及价态做出分析,对循环后的 FeNiS$_2$ 纳米片/rGO 做了 TEM 和 XPS 表征。图 10.10(b) 为稳定性测试后 FeNiS$_2$ 纳米片/rGO 的 TEM 图,结合上述内容对图 10.2(a) 和图 10.5(c) 进行了对比分析,在稳定性测试之后,对比新制的 FeNiS$_2$ 纳米片/rGO[图 10.5(c)],循环后的 FeNiS$_2$ 纳米片/rGO 二维薄片边缘结构略有模糊,说明在反应过程中也发生了自氧化过程,但是相对于图 10.2(a) 中单独的 FeNiS$_2$ 纳米片,其非晶化程度大大减弱,原始片状结构基本保留。对图 10.10(c)～(d) 的 XPS 结果进行分析,反应前后 Fe 的 2p 谱中信号几乎无变化;Ni 的 2p 谱中位于～855.6 eV 的 Ni—OH 峰增强,而位于～853.7 eV 的 Ni(Ⅱ)减弱;S 的 2p 谱前后差别较大,反应后位于～169 eV 的金属硫酸盐的峰增强,而原本位于～161.5 eV 的金属硫化物峰则减弱很多;同时结合 O 的 1s 谱分析,位于～530 eV 的金属氧化物峰略有增强,因此,可以推断出,在 OER 循环过程中,Ni 元素仍然有部分被氧化,S 存在部分溶出,但是在同等条件下,FeNiS$_2$ 纳米片/rGO 的自氧化过程比单独的 FeNiS$_2$ 纳米片缓慢得多。

以上分析结果表明,FeNiS$_2$ 纳米片在与 rGO 进行原位复合之后,减慢了自氧化过程的速率,阻碍了有效成分的溶出,从而使得材料稳定性有了极大提升,可以在长时间 OER 反应过程中保持高度稳定。

对 FeNiS$_2$ 纳米片在碱性体条件下 OER 催化过程中的自氧化现象进行了研究,发现在该过程中,FeNiS$_2$ 纳米片会发生结构的非晶化转变、金属氧化物与氢氧化物的生成和催化剂有效组分的溶出,虽然自氧化过程在一定程度上可以增强材料的催化活性,但更多的是对材料的稳定性造成不可逆转的下降。为解决这一问题,设计了简单的一步油相法,实现了 FeNiS$_2$ 纳米片和 rGO 的原位复合,这不仅保持了 FeNiS$_2$ 纳米片原本的二维薄片结构,还使 FeNiS$_2$ 纳米片组成的花球均匀分散在 rGO 表面。在碱性条件下的性能测试表明,将两者原位复合之后,无论是材料的本征催化活性,还是材料的催化稳定性,都有极大的提升。长时间稳定性测试前后的详细表征表明 FeNiS$_2$ 纳米片与 rGO 紧密复合形成的良好三维结构可以促进电子在两者之间的有效传递,增强催化剂导电能力;FeNiS$_2$ 纳米片在 rGO 基底上均匀分散,增大了材料的电化学活性面积;rGO 的

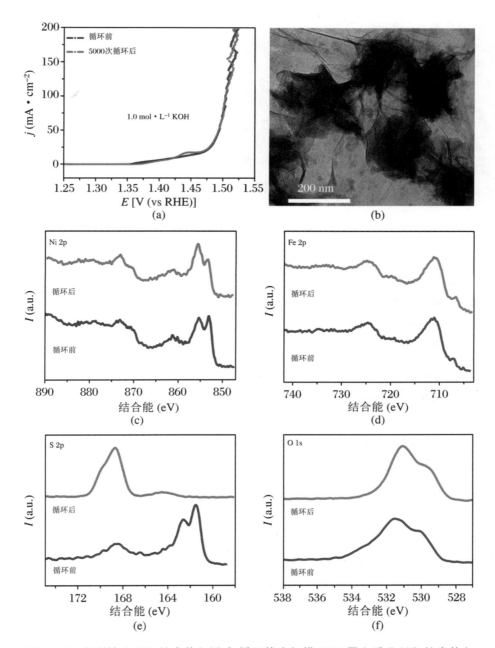

图 10.10 新制的 FeNiS₂ 纳米片/rGO 与循环伏安扫描 5000 圈之后 FeNiS₂ 纳米片/rGO 的 LSV 曲线对比图(a);循环伏安扫描 5000 圈之后 FeNiS₂ 纳米片/rGO 的 TEM 图(b);循环伏安扫描 5000 圈前后 FeNiS₂ 纳米片/rGO 中 Ni、Fe、S 和 O 元素的 XPS 图(c)~(f)

存在减缓了 FeNiS₂ 纳米片的自氧化速率,保留了无定形结构的优势,但有效阻止了活性成分的快速氧化和溶出,提升了材料的稳定性。基于此,不仅设计合成了一种适用于碱性条件下的性能优异的 OER 催化剂,还发展了一种油相反应原位复合rGO基底的方法,有效地优化了材料的催化活性和稳定性,有望应用于改善其他类似催化剂的性能。

参考文献

[1] Cheng F, Chen J. Metal-air batteries: from oxygen reduction electrochemistry to cathode catalysts [J]. Chemical Society Reviews, 2012, 41 (6): 2172-2192.

[2] Suen N T, Hung S F, Quan Q, et al., Electrocatalysis for the oxygen evolution reaction: recent development and future perspectives [J]. Chemical Society Reviews, 2017, 46 (2): 337-365.

[3] Munshi M, Tseung A, Parker J. The dissolution of iron from the negative material in pocket plate nickel-cadmium batteries [J]. Journal of Applied Electrochemistry, 1985, 15 (5): 711-717.

[4] Gong M, Dai H. A mini review of NiFe-based materials as highly active oxygen evolution reaction electrocatalysts [J]. Nano Research, 2015, 8 (1): 23-39.

[5] Trotochaud L, Young S L, Ranney J K, et al. Nickel-iron oxyhydroxide oxygen-evolution electrocatalysts: The role of intentional and incidental iron incorporation [J]. Journal of the American Chemical Society, 2014, 136 (18): 6744-6753.

[6] Ahn H S, Bard A J. Surface interrogation scanning electrochemical microscopy of $Ni_{1-x}Fe_xOOH$ ($0<x<0.27$) oxygen evolving catalyst: Kinetics of the 'fast' Iron sites [J]. Journal of the American Chemical Society, 2015, 138 (1): 313-318.

[7] Gong M, Li Y, Wang H, et al. An advanced Ni-Fe layered double hydroxide electrocatalyst for water oxidation [J]. Journal of the American Chemical Society, 2013, 135 (23): 8452-8455.

[8] Lu X, Zhao C. Electrodeposition of hierarchically structured three-dimensional nickel-iron electrodes for efficient oxygen evolution at high current densities [J]. Nature Communications, 2015, 6: 6616.

[9] Song F, Hu X. Exfoliation of layered double hydroxides for enhanced oxygen evolution catalysis [J]. Nature Communications, 2014, 5: 4477.

[10] Wang H, Lee H W, Deng Y, et al. Bifunctional non-noble metal oxide nanoparticle electrocatalysts through lithium-induced conversion for overall water splitting [J]. Nature Communications, 2015, 6: 7261.

[11] Smith R D, Prévot M S, Fagan R D, et al. Photochemical route for accessing amorphous metal oxide materials for water oxidation catalysis [J]. Science, 2013, 340 (6128): 60-63.

[12] Xu Q C, Jiang H, Zhang H X, et al. Heterogeneous interface engineered atomic configuration on ultrathin $Ni(OH)_2/Ni_3S_2$ nanoforests for efficient water splitting [J]. Applied Catalysis B-Environmental, 2019, 242: 60-66.

[13] Xu K, Chen P, Li X, et al. Metallic nickel nitride nanosheets realizing enhanced electrochemical water oxidation [J]. Journal of the American Chemical Society, 2015, 137 (12): 4119-4125.

[14] Zhu X, Dou X, Dai J, et al. Metallic nickel hydroxide nanosheets give superior electrocatalytic oxidation of urea for fuel cells [J]. Angewandte Chemie International Edition, 2016, 55 (40): 12465-12469.

[15] Yu X Y, Feng Y, Guan B, et al. Carbon coated porous nickel phosphides nanoplates for highly efficient oxygen evolution reaction [J]. Energy & Environmental Science, 2016, 9 (4): 1246-1250.

[16] Jiang J, Lu S, Gao H, et al. Ternary $FeNiS_2$ ultrathin nanosheets as an electrocatalyst for both oxygen evolution and reduction reactions [J]. Nano Energy, 2016, 27: 526-534.

[17] Chen G F, Ma T Y, Liu Z Q, et al. Efficient and stable bifunctional electrocatalysts $Ni/Ni_xM_y(M = P, S)$ for overall water splitting [J]. Advanced Functional Materials, 2016, 26(19):3314-3323.

[18] Wu Y, Liu Y, Li G-D, et al., Efficient electrocatalysis of overall water splitting by ultrasmall $Ni_xCo_{3-x}S_4$ coupled Ni_3S_2 nanosheet arrays [J]. Nano Energy, 2017, 35: 161-170.

[19] Friebel D, Louie M W, Bajdich M, et al. Identification of highly active Fe sites in (Ni, Fe)OOH for electrocatalytic water splitting [J]. Journal of the American Chemical Society, 2015, 137 (3): 1305-1313.

[20] Suryawanshi M P, Ghorpade U V, Shin S W, et al. Hierarchically coupled Ni:FeOOH nanosheets on 3D N-doped graphite foam as self-supported electrocatalysts for efficient and durable water oxidation [J]. ACS Catalysis, 2019, 9 (6): 5025-5034.

[21] Goörlin M, Ferreira de Araújo J, Schmies H, et al. Tracking catalyst redox states and reaction dynamics in Ni-Fe oxyhydroxide oxygen evolution reaction electrocatalysts: The role of catalyst support and electrolyte pH [J]. Journal of the American Chemical Society, 2017, 139 (5): 2070-2082.

[22] Hummers Jr W S, Offeman R E. Preparation of graphitic oxide [J]. Journal of the American Chemical Society, 1958, 80 (6): 1339.

[23] Huang Y X, Xie J F, Zhang X, et al. Reduced graphene oxide supported palladium nanoparticles via photoassisted citrate reduction for enhanced electrocatalytic activities [J]. ACS Applied Materials & Interfaces, 2014, 6 (18): 15795-15801.

[24] Indra A, Menezes P W, Sahraie N R, et al. Unification of catalytic water oxidation and oxygen reduction reactions: amorphous beat crystalline cobalt iron oxides [J]. Journal of the American Chemical Society, 2014, 136 (50): 17530-17536.

[25] Xie J, Zhang J, Li S, et al. Controllable disorder engineering in oxygen-incorporated MoS2 ultrathin nanosheets for efficient hydrogen evolution [J]. Journal of the American Chemical Society, 2013, 135 (47): 17881-17888.

[26] Youn D H, Han S, Kim J Y, et al. Highly active and stable hydrogen evolution electrocatalysts based on molybdenum compounds on carbon nanotube-graphene hybrid support [J]. ACS Nano, 2014, 8 (5): 5164-5173.

[27] Zhou X, Qiao J, Yang L, et al. A review of graphene-based nanostructural materials for both catalyst supports and metal-free catalysts in PEM fuel cell oxygen reduction reactions [J]. Advanced Energy Materials, 2014, 4 (8): 1301523.

[28] Long X, Li J, Xiao S, et al. A strongly coupled graphene and FeNi double hydroxide hybrid as an excellent electrocatalyst for the oxygen evolution reaction [J]. Angewandte Chemie International Edition, 2014, 53 (29): 7584-7588.

[29] Luo J, Im J H, Mayer M T, et al. Water photolysis at 12.3% efficiency via perovskite photovoltaics and Earth-abundant catalysts [J]. Science, 2014, 345 (6204): 1593-1596.

[30] Liu C, Kong D, Hsu P C, et al. Rapid water disinfection using vertically aligned MoS2 nanofilms and visible light [J]. Nature Nanotechnology, 2016, 11 (12): 1098.

[31] Deng S, Zhang K, Xie D, et al. High-index-faceted Ni_3S_2 branch arrays as bifunctional electrocatalysts for efficient water splitting [J]. Nano Micro Letters, 2019, 11 (1): 12.

[32] Li S, Wang Y, Peng S, et al. Co-Ni-based nanotubes/nanosheets as efficient water splitting electrocatalysts [J]. Advanced Energy Materials, 2016, 6 (3): 1501661.

[33] Cai P, Huang J, Chen J, et al. Oxygen-containing amorphous cobalt sulfide porous nanocubes as high-activity electrocatalysts for the oxygen evolution reaction in an alkaline/neutral medium [J]. Angewandte Chemie International Edition, 2017, 56 (17): 4858-4861.

[34] Gao X, Zhang H, Li Q, et al. Hierarchical $NiCo_2O_4$ hollow microcuboids as bifunctional electrocatalysts for overall water-splitting [J]. Angewandte Chemie International Edition, 2016, 55 (21): 6290-6294.

[35] Shi H, Liang H, Ming F, et al. Efficient overall water-splitting electrocatalysis using lepidocrocite VOOH hollow nanospheres [J]. Angewandte Chemie International Edition, 2017, 56 (2): 573-577.

[36] Wang Y, Zhang Y, Liu Z, et al. Layered double hydroxide nanosheets with multiple vacancies obtained by dry exfoliation as highly efficient oxygen evolution electrocatalysts [J]. Angewandte Chemie International Edition, 2017, 56 (21): 5867-5871.

[37] Xiao X, He C T, Zhao S, et al. A general approach to cobalt-based homobimetallic phosphide ultrathin nanosheets for highly efficient oxygen evolution in alkaline media [J]. Energy & Environmental Science, 2017, 10 (4): 893-899.

[38] Liu Y, Cheng H, Lyu M, et al. Low overpotential in vacancy-rich ultrathin $CoSe_2$ nanosheets for water oxidation [J]. Journal of the American Chemical Society, 2014, 136 (44): 15670-15675.

[39] Song F, Hu X. Ultrathin cobalt-manganese layered double hydroxide is an efficient oxygen evolution catalyst [J]. Journal of the American Chemical Society, 2014, 136 (47): 16481-16484.

[40] Zhao B, Zhang L, Zhen D, et al. A tailored double perovskite nanofiber catalyst enables ultrafast oxygen evolution [J]. Nature communications, 2017, 8: 14586.

[41] Chen X, Chen G. Stable $Ti/RuO_2\text{-}Sb_2O_5\text{-}SnO_2$ electrodes for O_2 evolution

[J]. Electrochimica Acta, 2005, 50 (20): 4155-4159.

[42] Jia J, Li X, Chen G. Stable spinel type cobalt and copper oxide electrodes for O_2 and H_2 evolutions in alkaline solution [J]. Electrochimica Acta, 2010, 55 (27): 8197-8206.

[43] Chen X, Chen G, Yue P L. Stable Ti/IrO_x-Sb_2O_5-SnO_2 anode for O_2 evolution with low Ir content [J]. The Journal of Physical Chemistry B, 2001, 105 (20): 4623-4628.

[44] Yang N, Tang C, Wang K, et al. Iron-doped nickel disulfide nanoarray: A highly efficient and stable electrocatalyst for water splitting [J]. Nano Research, 2016, 9 (11): 3346-3354.

[45] Zhang L, Zhang R, Ge R, et al. Facilitating active species generation by amorphous NiFe-Bi layer formation on NiFe-LDH nanoarray for efficient electrocatalytic oxygen evolution at Alkaline pH [J]. Chemistry-A European Journal, 2017, 23 (48): 11499-11503.

[46] Xie M, Yang L, Ji Y, et al. An amorphous Co-carbonate-hydroxide nanowire array for efficient and durable oxygen evolution reaction in carbonate electrolytes [J]. Nanoscale, 2017, 9 (43): 16612-16615.

第 11 章

多金属硫化物纳米复合材料界面电子传递过程中氧化/还原的双向调控

11.1
金属硫化物中多金属组合的优势

金属硫化物中由于存在多种金属共同形成化合物的可能性,为实现多金属位点的协同作用以及实现多种反应的催化提供了契机。同时,多金属硫化物界面较宽的可调电位,可以为金属空气电池的性能提升提供设计空间。锌空气电池以其能量密度高(理论能量密度可达 1084 W·h·kg^{-1})、放电电压平稳、环境友好、成本低廉和安全性高等优势,成为新型电气化交通工具和便携移动设备极具发展潜力的候选供能装置[1-2]。锌空气电池的充电方式包含两种,机械式充电和电化学式充电[3]。机械式充电即为定期更换消耗掉的锌片,然而该方式成本高、过程复杂,需要建立锌供应站网络,因此,很难被广泛采用。电化学充电则需要发生放电的逆过程,空气电极上发生水氧化反应生成氧气,锌电极表面的 ZnO 被还原为 Zn 在电极上沉积。因此,对于电化学式充电的锌空气电池,产氧反应(OER)和氧还原反应(ORR)是锌空气电池空气电极的两个主要电化学反应。这两个反应都为四电子转移过程,中间态复杂,动力学过程缓慢,因此,获得在相同电解质环境下具有 OER 和 ORR 双催化功能的稳定高效催化剂是推进锌空气电池实际应用的关键点[4-6]。

迄今为止,设计合理的适用于锌空气电池的 OER/ORR 双功能催化剂对于研究者们来说仍然是一个巨大的挑战,OER 和 ORR 的过电位过高是一个方面,同时催化剂还要能够抵抗高浓度碱溶液的腐蚀,在强氧化性的产氧电位区间和氧还原的强还原条件下稳定运行[3]。目前双功能催化剂的研究主要集中在过渡金属化合物(如氧化物、硫化物、氮化物和碳化物等)、碳基材料和前两者组成的复合物等材料上[3]。对于材料的设计和优化也已经有许多策略被提出,例如,晶体结构和组成成分调控[7-8],不同形态例如二维和三维结构的设计[9],催化剂尺寸的控制、晶体表面的选择性暴露[10]和缺陷的引入[11]等。

在目前的研究中,丰富且廉价的金属氧化物双功能电催化剂被大量报道[12-13],但金属氧化物作为氧电极催化剂仍然受到其不良固有电导率的阻碍。非贵金属硫化物中有大量的金属硫键,氧原子和硫原子之间的电负性和原子半径差异导致非贵金属氧化物和硫化物中共价键和金属键的比例不同,非贵金属硫化物中金属键的比例更多,导电率相对于氧化物有所增强,从该角度考虑,开发合适的非贵金属硫化物作为双功能催化剂具有可行性[14-16]。之前的研究表明,Co

金属硫化物 Co_9S_8 在催化 ORR 方面具有相对较低的过电位[17-18]。此外,在碱性条件下,引入少量 Fe 可使 Co 和 Ni(氧)氢氧化物的 OER 活性分别提高约 30 倍和 1000 倍[19]。受此启发,可以推测基于 Co_9S_8 的 FeCo 双金属硫化物可以作为高效的双功能氧电极催化剂,在同一电解质中表现出良好的 OER 和 ORR 活性。

基于上述假设,在本章中利用一步油相法及后续 NH_3 退火过程合成一系列 FeCo 双金属或单金属硫化物与 rGO 原位复合的纳米材料,利用各项表征分析引入 Fe 之后材料在物相和组成上的差异以及电化学测试评估材料在双功能上的催化性能。结合 EXAFS 测试和 DFT 理论计算对材料的原子配位环境和价态以及反应过程中生成中间态的路径进行分析,从原子尺度和理论角度阐述 Fe 的引入对材料本征性能的影响,并通过实际对可充电锌空气电池器件的测试,分析所制备材料作为双功能氧电极的潜在应用价值。

11.2 多金属硫化物纳米复合材料的制备与研究方法

11.2.1

催化剂的制备

采用第 3 章所述的一步油相法调控合成了一系列 Fe、Co 二维金属硫化物纳米片与 rGO 原位复合的催化剂材料,包括 Co_9S_8/rGO、$FeCo_8S_8$/rGO 和 $FeCoS_2$/rGO。具体步骤示意图如图 11.1 所示,在经过油相反应之后生成金属硫化物与氧化石墨烯的复合结构,随后在 NH_3 条件下进行退火处理将 GO 还原为 rGO。具体实验步骤陈述如下:

$FeCo_8S_8$/rGO 的制备:在洁净的 50 mL 三颈圆底烧瓶中,加入 0.0078 g (0.022 mmol)乙酰丙酮铁和 0.0458 g(0.178 mmol)乙酰丙酮钴,随后称取 5 mg

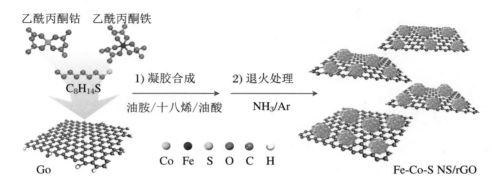

图 11.1 合成 Co_9S_8/rGO、$FeCo_8S_8$/rGO 和 $FeCoS_2$/rGO 的步骤示意图

冷冻干燥后的 GO 置于 1 mL 油胺中，超声分散，加入烧瓶中，再加入 5 mL 十八烯、0.2 mL 油酸和 0.2 mL 辛硫醇。在 N_2 保护下，在 20 min 内将反应体系升温至 150 ℃，并保持 30 min，然后以 10 ℃·min^{-1} 的升温速率升温至 260 ℃，保持反应 30 min 后自然冷却降至室温。取出离心后得到的黑色产物，然后用乙醇超声洗涤离心 2 遍，再用正己烷超声洗涤离心 2 遍，最后离心后得到的产物置于 80 ℃ 真空干燥箱 8 h 后取出。烘干后的产物研磨后置于刚玉瓷舟中，放入管式炉，通入 NH_3/Ar 混合气（NH_3 占混合气的体积百分数为 10%），以 5 ℃·min^{-1} 的速率升温至 400 ℃，在该温度下保持 2 h，随后自然降温冷却，得到最终产物 $FeCo_8S_8$/rGO。

Co_9S_8/rGO 的制备：制备方法和步骤与 $FeCo_8S_8$/rGO 相同，区别在于将金属前驱换成 0.0385 g (0.15 mmol) 乙酰丙酮钴。

$FeCoS_2$/rGO 的制备：制备方法和步骤与 $FeCo_8S_8$/rGO 相同，区别在于将金属前驱换成 0.0353 g (0.10 mmol) 乙酰丙酮铁和 0.0257 g (0.10 mmol) 乙酰丙酮钴。

11.2.2
材料表征和电化学测试

XRD 测试采用 Philips X′Pert PRO SUPER 衍射仪，阳极靶为铜靶，衍射仪装配有石墨单色化的 Cu-K 放射源（λ = 1.541874 Å，Rigaku TTR-Ⅲ，PHLIPS Co.，the Netherlands），用以分析样品的纯度、物相和晶格参数等信息。样品的形貌分析采用场发射扫描电子显微镜（SEM，Supra 40，Zeiss Co.，Germany）和透射电子显微镜（TEM，H-7650，Hitachi Co.，Japan，加速电压 100 kV）。高分辨透射电镜（HRTEM，JEM. ARM 200F，工作电压 200 kV）则

用以分析材料的形貌、晶格结构和元素组成等信息，得到包括高分辨透射显微(HRTEM)图、高角度环型暗场扫描透射(HAADF-STEM)图和对应的元素分布(EDS)图等。样品的化学组成和元素价态分析由 X-射线光电子能谱仪(XPS，Thermo Fisher ESCALAB250 型)测试得到。拉曼谱图则由共聚焦激光显微拉曼光谱仪(LABRAM-HR，Jobin-Yvon Co.，France)测得，设置激发波长为 514.5 nm。材料的比表面积分析采用 BET 分析方法，利用 Builder 4200 instrument (Tristar Ⅱ 3020M，Microme-ritics Co.，USA)测试得到结果。

所有电化学测试均在旋转圆盘装置(Pine Research Instrumentation Inc.，USA)上进行，参数设定和数据采集使用 CHI 760e 电化学工作站(Chenhua Instrument Co.，China)，测试采用三电极系统。测试使用的玻碳电极有效直径为 5.0 mm，几何面积为 0.196 cm^2，Ag/AgCl 电极(内装饱和 KCl 溶液)和铂丝电极分别作为参比电极和对电极。

对催化材料进行测试准备：分别称取 3 mg FeCo$_8$S$_8$ 纳米片/rGO、Co$_9$S$_8$ 纳米片、Co$_9$S$_8$ 纳米片/rGO、FeCoS$_2$ 纳米片/rGO 和商业 RuO$_2$，分别分散至 1.0 mL 异丙醇和水的混合溶剂中(体积比为 1∶3)，随后加入 20 μL Nafion 膜溶液，超声 60 min 使其混合分散均匀形成墨水状溶液。用移液管移取 10 μL 上述墨水状催化剂溶液滴到玻碳电极表面，待其自然风干成膜，催化剂在电极上负载量为 0.15 mg·cm^{-2}。测试得到的 LSV 曲线中的电位都经过溶液内阻校正，电流密度都经过电极面积归一化，并且所有的电极电势都统一转化为标准氢电极电势，通过以下公式进行转换：

$$E_{\text{vs. RHE}} = E_{\text{vs. Ag/AgCl}} + E_{\text{Ag/AgCl}}^{\theta} + 0.059 \times \text{pH} \tag{11.1}$$

催化剂的 OER 性能测试分别在 O$_2$ 饱和的 0.1 mol·L^{-1} KOH 溶液和 1 mol·L^{-1} KOH 溶液中进行。LSV 测试电压范围为 1.1～1.8 V(vs RHE)，扫速为 5 mV·s^{-1}，旋转圆盘转速为 1600 r·min^{-1}。电化学阻抗谱(EIS)的测试在 0.6 V(vs Ag/AgCl)的电压下进行，频率范围为 0.1 Hz～100 kHz，外加电压振幅为 10 mV。对于稳定性测试，采用循环伏安法(CV)，连续循环 3000 圈以衡量材料的稳定性。

催化剂的 ORR 性能测试在 N$_2$ 或 O$_2$ 饱和的 0.1 mol·L^{-1} KOH 溶液中进行，LSV 测试电压范围为 0.3～1.1 V(vs RHE)，扫描速度为 5 mV·s^{-1}，旋转圆盘转速为 1600 r·min^{-1}，将 O$_2$ 条件下得到的电流减去 N$_2$ 条件下得到的电流，得到最终 LSV 曲线。起始电位以达到 0.1 mA·cm^{-2} 的电流密度为准，电化学阻抗谱(EIS)的测试在 −0.2 V(vs Ag/AgCl)的电压下进行，频率范围为 0.1 Hz～100 kHz，外加电压振幅为 10 mV。同样地，稳定性测试采用循环伏安法(CV)连续循环 3000 圈以衡量材料的稳定性。反应中电子转移数和过氧化氢产率通过旋转环盘测试获得的数据计算得到，计算方法如下：

$$n = \frac{4i_d}{i_d + \frac{i_r}{N}} \tag{11.2}$$

$$H_2O_2\% = \frac{200i_r}{i_r + Ni_d} \tag{11.3}$$

其中，n 表示电子转移数，N 为旋转环盘电极 Pt 环的电流采集效率($N=0.4$)，i_r 和 i_d 分别是测试得到的环电流和盘电流数据。在进行 RRDE 测试过程中，设置环盘扫速为 5 mV·s^{-1}，环电压始终保持在 1.0 V(vs. RHE)。

11.2.3
锌空气电池的组装和测试方法

锌空气电极正极包含三个部分：泡沫镍支撑骨架、空气扩散层和催化剂层。对空气扩散层，先将 0.1 g 乙炔黑、0.2 g 炭黑混合分散在乙醇中，超声 10 min，随后加入 0.7 g PTFE 溶液(聚四氟乙烯溶液)，再均匀搅拌 60 min，搅拌之后置于 80 ℃烘箱内去除多余的乙醇，直至混合物变成面团状之后取出，在辊压机下辊压成 300 nm 厚的膜。对于催化剂层，将 15 mg FeCo$_8$S$_8$/rGO 和 5 mg 炭黑混合并分散在乙醇中，同样超声 10 min，随后加入 15 mg PTFE 溶液，在搅拌 60 min 之后，置于 80 ℃烘箱内去除多余的乙醇，直至混合物变成面团状，取出并辊压成膜且保证催化剂的负载量为 3 mg·cm^{-2}。完成上述操作之后，将辊压之后的空气扩散层和催化层分别贴合在泡沫镍的两侧，再经过辊压，使空气扩散层和催化层完全覆盖泡沫镍并压进泡沫镍骨架的缝隙中，辊压之后厚度为 0.4 mm。最后将电极片置于马弗炉中，在 320 ℃的空气中加热 20 min，冷却后取出待用。

最后为电池的组装，顺序如下：下层板，打磨后的锌片，厚度 0.8 mm 的留有与中层板圆孔大小一致圆孔的硅胶膜，中层板(侧面留有注入电解液的小孔)，厚度 0.8 mm 的留有与中层板圆孔大小一致圆孔的硅胶膜，正极电极片，上层板，随后整个组件用螺丝在四角固定。电解液为浓度 6 mol·L^{-1} KOH 和 0.2 mol·L^{-1} Zn(Ac)$_2$ 的混合溶液，由侧面的小孔注入，待排除气泡后用橡胶塞堵住。电池外壳采用PMMA材料，上层板有面积 3 cm^2 的圆形孔，便于正极与空气接触，使氧气参与反应。

在电池测试过程中，采用三电极系统，工作电极为负载催化剂的正极电极片，参比电极和对电极都为锌片，利用 CHI 760d 电化学工作站进行测试。对于充放电 LSV 测试，扫速为 5 mV·s^{-1}，正向扫描，电压范围为 0.6～2.2 V。

11.2.4
界面电子传递理论解析方法

计算方法：利用 CASTEP 软件[20]进行密度泛函理论计算。在 340 eV 平面波截断能下，采用超软赝势和广义梯度近似（GGA）的 perdewey-burke-ernzerhof（PBE）交换相关泛函进行计算[21-22]。考虑自旋极化和弛豫，采用 Broyden、Fletcher、Goldfarb 和 Shannom（BFGS）算法进行几何结构优化[23]。自洽场（SCF）设为 1×10^{-4} eV·cell^{-1}。这种设定是为了在速度和准确度之间取得平衡，并且计算结果对于探索性研究来说足够精确。利用 Monkhorst-Pack 算法生成的 k 点数[24]，对不同大小和形状晶胞的布里渊区进行积分。$FeCo_8S_8$、Co_9S_8 和 $FeCoS_2$ 的 slab 模型的真空层至少为 10 Å。在 slab 模型中，Co/Fe 位点被认为是 Co/Fe-S 表面层的活性中心。

自由能计算：OER/ORR 途径中所有中间吸附体都与活性位点结合，以计算自由能谱。在碱性电解液中，OER 反应有 4 步基本反应[25-26]：

$$\text{反应 1：}^*OH + OH^- \longrightarrow {}^*O + H_2O + e^- \tag{11.4}$$

$$\text{反应 2：}^*O + OH^- \longrightarrow {}^*OOH + e^- \tag{11.5}$$

$$\text{反应 3：}^*OOH + OH^- \longrightarrow {}^*OO + H_2O + e^- \tag{11.6}$$

$$\text{反应 4：}^*OO + OH^- \longrightarrow {}^*OH + O_2 + e^- \tag{11.7}$$

在 ORR 反应中，根据电化学测试和以往报道的机制，过程包含以下 6 步[27]：

$$\text{反应 1：} O_2 + {}^* \longrightarrow {}^*O_2 \tag{11.8}$$

$$\text{反应 2：}^*O_2 \longrightarrow {}^*O + {}^*O \tag{11.9}$$

$$\text{反应 3：}^*O + {}^*O + H_2O + e^- \longrightarrow {}^*O + {}^*OH + OH^- \tag{11.10}$$

$$\text{反应 4：}^*O + {}^*OH + H_2O + e^- \longrightarrow {}^*O + H_2O + OH^-$$
$$({}^*O + {}^*OH + e^- \longrightarrow {}^*O + OH^-) \tag{11.11}$$

$$\text{反应 5：}^*O + H_2O + H_2O + e^- \longrightarrow {}^*OH + H_2O + OH^-$$
$$({}^*O + H_2O + e^- \longrightarrow {}^*OH + OH^-) \tag{11.12}$$

$$\text{反应 6：}^*OH + H_2O + H_2O + e^- \longrightarrow {}^* + 2H_2O + OH^-$$
$$({}^*OH + e^- \longrightarrow {}^* + OH^-) \tag{11.13}$$

OER/ORR 在电化学反应路径中的中间物自由能的计算是基于计算标准

氢电极(CHE)模型[28]的,因此,在 pH 不为零的情况下,$G(H^+) = 1/2 G(H_2) - 2.303 k_B T \times pH$($p = 100$ kPa, $T = 298.15$ K)。根据中性溶液条件下的反应 $H_2O(l) \leftrightarrow H^+(aq) + OH^-(aq)$,$G(OH^-)$ 可以表示成 $G(H_2O) - G(H^+)$。除此之外,产生两个水分子的实验 ΔG 为 -4×1.23 eV($= -4.92$ eV),因此 O_2 的自由能为 $G(O_2) = 2G(H_2O) - 2G(H_2) + 4.92$。例如,OER 中反应 1 的 ΔG 可以表示为

$$\Delta G_1 = G(^*O) - G(^*OH) + G(H_2O) - G(OH^-) - eU \quad (11.14)$$

可以改写为

$$\Delta G_1 = G(^*O) - G(^*OH) + \frac{1}{2} G(H_2) - eU - 2.303 k_B T \times pH \quad (11.15)$$

其中,$G(\text{adsorbate}) = E_t + ZPE - TS$,$E_t$ 为含吸附物几何结构的 DFT 计算得到的总能量,ZPE 是零点能,T 为温度,S 为熵。ZPE 和 TS 远小于 ΔE,所以对 ΔG 的贡献很小,因此,ΔG 可以由 ΔE 进行估算[29-30]。e 为反应中转移的电子的电量,U 为实验中外加电压(平衡电压,计算中 $U = 1.23$ V)。对于氧的还原过程,在方程中添加 $+eU$ 来优化反应的 ΔG。例如,ORR 过程的反应 3 可以写为

$$\Delta G_3 = G(^*O + ^*OH) - G(^*O + ^*O) - \frac{1}{2} G(H_2) + eU + 2.303 k_B T \times pH \quad (11.16)$$

11.3 多金属硫化物界面电子传递催化氧化/还原反应的双向调控

11.3.1 材料的合成优化以及形貌物相和元素化合态

在采用一步油相法合成 $FeCo_8S_8/rGO$ 时,为了探究能使 $FeCo_8S_8$ 保持二维

薄片结构且能够将 $FeCo_8S_8$ 与 rGO 进行有效原位复合的影响因素和最佳条件，对油相反应中的溶剂和表面活性剂的添加量以及比例进行可控性调节实验。图 11.2(a)～(d)即为控制加入油酸(OA)的量得到的形貌不同的 $FeCo_8S_8$ 的 SEM 图(溶剂总量为 6 mL)。对图像进行分析，当不加入油酸时，产物几乎不形成二维薄片结构，多为团聚的小颗粒；当加入油酸量为 0.1 mL 时，产物中有片状结构产生，但组成的片花大小尺寸不均，且仍然含有不少团聚颗粒；当加入 0.24 mL 油酸，产物形成大小较为均一的花球，且边缘有清晰的二维片状结构；但当加入油酸为 0.4 mL 时，花球尺寸变大，花球边缘的二维片状结构破裂成为碎片。因此，在溶剂合适的条件下，油酸的最佳投加量为 0.2 mL。图 11.3(a)～(h)为调整十八烯(ODE)和油胺(OAM)比例得到的 $FeCo_8S_8$/rGO 复合物的 SEM 图，不加入 ODE 时，生成的 $FeCo_8S_8$ 易团聚；而溶剂全部为 ODE 时，形成的二维片厚度较大，且组成的片花直径接近且非常大，接近 1 μm。因此，调整两者比例为 OAM∶ODE＝3∶3，此时产物仍然存在团聚现象，并且在石墨烯基底上分布不均匀，为此增加了溶剂中 ODE 所占的比例；当 OAM∶ODE＝1∶5 时，花球状的 $FeCo_8S_8$ 均匀负载在铺展开的 GO 上，两者的片状结构得到有效结合。

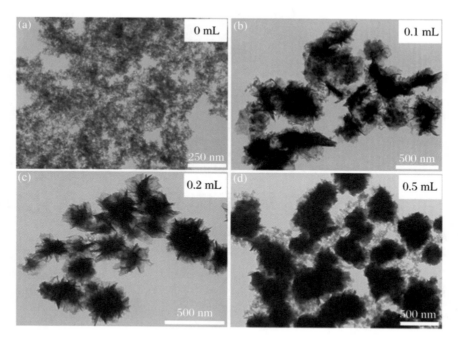

图 11.2　在添加不同量油酸情况下得到的 $FeCo_8S_8$/rGO，油酸添加量分别为 0 mL(a)；0.1 mL(b)；0.2 mL(c)；0.5 mL(d)

通过以上一步油相法合成条件的调控和后续的 NH_3 退火还原获得了 $FeCo_8S_8$ 与 rGO 的原位复合材料，图 11.4(a)为 $FeCo_8S_8$/rGO 的 SEM 图，二维纳米片组

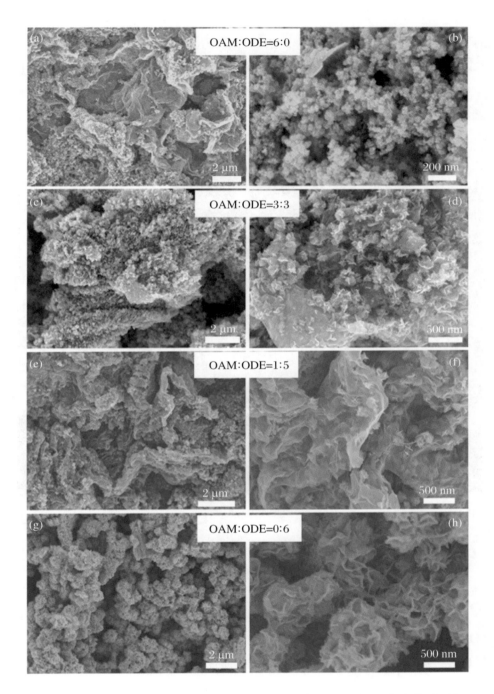

图 11.3 调整十八烯(ODE)和油胺(OAM)比例得到的 $FeCo_8S_8$/rGO 复合物:OAM：ODE=6：0(a);OAM：ODE=3：3(b);OAM：ODE=1：5(c);OAM：ODE=0：6(d)

成的花球状的 $FeCo_8S_8$ 均匀分布在 rGO 基底上,花球直径在 100 nm 左右,图 11.4(b)的 TEM 图进一步说明纳米花球均匀附着在石墨烯薄片上,并且 rGO

卷曲的边缘部分与 $FeCo_8S_8$ 的片状边缘重叠。图 11.4(c) 为 $FeCo_8S_8$/rGO 对应的元素含量图，Fe、Co、S 和 C 元素的原子含量比分别为 3.35%、26.48%、31.88% 和 38.29%，其中 Fe、Co 和 Co 的原子比例基本符合 $FeCo_8S_8$ 化合物的计量比。但是因为 C 的信号部分来自于铜网碳膜，因此，为了分析 $FeCo_8S_8$/rGO 复合物中 rGO 的具体含量，采用热重分析法(TG)。图 11.5 为 $FeCo_8S_8$/rGO 复合物的热重曲线，根据最终的失重百分数，得到其中碳含量质量比为 35.7%。

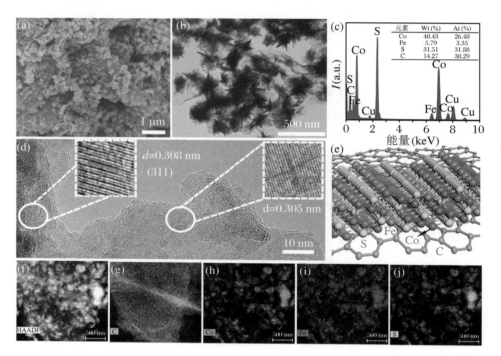

图 11.4　$FeCo_8S_8$ NS/rGO 形貌结构表征：SEM 图(a)；TEM 图(b)；元素分布和含量图(c)；HRTEM 图和对应的晶面条纹(d)；复合催化剂的原子排布示意图(e)；HAADF 图和对应的元素分布(f)～(j)，包括 C(靛青)、Co(绿色)、Fe(紫色)和 S(黄色)

为进一步了解 $FeCo_8S_8$/rGO 复合物的晶格结构信息，对其进行了高分辨(HRTEM)测试，并对结果进行了分析。如图 11.4(d)所示，对晶格条纹清晰的区域进行放大，通过测量条纹间距可以得出所选区域的晶格间距分别为 0.308 nm 和 0.305 nm，正好对应于立方 $FeCo_8S_8$ 的(311)晶面。根据 SEM、TEM 和 HRTEM 的表征结果，$FeCo_8S_8$ 纳米片与氧化还原石墨烯，因此，该复合物可以描述为两个二维结构的组合，其原子结构可以表示成图 11.4(e)。此外，还可以利用 HAADF 和对应的元素分布成像分析元素在 $FeCo_8S_8$/rGO 复合物中的具体分布，如图 11.4(f)～(j)所示。与图中标注的一样，Fe、Co 和 S 三种元素都在 $FeCo_8S_8$ 中均匀分布，而 C 元素则分布于 rGO 基底上，与图 11.4(e)所展示的模型非常符合。

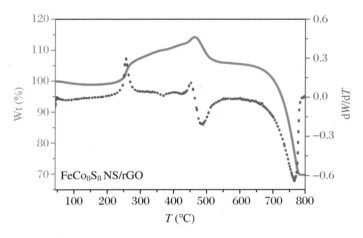

图 11.5 FeCo$_8$S$_8$/rGO 复合物的热重分析曲线

在对形貌进行分析之后,对制备的一系列产物做了物相、组成和化合态相关的表征。图 11.6(a) 为 FeCo$_8$S$_8$ NS/rGO、FeCoS$_2$ NS/rGO、Co$_9$S$_8$ NS/rGO 和 Co$_9$S$_8$ NS 的 XRD 谱图,与标准谱图进行对比,Co$_9$S$_8$ NS 和 Co$_9$S$_8$ NS/rGO 的主要衍射峰与立方 Co$_9$S$_8$ 的标准卡片(JCPDS Card,No. 29-0484)衍射峰对应,在晶格中引入少量 Fe 元素之后,产物的 XRD 衍射峰可以与立方 FeCo$_8$S$_8$ 的标准衍射峰匹配(JCPDS Card,No. 29-0484),当加入 Co 和 Fe 的前驱为 1∶1 时,得到的产物衍射峰可以归类为六方晶体 FeCoS$_2$(JCPDS Card,No. 75-0607),

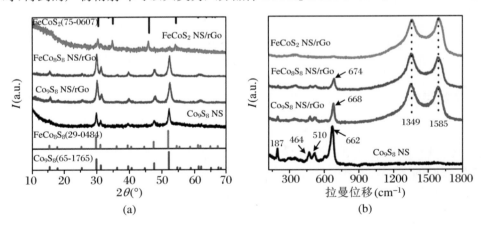

图 11.6 FeCo$_8$S$_8$ NS/rGO、FeCoS$_2$ NS/rGO、Co$_9$S$_8$ NS/rGO 和 Co$_9$S$_8$ NS 的 XRD 谱图 (a);Raman 谱图(b)和 XPS 图(c)(Co 2p);CoFe$_2$O$_4$、CoO、CoS、Co$_9$S$_8$ NS/rGO、FeCo$_8$S$_8$ NS/rGO 和 FeCoS$_2$ NS/rGO 的 Co 元素归一化的 K 边 XANES 图(d);Co 元素 K 边扩展的 XAFS 的 $k^3\chi(k)$ 振荡曲线(e)和 $k^3\chi(k)$ 振荡曲线对应的傅里叶变换曲线(f)

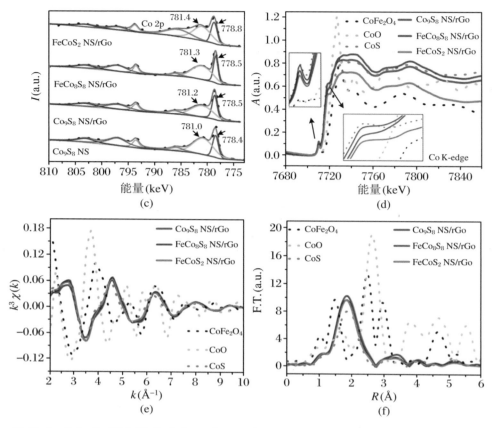

图 11.6 FeCo$_8$S$_8$ NS/rGO、FeCoS$_2$ NS/rGO、Co$_9$S$_8$ NS/rGO 和 Co$_9$S$_8$ NS 的 XRD 谱图(a)；Raman 谱图(b)和 XPS 图(c)(Co 2p)；CoFe$_2$O$_4$、CoO、CoS、Co$_9$S$_8$ NS/rGO、FeCo$_8$S$_8$ NS/rGO 和 FeCoS$_2$ NS/rGO 的 Co 元素归一化的 K 边 XANES 图(d)；Co 元素 K 边扩展的 XAFS 的 $k^3\chi(k)$ 振荡曲线(e)和 $k^3\chi(k)$ 振荡曲线对应的傅里叶变换曲线(f)(续)

以上结果都说明通过一步油相法成功合成了双金属 Fe-Co 为基础的纯相硫化物，并且 FeCo$_8$S$_8$ NS/rGO 与 Co$_9$S$_8$ NS/rGO 的 XRD 谱图非常相似，表明两者存在相似的晶格结构。同时，Co$_9$S$_8$ NS/rGO、FeCo$_8$S$_8$ NS/rGO 和 FeCoS$_2$ NS/rGO 的 XRD 谱图中都存在 $2\theta = 25°$ 的峰，该峰可以归属为 rGO 的衍射峰[31]，因此，在 Co$_9$S$_8$ 的 XRD 结果中并未出现。如图 11.6(b)所示的拉曼谱图也对该结果给出进一步证明，对于 Co$_9$S$_8$ NS/rGO、FeCo$_8$S$_8$ NS/rGO 和 FeCoS$_2$ NS/rGO，位于 1349 cm^{-1} 和 1585 cm^{-1} 处的峰对应于石墨烯的 D 峰和 G 峰，证明产物中存在石墨烯，而位于 187 cm^{-1}、464 cm^{-1}、510 cm^{-1}、662 cm^{-1} 处的峰则对应于化合物 Co$_9$S$_8$ 已知的拉曼峰，同时，在引入 Fe 元素之后，位于以上位置的峰都有所减弱，说明在引入 Fe 之后对产物的电子结构有所调整。

此外，采用 XPS 对催化剂表面的化学成分和价态进行了研究。图 11.7(a)

给出 Co_9S_8 NSs、Co_9S_8 NS/rGO、$FeCo_8S_8$ NS/rGO 和 $FeCoS_2$ NS/rGO 的 XPS 总谱图,除了材料中已有的元素 Co、S 和 C 元素的轨道峰外(Fe 元素峰太小),还出现了 O 和 N 元素的峰。N 的 1s 峰存在于含有 rGO 的产物中,可以归因于在 NH_3/Ar 气氛中退火还原 GO 时部分 N 元素掺杂进入 rGO 的结构中。图 11.7(b) 为 S 的 2p 谱,其自旋分裂的 $2p_{1/2}$ 和 $2p_{3/2}$ 峰分别位于 161.3 eV 和 162.5 eV,与 S^{2-} 的峰位相匹配,而在 166~170 eV 高结合能之间也存在微小的峰,归属于硫氧化物峰[16]。图 11.7(c) 为 Fe 的 2p 谱,因为 $FeCo_8S_8$ NS/rGO 中 Fe 含量较少,所以只在 710 eV 处有一个宽峰[32],但对于 $FeCoS_2$ NS/rGO,Fe 的 $2p_{3/2}$ 和 Fe $2p_{1/2}$ 两个裂分轨道在 710.9 eV 和 724.7 eV 处有两个明显的峰,结合能值与 Fe^{3+} 的结合能值匹配[33],因此,在 $FeCoS_2$ NS 中,Fe 处于 +3 价。图 11.7(d) 为 C 元素的 1s 谱图,位于 284.8 eV 的峰可以归属为 rGO 中的 C=C 键[31]。此外,在图 11.7(e) 的 N 1s 谱图中,位于 399.3 eV 的峰属于 C—N 峰,说明样品在 NH_3 气氛退火过程中,N 进入 rGO 的缺陷结构与 C 原子成键。

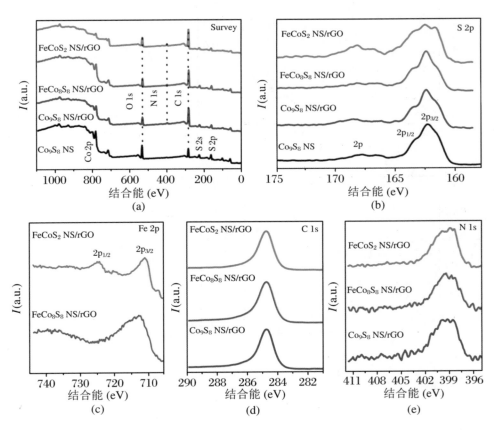

图 11.7 Co_9S_8 NSs、Co_9S_8 NS/rGO、$FeCo_8S_8$ NS/rGO 和 $FeCoS_2$ NS/rGO 的 XPS 总谱图 (a);对应的 S 2p 轨道、Fe 2p 轨道、C 1s 轨道和 N 1s 轨道的 XPS 谱图(b)~(e)

为了研究引入 Fe 元素前后产物中 Co 的配位和价态变化,对 Co 的 2p XPS 谱图做出分析。如图 11.8(a)所示,Co 的 $2p_{3/2}$ 和 $2p_{1/2}$ 两个轨道的峰位置结合能分别为 778.5 eV 和 793.5 eV,与 Co^{2+} 的结合能对应[34],因此在所有产物中,Co 的主要价态为 +2 价,同时谱图中还有两个分别位于 781.1 eV 和 797.0 eV 的小峰,说明样品表面有少量 CoO 存在[35],另外在引入 Fe 和 rGO 后,Co $2p_{3/2}$ 和 $2p_{1/2}$ 的峰位置明显向更高结合能的方向偏移。为了对此做出更精细的解析,采用 X 射线吸收精细结构光谱(XAFS)对产物(Co_9S_8 NS/rGO、$FeCo_8S_8$ NS/rGO 和 $FeCoS_2$ NS/rGO)和标准对比样($CoFe_2O_4$、CoO、CoS)进行测试,以进一步了解产物中 Co 元素的精细化学状态和局部原子排列。如图 11.8(b)所示,Co_9S_8 NS/rGO、$FeCo_8S_8$ NS/rGO 和 $FeCoS_2$ NS/rGO 在边前区域都有一个 7710.5 eV 左右的小峰,形状和位置都与标准样 CoO 和 CoS 在该区域的峰非

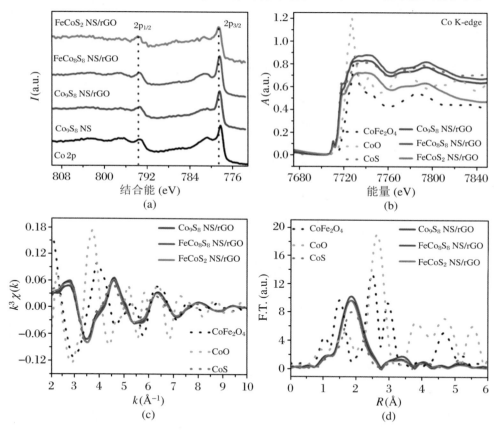

图 11.8　$FeCo_8S_8$ NS/rGO、$FeCoS_2$ NS/rGO、Co_9S_8 NS/rGO 和 Co_9S_8 NS 的 Co 2p XPS 谱图 (a);$CoFe_2O_4$、CoO、CoS、Co_9S_8 NS/rGO、$FeCo_8S_8$ NS/rGO 和 $FeCoS_2$ NS/rGO 的 Co 元素归一化的 K 边 XANES 谱图(b);Co 元素 K 边扩展的 XAFS 的 $k^3\chi(k)$ 振荡曲线(c)和 $k^3\chi(k)$ 振荡曲线对应的傅里叶变换曲线(d)

常匹配,并且产物的吸收边的特征与 CoS 谱一致,表明产物中 +2 价 Co 占主导且 Co—S 成键状态与 CoS 非常相似。此外,相对于 Co_9S_8 NS/rGO、$FeCo_8S_8$ NS/rGO 和 $FeCoS_2$ NS/rGO 的吸收边略微向高能方向偏移,该结果与 XPS 的分析结果一致。以上这些结果表明 rGO 与 Fe—Co 硫化物之间存在相互作用,而且适当引入 Fe 可以进一步调节硫化物中 Co 的电子结构。基于此,为了更细致地研究样品中引入 Fe 前后的原子近邻结构,进一步分析了产物 Co 元素 K 边扩展的 EXAFS 的 $k^3\chi(k)$ 振荡曲线和对应的傅里叶变换曲线。如图 11.8(c)、(d)所示,从 $k^3\chi(k)$ 振荡曲线的振幅来看,Co_9S_8 NS/rGO、$FeCo_8S_8$ NS/rGO 和 $FeCoS_2$ NS/rGO 的振幅几乎重合,而且与 CoS 的振幅相似度非常高,表明三个样品的原子排列差异较小,并与标样 CoS 的局域原子排列有相同的地方。但是经过傅里叶变换之后就能发现三个样品以及标样之间存在细微差异,因此,后续还将对傅里叶变换后的曲线进行拟合以获得具体的壳层半径和配位数信息。

图 11.9 即为 $FeCo_8S_8$ NS/rGO、$FeCoS_2$ NS/rGO 和 Co_9S_8 NS/rGO 的 Co 元素 K 边 EXAFS 的 $k^3\chi(k)$ 振荡曲线对应的傅里叶变换曲线及其拟合曲线,结合表 11.1 的拟合结果进行分析,在 $FeCo_8S_8$ NS/rGO 中,Co—S、Co—Co 和 Co—Fe 键的配位数分别约为 2.3、2.63 和 2.15。并且 Co—Co 和 Co—Fe 的配位数之和为 4.78,接近 $FeCo_8S_8$ 的理论配位数 6。此外,Co—S、Co—Co 和 Co—Fe 的配位数比块材 $FeCo_8S_8$ 的配位数小,表明片状 $FeCo_8S_8$ NS/rGO 在合成过程中生成局域原子排列的纳米结构。

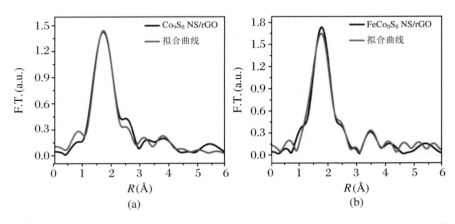

图 11.9 $FeCo_8S_8$ NS/rGO、$FeCoS_2$ NS/rGO 和 Co_9S_8 NS/rGO Co 元素 K 边 EXAFS 的 $k^3\chi(k)$ 振荡曲线对应的傅里叶变换曲线及其拟合曲线(a)~(f)

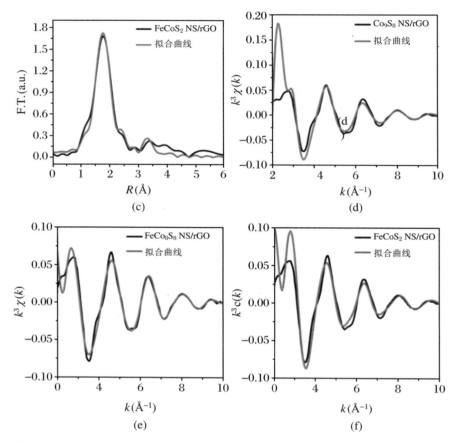

图 11.9 $FeCo_8S_8$ NS/rGO、$FeCoS_2$ NS/rGO 和 Co_9S_8 NS/rGO Co 元素 K 边 EXAFS 的 $k^3\chi(k)$ 振荡曲线对应的傅里叶变换曲线及其拟合曲线(a)~(f)(续)

表 11.1 $FeCo_8S_8$ NS/rGO、$FeCoS_2$ NS/rGO 和 Co_9S_8 NS/rGO 的理论键长和配位数以及 EXAFS 测试拟合键长和配位数

样品	壳层	键长(Å)	配位数	$\sigma^2(Å^2)$
$FeCo_8S_8$(理论上)	Co—S	2.235	4	
	Co—S	2.363	6	
	Co—Co/Fe	2.518	3	
	Co—Co/Fe	3.417	3	
Co_9S_8(理论上)	Co—S	2.127	4	
	Co—S	2.359	5	
	Co—Co	2.505	3	
	Co—Co	4.110	24	

续表

样品	壳层	键长(Å)	配位数	σ^2(Å^2)
FeCoS$_2$(理论上)	Co—S	2.348	6	
	Co—Co/Fe	2.645	2	
	Co—Co/Fe	3.360	4	
FeCo$_8$S$_8$/rGO	Co—S$_1$	2.1709±0.0001	2.32	0.00523
	Co—Co$_1$	2.4427±0.0009	2.63	0.01405
	Co—Fe$_1$	2.4305±0.0003	2.15	0.01481
	Co—Co$_2$	3.4967±0.0002	4.76	0.01282
	Co—Co$_2$	3.5546±0.0002	1.85	0.03352
	Co—Fe$_2$	3.6173±0.0003	10.05	0.07625
Co$_9$S$_8$/rGO	Co—S$_1$	2.3626±0.0009	4.77	0.00624
	Co—Co$_2$	4.1302±0.0003	19.53	0.04090
FeCoS$_2$/rGO	Co—S$_1$	2.2471±0.0006	2.94	0.00494
	Co—Co$_1$	2.8144±0.0004	1.82	0.01300
	Co—Co$_2$	3.5753±0.0004	16.77	0.03302
	Co—Fe$_2$	3.9141±0.0007	5.28	0.03827

11.3.2
材料的 OER/ORR 电化学性能

为评估材料的 OER 催化性能,采用三电极系统对 Co$_9$S$_8$ NS、Co$_9$S$_8$ NS/rGO、FeCo$_8$S$_8$ NS/rGO 和 FeCoS$_2$ NS/rGO 的 OER 性能进行测试,并在相同实验条件下对商业 RuO$_2$ 也进行同样的测试以便进行性能对比。图 11.10(a)即为以上催化剂在 O$_2$ 饱和的 0.1 mol·L^{-1} KOH 电解液中测试得到的 LSV 曲线,溶液内阻导致的电压降已经补偿校正,根据曲线可以发现,FeCo$_8$S$_8$ NS/rGO 和 RuO$_2$ 的起始电位相当,且明显小于 Co$_9$S$_8$ NS、Co$_9$S$_8$ NS/rGO 和 FeCoS$_2$ NS/rGO 的起始电位,当电势逐渐增大时,FeCo$_8$S$_8$ NS/rGO 的电流增长要比 RuO$_2$ 更快,因此在达到 10 mA·cm^{-2} 电流密度时 FeCo$_8$S$_8$ NS/rGO 的过电位为 1.56 V,而 RuO$_2$ 的过电位为 1.58 V,同时 Co$_9$S$_8$ NS、Co$_9$S$_8$ NS/rGO 和 FeCoS$_2$ NS/rGO 的过电位分别为 1.73 V、1.61 V 和 1.67 V。以上结果说明 FeCo$_8$S$_8$ NS/rGO 在 OER 催化活性上要优于同类催化剂 Co$_9$S$_8$ NS、Co$_9$S$_8$ NS/rGO 和 FeCoS$_2$ NS/rGO,甚至与

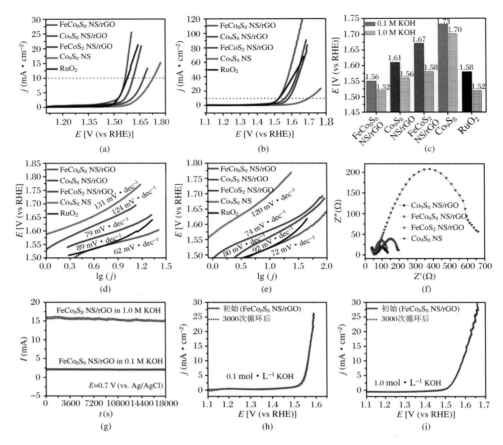

图 11.10 Co_9S_8 NS、Co_9S_8 NS/rGO、$FeCo_8S_8$ NS/rGO、$FeCoS_2$ NS/rGO 和 RuO_2 的 OER 性能测试:在 0.1 mol·L^{-1} KOH 溶液中的 LSV 曲线(a);在 1.0 mol·L^{-1} KOH 溶液中的 LSV 曲线(b);在 0.1 mol·L^{-1} KOH 和 1.0 mol·L^{-1} KOH 溶液中达到 10 mA cm^{-2} 电流密度所需要的电势(c);在 0.1 mol·L^{-1} KOH 溶液中的塔菲尔斜率(d);在 1 mol·L^{-1} KOH 溶液中的塔菲尔斜率(e);在 0.1 mol·L^{-1} KOH 溶液中的电化学阻抗谱(f);$FeCo_8S_8$ NS/rGO 在 0.1 mol·L^{-1} KOH 溶液和 1.0 mol·L^{-1} KOH 溶液中的稳定性测试(g)~(i)

商业 RuO_2 相比也略占优势,除此之外,塔菲尔斜率通常被用来分析催化过程的动力学速率,图 11.10(d)为上述几种催化剂的塔菲尔斜率,$FeCo_8S_8$ NS/rGO、Co_9S_8 NS、Co_9S_8 NS/rGO、$FeCoS_2$ NS/rGO 和 RuO_2 的塔菲尔斜率具体值分别为 62 mV·dec^{-1}、131 mV·dec^{-1}、79 mV·dec^{-1}、124 mV·dec^{-1} 和 89 mV·dec^{-1},$FeCo_8S_8$ NS/rGO 的塔菲尔斜率值最小,说明该催化剂在同样条件下的动力学反应速率最快,性能最好。为了评估材料在碱性更强条件下的性能,也在 1.0 mol·L^{-1} KOH 溶液中对以上材料进行了 OER 性能测试。如图 11.10(b)所示,$FeCo_8S_8$ NS/rGO 依然表现出最优的 OER 催化性能,在电势为 1.52 V 处达到 10 mA·cm^{-2}

电流密度，塔菲尔斜率为 72 mV·dec^{-1}，对于 Co$_9$S$_8$ NS、Co$_9$S$_8$ NS/rGO、FeCoS$_2$ NS/rGO 和 RuO$_2$，塔菲尔斜率值分别为 120 mV·dec^{-1}、80 mV·dec^{-1}、74 mV·dec^{-1} 和 92 mV·dec^{-1}，达到 10 mA·cm^{-2} 电流密度时的电势如图 12.11(c)所示，分别为 1.70 V、1.56 V、1.58 V 和 1.52 V。基于以上实验结果可以得出结论，FeCo$_8$S$_8$ NS/rGO 晶格结构中的少量 Fe，在 Co$_9$S$_8$ NS/rGO 的基础上对产物的 OER 性能有非常显著的提升。图 11.10(f)的电化学阻抗谱也从动力学角度说明了 FeCo$_8$S$_8$ NS/rGO 的性能优越性，Nyquist 曲线表明，FeCo$_8$S$_8$ NS/rGO 的电荷转移电阻与 Co$_9$S$_8$ NS/rGO 的电阻相当，但明显小于 FeCoS$_2$ NS/rGO 和 Co$_9$S$_8$ NS，说明 FeCo$_8$S$_8$ 在与 rGO 原位复合后，rGO 极大地增强了材料的电子转移性能。

稳定性也是衡量催化剂性能好坏的重要因素，采用计时电流法(i-t)和多次循环伏安法(CV)对 FeCo$_8$S$_8$ NS/rGO 的稳定性进行测试。如图 11.10(g)所示，设定反应电压为 0.7 V(Ag/AgCl)，在 1 mol·L^{-1} KOH 溶液中，反应 5 h 后 FeCo$_8$S$_8$ NS/rGO 的电流只出现微小的下降，而在 0.1 mol·L^{-1} KOH 溶液中反应 5 h，电流密度一直保持恒定。随后在电位 1.1～1.7 V 之间进行 CV 循环实验，实验过程中工作电极保持 1000 r·min^{-1} 转速以加快产生的氧气在电极上的脱离。如图 11.10(h)所示在 0.1 mol·L^{-1} KOH 溶液中进行 CV 循环 3000 次后，FeCo$_8$S$_8$ NS/rGO 的线性扫描伏安曲线与循环之前完全重合，说明性能未出现衰减。同样地，在 1.0 mol·L^{-1} KOH 中也进行了 3000 次 CV 循环，循环前后的 LSV 曲线也几乎完全重合，表明 FeCo$_8$S$_8$ NS/rGO 具有良好的稳定性，能在碱性环境中长时间稳定进行催化反应。在稳定性测试前后，对产物进行 TEM 表征以判断 FeCo$_8$S$_8$ NS/rGO 在该过程中是否出现形貌结构的改变。图 11.11(a)和(b)分别是 FeCo$_8$S$_8$ NS/rGO 在 0.1 mol·L^{-1} KOH 溶液中进行稳定性测试前和测试后的 TEM 图，对比可以发现，长时间循环反应后的 FeCo$_8$S$_8$ NS/rGO 依然保持原本的形貌特征，没有明显的分离和团聚。此外，也采用 XPS 对反应后的 FeCo$_8$S$_8$ NS/rGO 进行表征并与反应前的 XPS 谱图对比，如图 11.11(c)～(f)所示，图 11.11(d)的 Co 2p 谱图显示位于 781.3 eV 和 797.1 eV 的峰均有减弱，说明 Co 元素在反应过程中有少量原子脱离材料表面进入溶液中，图 11.11(e)的 Fe 2p 几乎没有差异，而图 11.11(f)中 S 的 2p 位于 168.9 eV 的峰明显减弱，说明在反应过程中有部分 S 元素也从材料表面溶出，进入溶液中。

综合以上结果分析，在循环过程中除了存在少量 Co 和 S 元素的溶出，FeCo$_8$S$_8$ NS/rGO 只出现轻微的活性下降，并在反应过程中保持良好的形貌和结构，表明 FeCo$_8$S$_8$ NS/rGO 具有良好的稳定性。

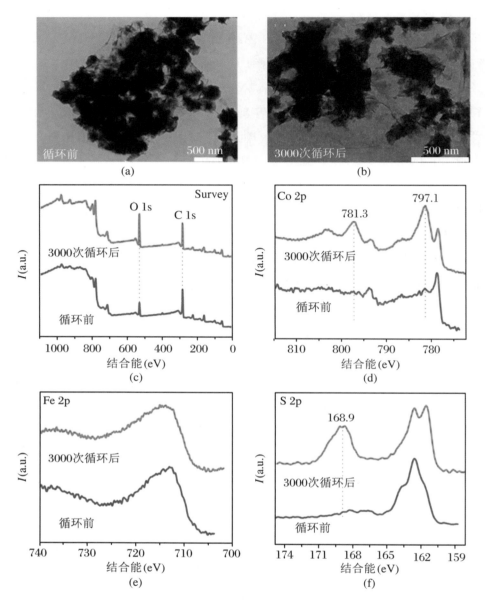

图 11.11　FeCo₈S₈ NS/rGO 在稳定性循环测试前的 TEM 图(a);FeCo₈S₈ NS/rGO 在稳定性循环测试后的 TEM 图(b);FeCo₈S₈ NS/rGO 在稳定性循环测试前后的 XPS 图(c)~(f)

随后为了评估引入 Fe 之后的 $FeCo_8S_8$ NS/rGO 是否保留 Co_9S_8 NS/rGO 的 ORR 催化优势,在 $0.1\ mol\cdot L^{-1}$ KOH 溶液中对 $FeCo_8S_8$ NS/rGO 进行 ORR 性能测试,为了进行性能对比,Co_9S_8 NSs、Co_9S_8 NS/rGO、$FeCoS_2$ NS/rGO 和商业 Pt/C 也在相同的条件下进行测试。图 11.12(a)为以上材料进行溶液内阻补偿后的 LSV 曲线,$FeCo_8S_8$ NS/rGO 与 Co_9S_8 NS/rGO 的催化性能相当,说

明引入少量 Fe 之后，$FeCo_8S_8$ NS/rGO 仍然保持了 Co_9S_8 NS/rGO 优异的 ORR 催化性能。图 11.12(b) 展示了以上材料的半波电位 $E_{1/2}$、$FeCo_8S_8$ NSs/rGO、Co_9S_8 NSs、Co_9S_8 NS/rGO、$FeCoS_2$ NS/rGO 和 Pt/C 的半波电位分别为 0.79 V、0.78 V、0.78 V、0.69 V 和 0.85 V，从半波电位判断，$FeCo_8S_8$ NSs/rGO 要优于同类型材料 Co_9S_8 NSs、Co_9S_8 NS/rGO 和 $FeCoS_2$ NS/rGO，但稍逊于商业 Pt/C。图 11.12(b) 为利用旋转环盘电极（RRDE）测试得到的盘电流和环电流，以此计算材料的电子转移数和过氧化氢产率（根据式 11.2 和式 11.3）来衡量材料的催化选择性，在 0.5 V（Ag/AgCl）电势下，$FeCo_8S_8$ NSs/rGO 的电子转移数和过氧化氢产率分别为 3.94% 和 3.0%，同样条件下，Co_9S_8 NSs 的为 3.92% 和 3.7%，Co_9S_8 NS/rGO 的为 3.94% 和 2.9%，$FeCoS_2$ NS/rGO 的为 3.93% 和 3.3%，

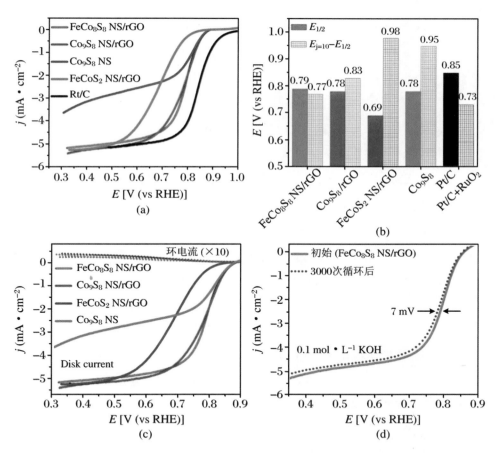

图 11.12　Co_9S_8 NSs、Co_9S_8 NS/rGO、$FeCo_8S_8$ NS/rGO、$FeCoS_2$ NS/rGO 和 Pt/C 在 0.1 mol·L^{-1} KOH 溶液中的 ORR 极化曲线(a)；ORR 半波电位和 OER 中电流密度为 10 mA cm^{-2} 时的电位差值 ΔE(b)；旋转环盘电极（RRDE）测试得到的盘电流和环电流(c)；0.1 mol·L^{-1} KOH 溶液中 $FeCo_8S_8$ NS/rGO 的 ORR 稳定性(d)

Pt/C 的为 3.97% 和 2.9%。以上结果清晰地表明，所测试的催化剂在催化 ORR 过程中均为四电子转移路径，且过氧化氢产率都处于较低的水平。值得一提的是，虽然 FeCo$_8$S$_8$ NS/rGO 在性能上依然不及商业 Pt/C，但与之前已报道的其他以 Fe-Co 为基础的催化剂相比仍具有较大的优势，如表格 11.2 所示。表中列举了部分 Co$_9$S$_8$ 与碳基材料复合的 ORR 催化剂和 Fe-Co-S 与碳材料复合的催化剂[36-40]。

表 11.2 已报道的 Co$_9$S$_8$ 类 ORR 催化材料与本工作中材料的催化性能对比

ORR 催化剂	起始电位（V）	半波电位（V）	电子转移数	电解液	参考文献
Co$_{0.5}$Fe$_{0.5}$S@N-MC	0.913	~0.808	3.8~4.0	0.1 mol·L^{-1} KOH	36
Co$_9$S$_8$/N-C hybrid	0.914	~0.76	3.63~3.88	0.1 mol·L^{-1} KOH	37
N-Co$_9$S$_8$/G	0.941	~0.76	3.7~3.9	0.1 mol·L^{-1} KOH	38
Co-C@Co$_9$S$_8$ DSNCs	0.96	—	~3.8	0.1 mol·L^{-1} KOH	39
Co$_9$S$_8$/NSPC9-45	~0.9	0.79	3.94	0.1 mol·L^{-1} KOH	40
FeCo$_8$S$_8$ NS/rGO	0.92	0.79	3.94	0.1 mol·L^{-1} KOH	本书

为了测试材料的 ORR 催化稳定性，比较了 FeCo$_8$S$_8$ NS/rGO 在 3000 圈 CV 循环前后的 LSV 曲线，如图 11.12(d) 所示。循环之后的 LSV 曲线在半波电位上较循环之前的 LSV 曲线轻微衰减 7 mV，表明 FeCo$_8$S$_8$ NS/rGO 具有良好的循环稳定性。

基于以上的电化学测试结果，在碱性条件下，FeCo$_8$S$_8$ NS/rGO 对 OER 和 ORR 均表现出优异的催化活性和稳定性。因此，通过 ORR 半波电位 $E_{1/2}$ 与 OER 中电流密度为 10 mA·cm^{-2} 时的电位 $E_{j=10}$ 之间的差值 ΔE 来评估材料的双功能性能，ΔE 值越小，说明材料的双功能性能越好。如图 11.12(b) 所示，FeCo$_8$S$_8$ NS/rGO 的 ΔE 值为 0.77 V，该值要远小于 Co$_9$S$_8$ NSs、Co$_9$S$_8$ NS/rGO 和 FeCoS$_2$ NS/rGO 的 ΔE 值，分别为 0.95 V、0.83 V 和 0.98 V，说明 FeCo$_8$S$_8$ NS/rGO 在上述材料中具有最优的双功能性能。同样地，FeCo$_8$S$_8$ NS/rGO 与之前已报道的以 Co 为基础的双功能催化剂相比也具有性能上的优势[36-38,40-45]，如表 11.3 所示。

表 11.3　FeCo$_8$S$_8$ NS/rGO 与已报道的以 Co 为基础的双功能催化剂性能比较

OER 和 ORR 的双功能催化剂	ΔE(V)	电解液	参考文献
Co$_4$N/CNW/CC	0.74	0.1 mol·L^{-1} KOH	41
Co$_3$O$_4$/NPGC	0.838	0.1 mol·L^{-1} KOH	42
Co$_3$O$_4$/NC	0.87	0.1 mol·L^{-1} KOH	43
NiCo$_2$O$_4$	0.84	0.1 mol·L^{-1} KOH	44
Co$_{0.51}$Mn$_{0.49}$O/NCNT	0.73	1 mol·L^{-1} KOH	45
Co$_9$S$_8$/N-C hybrid	0.84	0.1 mol·L^{-1} KOH	37
N-Co$_9$S$_8$/G	0.88	0.1 mol·L^{-1} KOH	38
Co$_9$S$_8$/NSPC9-45	0.75	0.1 mol·L^{-1} KOH	40
Co$_{0.5}$Fe$_{0.5}$S@N-MC	0.84	0.1 mol·L^{-1} KOH	36
FeCo$_8$S$_8$ NS/rGO	0.77	0.1 mol·L^{-1} KOH	本书

11.3.3

理论计算分析结果

为深入了解制备的 OER 和 ORR 双功能催化剂在原子水平上的合金化效果,采用密度泛函理论(DFT)进行理论计算分析。基于之前的 HRTEM 表征结果,设定 FeCo$_8$S$_8$、Co$_9$S$_8$ 和 FeCoS$_2$ 在 OER/ORR 催化过程中暴露的晶面分别为(311)、(311)和(102)。在 FeCo$_8$S$_8$ 和 Co$_9$S$_8$ 的晶格结构中,Co 和 Fe 两种原子占据相同的晶格位点,并作为反应的活性中心,在 slab 模型中,FeCo$_8$S$_8$ 和 Co$_9$S$_8$ 在端面上各有三种可能的 OER/ORR 活性位点,图 11.13(a)~(b)中用虚线圆圈标出。但是 FeCoS$_2$ 的 slab 模型中只有一种可能的反应活性位点,如图 11.13(c)所示。根据之前的工作[46-47],碱性介质中的 OER 过程,OH$^-$ 吸附并富集在催化剂表面之后还有接下来的 4 步反应,如图 11.16(a)所示。在四电子转移反应路径中,*OH、*O、*OOH 和 *OO 等不同的吸附物种都吸附在 Co/Fe 所在的位点,如图 11.14(d)~(i)所示。整个体系中反应的 Gibbs 自由能是在平衡电势($U=1.23$ eV)条件下通过 DFT 计算得到[28,48]。如图 11.15 所示,FeCo$_8$S$_8$ 和 Co$_9$S$_8$ 表面发生的 OER 反应在不同的起始吸附位点上表现出不同的自由能分布。对于 FeCo$_8$S$_8$ 的 Co/Fe—S 位点,如图 11.14(a)~(b)所示,步骤 1 从 *OH 到 *O 和步骤 2 从 *O 到 *OOH 在位点 1(最初的吸附位置)都表现出上升的自

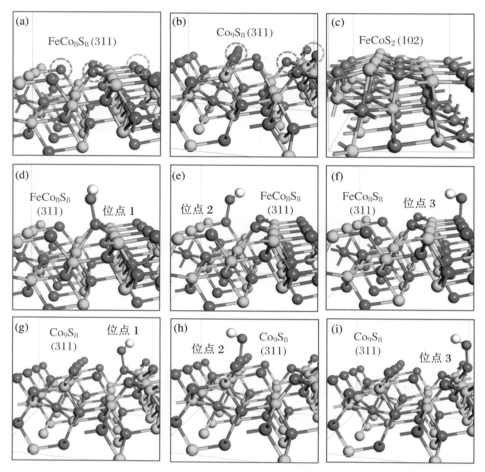

图 11.13　制备的三种材料的能量最小化 slab 模型和 OER 起始时催化剂吸附 *OH 的结构：$FeCo_8S_8$ 的(311)表面三种可能的 OER/ORR 活性位点(a)；Co_9S_8 的(311)表面三种可能的 OER/ORR 活性位点(b)；$FeCoS_2$ 的(102)表面一种 OER/ORR 活性位点(c)；OER 反应起始时 *OH 在 $FeCo_8S_8$ (311)表面不同位点的吸附形态(d)～(f)；OER 反应起始时 *OH 在 Co_9S_8 (311)表面不同位点的吸附形态(g)～(i)

由能变，并且 *OOH 的形成(步骤 2)需要更高的 ΔG，说明该步骤是 $FeCo_8S_8$ 催化产氧过程中的决速步骤。同样地，*OOH 的形成也是位点 2 和位点 3 作为 *OH 起始吸附位点的限速步，并且需要比从位点 1 开始的反应更多的自由能。因此，$FeCo_8S_8$ 表面的 OER 反应主要发生在活性位点 1。对 Co_9S_8 也采取了同样的分析过程，如图 11.14(c)～(d)所示。结果显示在 Co_9S_8 (311)表面，活性位点 2 为最佳的反应起始位点，对降低 OER 活化能最有利。上述 3 种 Fe—Co—S 纳米化合物的产氧性能在图 11.15(b)中进行对比，结果表明催化剂的 Fe/Co 比

例和材料表面原子排列的不同对 OER 的自由能分布有显著影响。$FeCo_8S_8$
(311)表面的位点 1 作为活性中心需要最少的能量促使反应发生，具体为：步骤
2 中从 *O 到 *OOH 需要 0.49 eV 使反应发生。吸附位点上一旦生成 *OOH，
反应就会放热，并自发地将 *OO 中间体转化生成 O_2。因此，理论计算结果表
明，$FeCo_8S_8$ 独特的结构提高了其 OER 活性，与实验结果吻合。

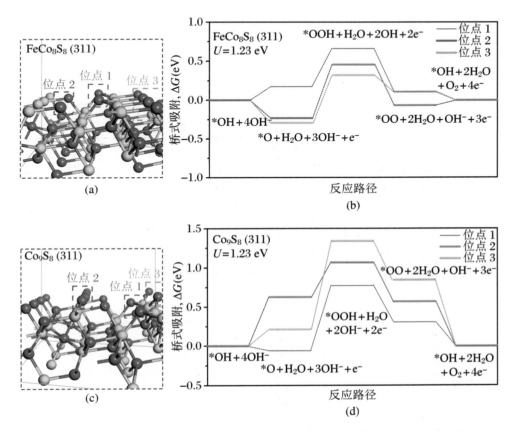

图 11.14　OER 反应在 $FeCo_8S_8$ 和 Co_9S_8 表面的理论计算信息：$FeCo_8S_8$(311)表面吸附物进
行结合的活性位点(a)；$FeCo_8S_8$(311)表面不同位点作为 OER 活性中心计算所
得的自由能数据(b)；Co_9S_8(311)表面吸附物进行结合的活性位点(c)；Co_9S_8
(311)表面不同位点作为 OER 活性中心计算所得的自由能数据(d)

对于 ORR 过程，根据电化学实验结果，O_2 在催化剂表面通过四电子过程直
接被还原为 H_2O。结合之前已报道工作中提出的 ORR 反应机制[27]，碱性介质
中 Fe—Co—S 化合物催化氧还原的基本步骤如图 11.15(c)所示[O_2 解离的机制
详见实验部分的式(11.8)～式(11.13)]。事实上，反应初始 O_2 的吸附构型对中
间体结构和整个 ORR 过程的机制都有影响[49]。在直接的四电子转化路径中，
O_2 倾向于以桥式构型与催化剂表面结合，如图 11.16 所示。O_2 的两个 O 原子各

结合催化剂表面的一个活性原子,考虑到Fe/Co活性原子之间的距离,$FeCo_8S_8$的位点3和Co_9S_8的位点2是O_2进行桥式吸附的最佳位点,如图11.16(a)~(b)所示。O_2吸附是ORR过程中的关键反应,根据图11.15(d)所示的反应路径中O_2的吸附能(ORR过程的步骤1),在$FeCo_8S_8$位点3、Co_9S_8位点2和$FeCoS_2$的位点中,$FeCo_8S_8$位点3对吸附O_2最有利,但是$FeCoS_2$的位点却不利于O_2吸附,这可能是导致其ORR效率相对较低的主要因素。在接下来的质子/电子转移步骤中,OH^-脱附的两个步骤(步骤4和步骤6)是吸热的,但是$FeCo_8S_8$使反应发生所需能量要少于Co_9S_8。因此,理论计算与实验结果非常吻合,这说明通过优化催化剂表面的Fe/Co组成可以有效提升其OER/ORR催化能力。

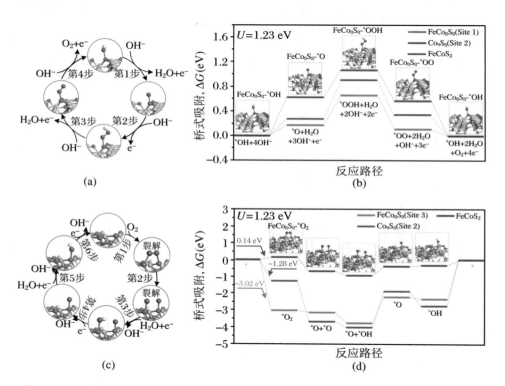

图11.15 碱性条件下OER在$FeCo_8S_8$、Co_9S_8和$FeCoS_2$表面的反应路径(a);材料活性中心的OER各步反应自由能分布,插入图为OER反应各步材料材料活性中心与吸附物结合的原子结构示意图(b);碱性条件下的ORR四电子反应路径(c);ORR各步反应的自由能分布,插入图为ORR反应各步材料活性中心与吸附物结合的原子结构示意图(d)

除了调控表面Fe/Co原子组成导致的催化剂本征性能的提升,rGO基底对催化剂活性和稳定性的提升也起着至关重要的作用。为了评估rGO对催化剂的贡献,采用电化学阻抗谱(EIS)对制备的催化剂在OER和ORR过程中的电

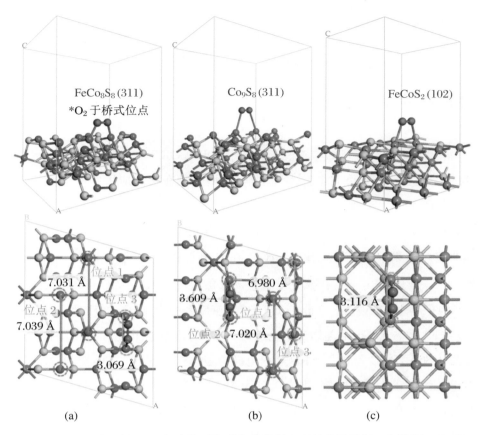

图 11.16　ORR 过程中 *O_2 以桥式构型吸附在催化剂表面的能量最低几何构型：*O_2 吸附在 $FeCo_8S_8$(311)面位点 3 上的三维视图和俯视图(a)；*O_2 吸附在 Co_9S_8 (311)面位点 2 上的三维视图和俯视图(b)；*O_2 吸附在 $FeCoS_2$(102)面上的三维视图和俯视图(c)

极动力学进行了研究。如图 11.10(f)所示，半圆与实阻抗轴的第一个交点为电极/电解质界面上的溶液内阻(R_s)，半圆直径为催化剂的电荷转移电阻(R_{ct})[50]，根据图中 EIS 谱，溶液内阻接近 60 Ω，$FeCo_8S_8$ NS/rGO 的电荷转移电阻约为 65 Ω，Co_9S_8 NS/rGO、$FeCoS_2$ NS/rGO 和 Co_9S_8 NS 的电荷转移电阻分别约为 70 Ω、140 Ω 和 600 Ω。由此可见，原位负载 rGO 极大程度地降低了催化剂的电荷转移电阻，增强了电子传导效率。

综合来说，以上结果表明 $FeCo_8S_8$ NS/rGO 优越的双功能催化活性和稳定性可以归因于以下两点：① 适量 Fe 原子的引入优化了 Co_9S_8 表面的活性原子位点，提升了材料的本征催化活性；② $FeCo_8S_8$ NS 与 rGO 的原位复合结构增强了材料整体的导电能力和长时间循环稳定性。

11.3.4 材料在锌空气电池中的应用

在 FeCo$_8$S$_8$ NS/rGO 具有良好的催化 OER/ORR 双功能活性和稳定性的基础上,将其应用到可充电锌空气电池中,研究其在实际应用中的活性和稳定性。如图 11.17(a)和(b)分别为锌空气电池的示意图和实物图,图 11.17(c)为 FeCo$_8$S$_8$ NS/rGO、Pt/C 和 Pt/C + RuO$_2$/C 分别作为电池空气电极的充放电极化曲线。结果显示 FeCo$_8$S$_8$ NS/rGO 在充电过程中相对于 Pt/C 和 Pt/C + RuO$_2$/C 更有优势,尽管在放电过程中需要的过电势要略大于 Pt/C 和 Pt/C + RuO$_2$/C。具体来说,在 10 mA·cm^{-2} 电流密度下进行充放电循环时,以 FeCo$_8$S$_8$ NS/rGO 为空气电极的起始充电电压为 2.00 V,放电电压为 1.14 V,因此,电压差为 0.86 V,Pt/C 和 Pt/C + RuO$_2$/C 电极的充电电压分别为 2.15 V 和 1.95 V,放电电压为 1.20 V,电压差分别为 0.95 V 和 0.75 V。如图 11.17(d)所示,在充放电各循环 200 圈之后,FeCo$_8$S$_8$ NS/rGO 电极的充放电电压差上升 0.11 V 至 0.97 V,而 Pt/C 和 Pt/C + RuO$_2$/C 的电压差分别上升 0.63 V 和 0.15 V,由此可见,以其作为锌空气电池的电极在稳定性上要远优于 Pt/C,而与 Pt/C + RuO$_2$/C 的性能相当。

综上,设计并制备了一种新型 FeCo$_8$S$_8$ NS/rGO 纳米复合材料,该材料具有优异的双功能 OER/ORR 催化活性,可作为电极材料应用于可充电锌空气电池。对于该三元硫化物,其组成中适当的铁含量是其双功能性能的关键。制备的 FeCo$_8$S$_8$ NS/rGO 纳米复合材料在 0.1 mol·L^{-1} KOH 溶液中展示出极低的双功能电位差($\Delta E_j = E^{-2}_{10\,\text{mA·cm}} - E_{1/2} = 0.77\text{ V}$),在制备得到的 FeCo 为基础的催化剂中表现出最佳的双功能电催化活性,在实际的锌空气电池试验中具有较低的充放电电压差和良好的耐久性。DFT 计算结果进一步表明,适当引入 Fe 含量可以降低 OH$^-$ 氧化为 O$_2$ 的活化能,维持 O$_2$ 还原为 OH$^-$ 的活化能。因此,FeCo$_8$S$_8$ NS/rGO 纳米复合材料作为可充电锌空气电池的空气电极具有很大的潜力,也可以为多种氧化还原反应提供界面催化作用,增强有机污染物的降解效率。

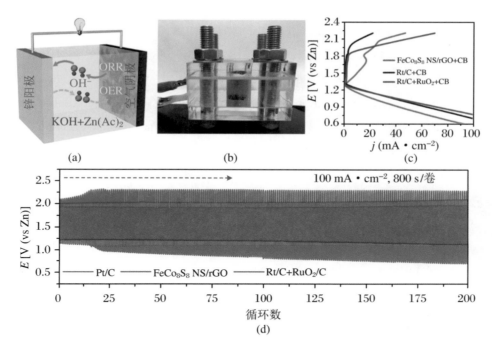

图11.17 锌空气电池装置示意图(a)和实物图(b);锌空气电池充放电极化曲线图(c);锌空气电池在10 mA·cm^{-2}电流密度下充放电循环稳定性图(d)

参考文献

[1] Zhang X, Li L, Fan E, et al. Toward sustainable and systematic recycling of spent rechargeable batteries [J]. Chemical Society Reviews, 2018, 47 (19): 7239-7302.

[2] Berg E J, Villevieille C, Streich D, et al. Rechargeable batteries: grasping for the limits of chemistry [J]. Journal of the Electrochemical Society, 2015, 162 (14): A2468-A2475.

[3] Fu J, Cano Z P, Park M G, et al. Electrically rechargeable zinc-air batteries: progress, challenges, and perspectives [J]. Advanced Materials, 2017, 29 (7): 1604685.

[4] Li G, Wang X, Fu J, et al. Pomegranate-inspired design of highly active and durable bifunctional electrocatalysts for rechargeable metal-air batteries [J]. Angewandte Chemie International Edition, 2016, 55 (16): 4977-4982.

[5] Liu Q, Chang Z, Li Z, et al. Flexible metal-air batteries: progress, challenges, and perspectives [J]. Small Methods, 2018, 2 (2): 1700231.

[6] Amiinu I S, Liu X B, Pu Z H, et al. From 3D ZIF nanocrystals to Co-N-x/C nanorod array electrocatalysts for ORR, OER, and Zn-air batteries [J]. Advanced Functional Materials, 2018, 28 (5).

[7] Gorlin Y, Lassalle-Kaiser B, Benck J D, et al. In situ X-ray absorption spectroscopy investigation of a bifunctional manganese oxide catalyst with high activity for electrochemical water oxidation and oxygen reduction [J]. Journal of the American Chemical Society, 2013, 135 (23): 8525-8534.

[8] Cheng F, Su Y, Liang J, et al. MnO_2-based nanostructures as catalysts for electrochemical oxygen reduction in alkaline media [J]. Chemistry of Materials, 2009, 22 (3): 898-905.

[9] Peng S, Li L, Tan H, et al. MS_2 (M = Co and Ni) hollow spheres with tunable interiors for high-performance supercapacitors and photovoltaics [J]. Advanced Functional Materials, 2014, 24 (15): 2155-2162.

[10] Yang H G, Sun C H, Qiao S Z, et al. Anatase TiO_2 single crystals with a large percentage of reactive facets [J]. Nature, 2008, 453 (7195): 638-641.

[11] An L, Huang B L, Zhang Y, et al. Interfacial defect engineering for improved portable Zinc-air batteries with a broad working temperature [J]. Angewandte Chemie-International Edition, 2019, 58 (28): 9459-9463.

[12] Liu X, Liu W, Ko M, et al. Metal (Ni, Co)-Metal oxides/graphene nanocomposites as multifunctional electrocatalysts [J]. Advanced Functional Materials, 2015, 25 (36): 5799-5808.

[13] Mei J, Liao T, Kou L, et al. Two-dimensional metal oxide nanomaterials for next-generation rechargeable batteries [J]. Advanced Materials, 2017, 29 (48): 1700176.

[14] Gao Q, Huang C Q, Ju Y M, et al. Phase-selective syntheses of cobalt telluride nanofleeces for efficient oxygen evolution catalysts [J]. Angewandte Chemie International Edition, 2017, 56 (27): 7769-7773.

[15] Zhang Y, Zhou Q, Zhu J, et al. Nanostructured metal chalcogenides for energy storage and electrocatalysis [J]. Advanced Functional Materials, 2017, 27 (35): 1702317.

[16] Yan B, Krishnamurthy D, Hendon C H, et al. Surface restructuring of nickel sulfide generates optimally coordinated active sites for oxygen reduction catalysis [J]. Joule, 2017, 1 (3): 600-612.

[17] Li L H, Song L, Guo H, et al. N-Doped porous carbon nanosheets decora-

ted with graphitized carbon layer encapsulated Co_9S_8 nanoparticles: an efficient bifunctional electrocatalyst for the OER and ORR [J]. Nanoscale, 2019, 11 (3): 901-907.

[18] Li S S, Hao X G, Abudula A, et al. Nanostructured Co-based bifunctional electrocatalysts for energy conversion and storage: current status and perspectives [J]. Journal of Materials Chemistry A, 2019, 7 (32): 18674-18707.

[19] Stevens M B, Trang C D, Enman L J, et al. Reactive Fe-sites in Ni/Fe (oxy) hydroxide are responsible for exceptional oxygen electrocatalysis activity [J]. Journal of the American Chemical Society, 2017, 139 (33): 11361-11364.

[20] Segall M, Lindan P J, Probert M A, et al. First-principles simulation: ideas, illustrations and the CASTEP code [J]. Journal of Physics: Condensed Matter, 2002, 14 (11): 2717.

[21] Perdew J P, Burke K, Ernzerhof M. Generalized gradient approximation made simple [J]. Physical Review Letters, 1996, 77 (18): 3865.

[22] Vanderbilt D. Soft self-consistent pseudopotentials in a generalized eigenvalue formalism [J]. Physical Review B, 1990, 41 (11): 7892.

[23] Pfrommer B G, Côté M, Louie S G, et al. Relaxation of crystals with the quasi-Newton method [J]. Journal of Computational Physics, 1997, 131 (1): 233-240.

[24] Monkhorst H J, Pack J D. Special points for Brillouin-zone integrations [J]. Physical Review B, 1976, 13 (12): 5188.

[25] Ling T, Yan D Y, Jiao Y, et al. Engineering surface atomic structure of single-crystal cobalt (Ⅱ) oxide nanorods for superior electrocatalysis [J]. Nature Communications, 2016, 7: 12876.

[26] Suntivich J, May K J, Gasteiger H A, et al. A perovskite oxide optimized for oxygen evolution catalysis from molecular orbital principles [J]. Science, 2011, 334 (6061): 1383-1385.

[27] Lim D H, Wilcox J. Mechanisms of the oxygen reduction reaction on defective graphene-supported Pt nanoparticles from first-principles [J]. The Journal of Physical Chemistry C, 2012, 116 (5): 3653-3660.

[28] Man I C, Su H Y, Calle-Vallejo F, et al. Universality in oxygen evolution electrocatalysis on oxide surfaces [J]. ChemCatChem, 2011, 3 (7):

1159-1165.

[29] Roudgar A, Eikerling M, van Santen R. Ab initio study of oxygen reduction mechanism at Pt_4 cluster [J]. Physical Chemistry Chemical Physics, 2010, 12 (3): 614-620.

[30] Chen J J, Wang W K, Li W W, et al. Roles of crystal surface in Pt-loaded titania for photocatalytic conversion of organic pollutants: a first-principle theoretical calculation [J]. ACS Applied Materials & Interfaces, 2015, 7 (23): 12671-12678.

[31] Singh V K, Shukla A, Patra M K, et al. Microwave absorbing properties of a thermally reduced graphene oxide/nitrile butadiene rubber composite [J]. Carbon, 2012, 50 (6): 2202-2208.

[32] Wang D Y, Gong M, Chou H L, et al. Highly active and stable hybrid catalyst of cobalt-doped FeS_2 nanosheets-carbon nanotubes for hydrogen evolution reaction [J]. Journal of the American Chemical Society, 2015, 137 (4): 1587-1592.

[33] Deng J, Lv X, Gao J, et al. Facile synthesis of carbon-coated hematite nanostructures for solar water splitting [J]. Energy & Environmental Science, 2013, 6 (6): 1965-1970.

[34] Huo J, Wu J, Zheng M, et al. High performance sponge-like cobalt sulfide/reduced graphene oxide hybrid counter electrode for dye-sensitized solar cells [J]. Journal of Power Sources, 2015, 293: 570-576.

[35] Zhang P, Wang R, He M, et al. 3D hierarchical Co/CoO-graphene-carbonized melamine foam as a superior cathode toward long-life lithium oxygen batteries [J]. Advanced Functional Materials, 2016, 26 (9): 1354-1364.

[36] Shen M, Ruan C, Chen Y, et al. Covalent entrapment of cobalt-iron sulfides in N-doped mesoporous carbon: Extraordinary bifunctional electrocatalysts for oxygen reduction and evolution reactions [J]. ACS Applied Materials & Interfaces, 2015, 7 (2): 1207-1218.

[37] Cao X, Zheng X, Tian J, et al. Cobalt sulfide embedded in porous nitrogen-doped carbon as a bifunctional electrocatalyst for oxygen reduction and evolution reactions [J]. Electrochimica Acta, 2016, 191: 776-783.

[38] Dou S, Tao L, Huo J, et al. Etched and doped Co_9S_8/graphene hybrid for oxygen electrocatalysis [J]. Energy & Environmental Science, 2016, 9 (4): 1320-1326.

[39] Hu H, Han L, Yu M, et al. Metal-organic-framework-engaged formation of Co nanoparticle-embedded carbon@ Co_9S_8 double-shelled nanocages for efficient oxygen reduction [J]. Energy & Environmental Science, 2016, 9 (1): 107-111.

[40] Zhong H X, Li K, Zhang Q, et al. In situ anchoring of Co_9S_8 nanoparticles on N and S co-doped porous carbon tube as bifunctional oxygen electrocatalysts [J]. NPG Asia Materials, 2016, 8 (9): e308.

[41] Meng F, Zhong H, Bao D, et al. In situ coupling of strung Co_4N and intertwined N-C fibers toward free-standing bifunctional cathode for robust, efficient, and flexible Zn-air batteries [J]. Journal of the American Chemical Society, 2016, 138 (32): 10226-10231.

[42] Liang Y, Li Y, Wang H, et al. Co_3O_4 nanocrystals on graphene as a synergistic catalyst for oxygen reduction reaction [J]. Nature Materials, 2011, 10 (10): 780.

[43] Masa J, Xia W, Sinev I, et al. Mn_xO_y/NC and Co_xO_y/NC nanoparticles embedded in a nitrogen-doped carbon matrix for high-performance bifunctional oxygen electrodes [J]. Angewandte Chemie International Edition, 2014, 53 (32): 8508-8512.

[44] Prabu M, Ketpang K, Shanmugam S. Hierarchical nanostructured $NiCo_2O_4$ as an efficient bifunctional non-precious metal catalyst for rechargeable zinc-air batteries [J]. Nanoscale, 2014, 6 (6): 3173-3181.

[45] Liu X, Park M, Kim M G, et al. High-performance non-spinel cobalt-manganese mixed oxide-based bifunctional electrocatalysts for rechargeable zinc-air batteries [J]. Nano Energy, 2016, 20: 315-325.

[46] Hong W T, Risch M, Stoerzinger K A, et al. Toward the rational design of non-precious transition metal oxides for oxygen electrocatalysis [J]. Energy & Environmental Science, 2015, 8 (5): 1404-1427.

[47] Chen D, Chen C, Baiyee Z M, et al. Nonstoichiometric oxides as low-cost and highly-efficient oxygen reduction/evolution catalysts for low-temperature electrochemical devices [J]. Chemical Reviews, 2015, 115 (18): 9869-9921.

[48] Nørskov J K, Rossmeisl J, Logadottir A, et al. Origin of the overpotential for oxygen reduction at a fuel-cell cathode [J]. The Journal of Physical Chemistry B, 2004, 108 (46): 17886-17892.

[49] Anderson A B, Albu T V. Catalytic effect of platinum on oxygen reduction an ab initio model including electrode potential dependence [J]. Journal of the Electrochemical Society, 2000, 147 (11): 4229-4238.

[50] Li J, Yan M, Zhou X, et al. Mechanistic insights on ternary $Ni_{2-x}Co_xP$ for hydrogen evolution and their hybrids with graphene as highly efficient and robust catalysts for overall water splitting [J]. Advanced Functional Materials, 2016, 26 (37): 6785-6796.

第 12 章

光催化硝化与反硝化过程中的界面电子转移

12.1
光催化硝化现象与光电催化反硝化

人类活动持续地改变了全球氮循环,这反映在化石燃料消耗量的增加,农业和工业上氮需求的不断增长,而使用效率普遍较低。大量人工合成的含氮物质流失到空气、水和土壤中,造成越来越严重的环境和人类健康问题[1-2]。通常,生物固氮是氮元素从气态氮进入生物圈的主要途径,固氮反应是通过固氮酶的作用,固氮酶是一种金属酶,其中的金属可以是钼、钒或铁[3]。对氮循环的认识已经从如何提高氮的固定来促进粮食的生产和能源的获得,转移到如何实现活性氮水平的提升上来。Haber-Bosch 和 Ostwald 过程将 N_2 转化为化肥,是最知名的工业硝酸盐生产方法[4]。

Yuan 等人发现,在常压下空气中的 N_2 和 O_2 能够在 TiO_2 催化剂表面光催化形成硝酸盐[5],这一氧化过程是一种硝化途径。二氧化钛(TiO_2)是广泛使用且环境友好的光催化剂和添加剂[6-7]。而光催化氧化过程一般发生在 TiO_2 具有高反应活性的(001)晶面。在反应过程中,气体分子与 TiO_2 表面存在电子转移。可以采用第一性原理密度泛函理论(DFT)计算来阐明光催化形成硝酸盐的机制,分析中间产物在 TiO_2 晶面上的吸附构型,探索硝化过程的热力学性质和速率控制步骤。

由于 TiO_2 目前在涂料、油漆、塑料和橡胶等领域广泛使用,而过量的硝酸盐排放到水系统中,会对环境和人类健康造成影响[8],对淡水系统和河口造成蓝藻暴发。饮用水中高浓度的硝酸盐以及其还原中间产物——亚硝酸盐,会导致高铁血红蛋白症和癌症[9]。而商业化的 TiO_2 纳米晶体 P25 可以作为硝酸盐还原催化剂,同时可以使用微生物代谢产生的生物电子作为空穴牺牲剂,利用微生物的方法具有成本效益和可持续性[10-11]。还原反应一般发生在 TiO_2 纳米晶体的(101)晶面。在反硝化过程中,TiO_2 晶面与硝酸根离子之间存在界面电子转移。可以通过理论计算对光催化反硝化过程中水溶液中反应物、中间产物和最终产物进行解析,并通过对该反硝化过程的热力学性质和动力学性质分析来阐明硝酸盐的高选择性还原机制。

12.2 典型光催化剂界面硝化与反硝化机制解析方法

12.2.1 模型的建立

硝化过程由于是氧化反应,而且商业化的 TiO_2 P25(Degussa)中锐钛矿的比例大于金红石,(001)晶面是主要的活性位点,对 TiO_2 纳米颗粒的特殊性质有很大的贡献[12-13],对此构建了 TiO_2(001)晶面来进行计算。周期性(001)晶面的 slab 之间的间距是 10 Å,每个 slab 包含 8 个 Ti 原子和 16 个 O 原子。晶面(001)上在两个 Ti 原子之间存在 2-配位的桥氧键(Ti—O—Ti 角度是 156°),另外还有 5-配位的 Ti 原子(Ti-5c)[14]。N_2、O_2、H_2O、·OH 以及产物分子都只吸附在 slab 的一侧。反硝化过程则是还原反应,且(101)晶面是锐钛矿中较为稳定的晶面,周期性(101)晶面的 slab 之间的间距也定为 10 Å,每个 slab 包含 18 个 Ti 原子和 36 个 O 原子,Ti 原子有 5-、6-配位(Ti-5c、Ti-6c)两种情况,O 原子则是 2-、3-配位(O-2c、O-3c)[15]。硝酸根、水合质子、水分子和还原产物也只吸附在 slab 的一侧。

12.2.2 第一性原理计算

采用广义梯度近似(GGA)结合平面波基组和超软赝势[16]在 CASTEP 程序[17]中进行第一性原理 DFT 计算。交换相关能量和势能通过 Perdew-Wang 91(PW91)泛函[18]自洽描述。能量、应力和位移的收敛标准分别是 1×10^{-5} eV·atom^{-1}、0.05 GPa 和 0.001 Å。布里渊区积分通过 Monkhorst-Pack 算法产生的 k-points 进行,而 k-points 的取值则取决于晶胞的尺寸和形状[19]。采用线性同步度越（LST）/二次同步度越（QST）方法[20]进行过渡态搜索和能垒的计算。

光催化硝化与反硝化的热力学性质通过 DMol^3 程序[21]进行计算。采用 GGA 结合 Perdew、Burke 和 Ernzerhof（PBE）泛函[22]进行计算，双精度数值基组结合 d-轨道极化函数（DND）用于处理价电子，全电子核处理方法用于描述核电子。

12.3 典型光催化剂界面硝化与反硝化过程的电子转移

12.3.1 光催化硝化过程的电子转移

实验中发现 NO 可能是 TiO_2 晶面上光催化硝化过程中的初始产物。反应中 Cr(Ⅵ)的存在表明光生电子（e^-）在这个过程中起到了重要的作用，但光生空穴（h^+）的作用也不能被忽略。因此，提出了 TiO_2 晶面上光催化形成 NO 的反应。

$$N_2 + O_2 \longrightarrow 2NO \tag{12.1}$$

$$N_2 + 2H_2O + 4h^+ \longrightarrow 4H^+ + 2NO \tag{12.2}$$

对反应方程(12.1)、(12.2)中 NO 形成机制进行详细的分析，在 TiO_2(001)晶面（图 12.1）的导带（CB）上初始的反应可能是 O_2 的吸附以及光生电子的转移，而在价带（VB）上起始的步骤可能是 H_2O 与电子跃迁后留下的光生空穴发生氧化反应。可以通过热力学性质与反应动力学计算，对 CB 与 VB 上不同的 NO 光催化形成机制进行深入解析。

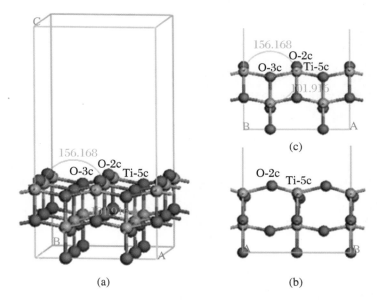

图 12.1 锐钛矿 TiO_2(001)的结构：视图(a)；前视图(b)；侧视图(c)

1. 导带(CB)上的光催化硝化机制

对 CB 机制中的关键中间产物的几何结构进行优化，如图 12.2 中(a)～(e)所示。基元反应步骤的热力学性质和动力学性质如表 12.1 所示。

CB 机制从 O_2 的吸附开始，O_2 吸附于两个 Ti-5c 位点形成[(Ti-5c)—O—O—(Ti-5c)]的结构，通过光生电子转移获得超氧负离子($O_2 \cdot ^-$)（反应 C1，图 12.3）。这一步骤的 Gibbs 自由能为负值（$\Delta_r G_m^{\ominus} = -2.605 \text{ kcal} \cdot \text{mol}^{-1}$），说明 O_2 的吸附并形成 $O_2 \cdot ^-$ 是热力学可行的。接着，通过 $O_2 \cdot ^-$ 与 N_2 相互作用形成如图 12.2(c)的结构，其中两个 N 原子之间通过双键链接（N=N）。之后，吸附物中的结构经过键类型的调整来降低整个体系的能量，得到图 12.2(d)，在 N 和 O 之间形成双键（N=O）。在结构调整过程中，O—O 和 N—N 键长不断地发生改变，最后形成 NO 气体分子，从锐钛矿 TiO_2(001)晶面解析，完成整个光催化反应。

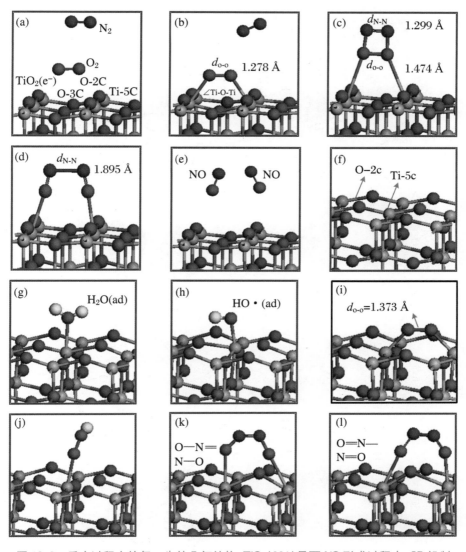

图 12.2 反应过程中的每一步的几何结构,TiO₂(001)晶面 NO 形成过程中,CB 机制 (a)～(e)和 VB 机制(f)～(l)的反应物、中间产物、最终产物的结构

表 12.1 NO 在锐钛矿 TiO₂(001)晶面的 CB 机制的热力学性质(298.15 K,1 atm)和能垒

步骤	$\Delta_r G_m^\ominus$ (kcal·mol^{-1})	$\Delta_r H_m^\ominus$ (kcal·mol^{-1})	$\Delta_r S_m^\ominus$ [cal·(mol·K^{-1})$^{-1}$]	E_a(eV)
(a) + O₂→(b)	−2.605	2.734	17.906	0.024
(b) + N₂→(c)	2.101	0.707	−4.676	3.132
(c)→(d)	−1.072	−0.226	2.838	0.561
(d)→(e) + 2NO	−1.108	0.688	6.024	0.087
Total	−2.684	3.903	22.092	—

表中(a)～(e)对应图 12.2。

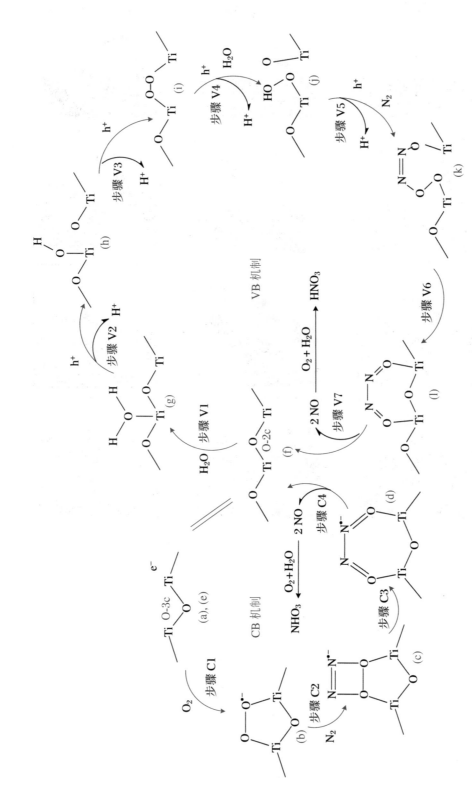

图 12.3 锐钛矿 TiO_2 (001) 晶面 NO 在导带 (CB) 和价带 (VB) 上的形成机制，(a)，(e) 和 (f) 分别是锐钛矿 TiO_2 (001) 晶面的前视图和侧视图

在光催化过程中，整个过程的 $\Delta_r G_m^{\ominus}$ 值（表12.1）表明在 1 atm, 298.15 K 的条件下，NO 在锐钛矿 TiO_2(001)晶面的 CB 机制是热力学可行的。

此外，还考虑了不同的温度对 CB 机制的形成 NO 的影响。随着温度的增加，CB 机制的 $\Delta_r G_m$ 变得更负，如图12.4所示。虽然(b)+N_2→(c)步骤的 $\Delta_r G_m$ 随着温度的增加，略有增大，但整体的 $\Delta_r G_m$ 是随着温度的升高而减小的，表明温度升高有利于光催化硝化过程进行。

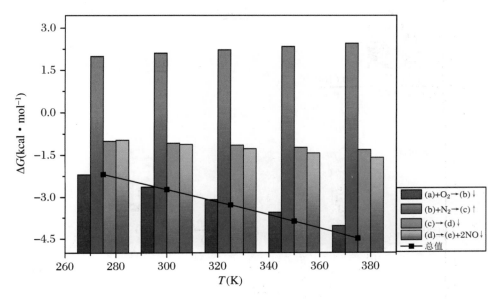

图12.4 CB 机制中反应 Gibbs 自由能随温度的变化，温度范围是 275～375 K，增幅是 25 K

整个反应的能量变化以及过渡态的结构，如图12.5(a)所示。可以发现速率控制步骤[(b)+N_2→(c)]的能垒(E_a)是 3.132 eV。这表明 O_2 吸附并形成的 $O_2 \cdot^-$ 作用于游离的 N_2 分子需要克服一个相对较高的 E_a，因为 N_2 分子中 N 原子之间是三键结构，具有较大的键能。但相对于后一节进行讨论的 VB 机制，可以发现 CB 机制具有较低的能垒。因此，光催化 NO 形成机制的分析表明，在实际体系中光催化硝化过程的进行可能主要是按照 CB 机制进行的，且具有较高的反应速率，但同时会存在 VB 机制。

2. 价带(VB)上的光催化硝化机制

在上文中提出了在 TiO_2(001)晶面上形成 NO 的 VB 机制，如图12.3所示。VB 机制存在 7 个步骤，所有中间产物的最低能量结构如图12.2(f)～(l)所示。表12.2给出了每一步基元反应的热力学性质和动力学性质，整体反应的能量变化如图 12.5(b)所示。

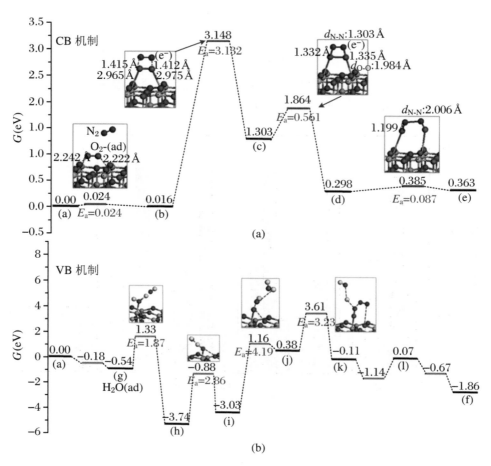

图 12.5　TiO$_2$(001)晶面上光催化形成 NO 的能量变化：CB 机制(a)；VB 机制(b)

表 12.2　NO 在锐钛矿 TiO$_2$(001)晶面的 VB 机制的热力学性质(298.15 K, 1 atm)和能垒

步骤	$\Delta G'$(eV)	$-|e|U$(eV)	ΔG(eV)	E_a(eV)
(f) + H$_2$O → (g)	−0.17	0	−0.17	−0.18
(g) + h$^+$ → (h) + H$^+$	−3.24	−2.56	−5.80	1.87
(h) + h$^+$ → (i) + H$^+$	0.92	−2.56	−1.64	2.86
(i) + H$_2$O + h$^+$ → (j) + H$^+$	3.77	−2.56	1.21	4.19
(j) + N$_2$ + h$^+$ → (k) + H$^+$	−0.80	−2.56	−3.36	3.23
(k) → (l)	0.10	0	0.10	−1.14
(l) → 6(f) + 2NO	−2.07	0	−2.07	−0.67

表中(f)~(l)对应图 12.2。

VB 机制的初始步骤是 H$_2$O 分子的吸附（反应 V1，图 12.3），其中 H$_2$O 分子是以 O 原子配位于 Ti-5c 位点的方式进行吸附。由于 Ti 的电子排布是[Ar]

$3d^24s^2$，在 3d 轨道上存在 4 个空轨道，能够接受 O 原子的成对电子进行配位。计算表明，这一步的 Gibbs 自由能变为负值（$\Delta G_1 = -0.17$ eV），所以 H_2O 以这种方式吸附在 TiO_2（001）晶面上是热力学可行的。接着，通过吸附的 H_2O 分子中 O—H 键的断裂[反应 V2，(g) + h^+ →(h) + H^+]形成吸附的·OH，同时导致晶面上 Ti—O 键的断裂(图 12.2(h))。

在光催化 NO 形成过程的 VB 机制中，质子和光生空穴参与的基元反应，ΔG 的计算中应该扣除光生电子-空穴对产生所需的能量[23]。比如，对于 VB 上的反应(g) + h^+ →(h) + H^+，可以写成(g) + h^+ + e^- →(h) + H^+ + e^-。因此，ΔG 可以通过下列方程进行计算：

$$\Delta G = G[(h)] + G(H^+ + e^-) - G[(g)] - G(h^+ + e^-) \quad (12.3)$$

标准氢电极(SHE)作为参比电极 $H^+ + e^- \to 1/2H_2$（pH = 0，$p = 100$ kPa，$T = 298.15$ K），$\Delta G_H = 1/2 G(H_2) - G(H^+ + e^-) = 0$，方程(13.3)可以转化为

$$\Delta G = G[(h)] + \frac{1}{2}G(H_2) - G[(g)] - |e|U \quad (12.4)$$

其中，U 是价带最高点空穴相对于 SHE（$U = 0$，$\Delta G_H = 0$）的电势，所以$|e|U$表示光生电子-空穴对产生所需要的能量。

依据图 12.3 的机制解析，TiO_2（001）晶面上光催化形成 NO 的整体反应可以描述为反应方程(12.2)：$N_2 + 2H_2O + 4h^+ \longrightarrow 4H^+ + 2NO$。所以，可以据此转化对氧化还原电对 NO/N_2 的质子耦合电子转移反应 $2NO + 4H^+ + 4e^- \longrightarrow N_2 + 2H_2O$。通过量化计算可以得到这个电对的标准电极电势。水溶液中反应物与产物的结构通过 PW91/DNP 基组[22]进行优化，并在计算过程中考虑连续溶剂模型(COSMO)[26]。表 12.3 给出了计算得到的热力学数据，半反应的标准 Gibbs 自由能($\Delta G^{\ominus}$)可以通过方程(12.5)进行计算：

表 12.3　反应 $2NO + 4H^+ + 4e^- \longrightarrow N_2 + 2H_2O$（$2NO + 4H_3O^+ + 4e^- \longrightarrow N_2 + 6H_2O$）中反应物和产物的热力学性质

	G(kcal·mol^{-1})	E_T(Hatree)	$\Delta G^{\ominus}$(kcal·mol^{-1})
NO	−9.440	−129.909	—
*H_3O^+	9.632	−76.857	
N_2	−8.587	−109.534	−631.367
H_2O	1.553	−76.448	—

*水溶液中质子一般以水合质子的形式存在，这里用 H_3O^+ 进行计算。

$$\Delta G^{\ominus} = (G_P + E_{T,P} \times 627.51) - (G_R + E_{T,R} \times 627.51) \quad (12.5)$$

其中，G_P 和 G_R 分别表示产物和反应物在 298.15 K 时的自由能，而 $E_{T,P}$ 和

$E_{T,R}$ 分别表示产物和反应物的总能量。

标准电极电势($E^\ominus$ vs SHE)的计算方法如方程(12.6)所示：

$$E^\ominus = -\frac{\Delta G^\ominus}{nF} - E_H \tag{12.6}$$

其中，n 是转移的电子数，F 是 Faraday 常数，为 23.06 kcal·mol^{-1}·V^{-1}，E_H 是标准氢电极的还原电势，为 4.28 V[27]。

计算得到的 $E^\ominus$(vs SHE)是 2.56 V，所以$|e|U$ 等于 2.56 eV，就足以驱动光催化反应步骤 V3[(h) + h$^+$ ⟶ (l) + H$^+$]的进行，因为此时该基元反应的 ΔG 值为负值(-1.64 kcal·mol^{-1}，表 12.2)。对于反应 V3，在去除第二个质子后，形成桥连的过氧化物[图 12.2(i)]。随后第二个 H$_2$O 分子攻击 O—O 键形成 Ti—O—OH O—Ti 结构(图 12.2(j))，随着质子的去除形成氧自由基[(i) + H$_2$O + h$^+$ ⟶ (j) + H$^+$]。反应 V4 是 VB 机制中的速率控制步骤，$|e|U$ 值为 2.56 eV 不足以克服该基元反应的能垒 E_a。考虑在(001)晶面施加 1.22 V 的过电势，反应 V4 的 ΔG 则为负值，此时$|e|U$ 等于 3.78 eV (2.56+1.22 eV，~330 nm)。在这种情况下，N$_2$ 与 O 自由基形成 N—O 键，可克服 3.23 eV 的能垒。之后，反应 V6 通过结构的调整进一步降低总能量。最后，形成 NO，从 TiO$_2$(001)晶面解吸(反应 V7)。对于图 12.2(f)，TiO$_2$ 表面的结构是相同的，但晶面真空层中的分子是不同的，图 12.2(f)中是反应物，而图 12.2(f)是产物，而且图 12.2(f)的总能量低于图 12.2(f)，如图 12.5(b)所示。这说明从 N$_2$ 和 H$_2$O 形成 NO，系统的整体能量降低，系统趋于稳定。TiO$_2$ 的导带最低点相对于 SHE 高了 0.2 eV，表明高于 3.98 eV(0.2+3.78 eV，~310 nm)的紫外光能量就足够驱使光催化形成 NO。

12.3.2
光催化反硝化的电子转移

由于光催化反硝化过程是还原反应，锐钛矿 TiO$_2$(101)晶面是最稳定的晶面，所以反硝化过程应该发生在(101)晶面。TiO$_2$ 表面进行光催化反硝化过程可以用一下反应式描述：

$$10\ e_{cb}^- + 2\ NO_3^- + 12\ H^+ \longrightarrow N_2 + 6\ H_2O \tag{12.7}$$

1. 生物电辅助的光催化反硝化过程

构建了生物光电催化反硝化系统，微生物产生的电子可以与光催化剂的光生空穴耦合,从而防止光生电子与空穴的复合,提高光催化效率。在该系统中,反硝化过程不会产生亚硝酸盐,因而可能与传统的反硝化过程的反应途径不同,如图 12.6 所示。对生物电子辅助过程以及传统反硝化过程都进行了计算,在生物光电催化系统中,反应中间产物的几何结构也都进行了优化(图 12.7)。

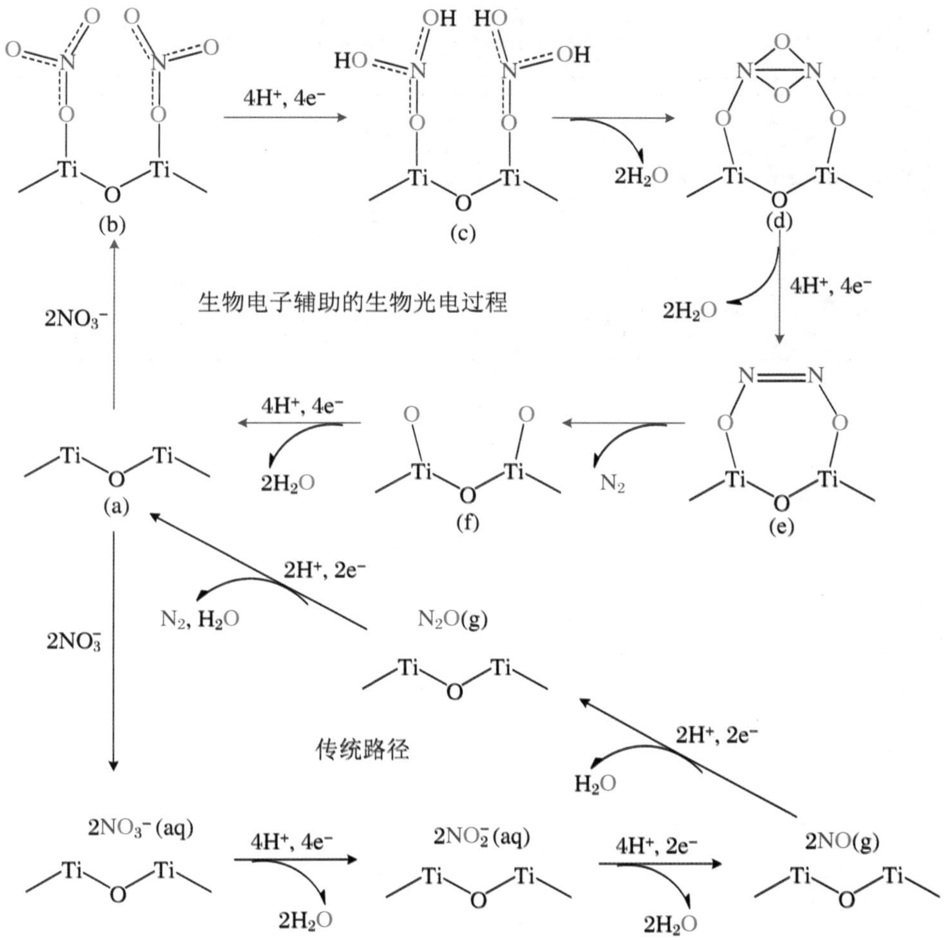

图 12.6　锐钛矿 TiO_2(101)晶面光催化反硝化的机制,蓝色箭头代表生物电子辅助的光电过程(Bioelectron-assisted way in bioelectron-photocatalytic system),黑色箭头代表传统反硝化过程(Conventional pathway)

表 12.4 中是基元反应的热力学和动力学数据,图 12.8 给出了每一步反应的能垒以及过渡态结构。光催化过程中的初始步骤是硝酸根离子[$(NO_3^-)_{ad}$]的吸附,以 NO_3^- 的 O 原子吸附于 TiO_2(101)晶面的 Ti 原子上,形成 Ti—O 键[图 12.7(b)]。吸附态的$(NO_3^-)_{ad}$ 的 O_{ad} 位于 TiO_2(101)晶面上一层的晶格氧的位置,如图 12.9 所示。第一步反应过程的 Gibbs 自由能为负值($\Delta_r G_m^\ominus$ = -6.908 kcal·mol^{-1}),表明这个过程在光催化过程中是热力学可行的。接着,

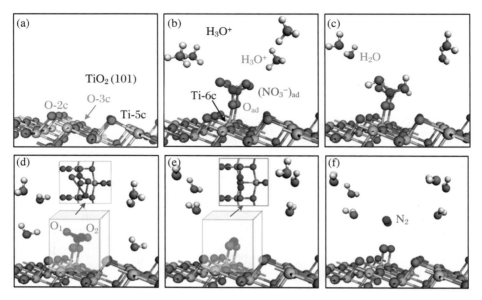

图 12.7 TiO$_2$(101)晶面光催化反硝化过程每一步反应中间产物的结构:TiO$_2$(101)晶面(a);NO$_3^-$的 O 原子吸附在 Ti 原子上,O 原子在晶格氧的位置(b);还原过程的中间产物,插图中是俯视图(c)～(e);N$_2$的生成(f)

4 个水合质子的 H$^+$ 分别添加到两个吸附态(NO$_3^-$)$_{ad}$另外 4 个 O 原子上,形成 OH 基团[图 12.7(c)],对应于图 12.6 中的(b)到(c)的过程,其中能垒 E_a 为 1.658 eV。水合质子是带正电的多原子离子,化学试为 H$_3$O$^+$ 是一类氧鎓离子。这种阳离子易于与溶剂结合,经常被用于代表水溶液中的质子。

表 12.4 硝酸盐在生物光电催化反硝化系统中还原的热力学和动力学性质,NO$_3^-$在锐钛矿 TiO$_2$(101)晶面发生还原(298.15 K,1 atm)

步骤	$\Delta_r G_m^{\ominus}$ (kcal·mol^{-1})	$\Delta_r H_m^{\ominus}$ (kcal·mol^{-1})	$\Delta_r S_m^{\ominus}$ (cal·mol·K^{-1})	E_a (eV)
(a) + 2NO$_3^-$ → (b)	−6.908	−11.145	−14.211	—
(b) + 4H$^+$ + 2e$_{cb}^-$ → (c)	−186.016	−178.117	26.494	1.658
(c) → (d) + 2H$_2$O	−613.227	−72.432	−14.105	2.265
(d) + 4H$^+$ + 4e$_{cb}^-$ → (e) + 2H$_2$O	−269.880	−254.843	50.432	−2.559
(e) → (f) + N$_2$	20.949	19.797	−3.862	3.115
(f) + 4H$^+$ + 4e$_{cb}^-$ → 1 + 2H$_2$O	−397.768	−391.597	20.697	−0.199
总计	−907.850	−8813.337	65.445	—

表中(a)～(f)对应图 12.7。

在反应 II 之后,在两个相邻的 OH 之间进行分子间脱水反应,形成图 12.7(d)。之后,4 个 H$_3$O$^+$ 的质子与 O$_1$ 和 O$_2$ 原子相结合形成 6 个 H$_2$O,从而 N 原

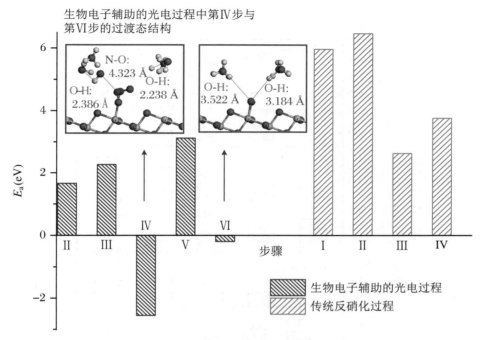

图12.8 光催化反硝化过程中生物电辅助路径与传统路径的能垒(E_a)的比较以及生物电辅助路径中步骤Ⅳ和Ⅵ的过渡态结构信息,且反应Ⅰ是NO_3^-在TiO_2(101)晶面的吸附过程

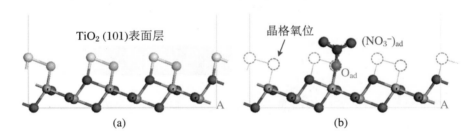

图12.9 吸附态NO_3^-的O_{ad}原子位于锐钛矿TiO_2(101)晶面上一层的晶格氧位:TiO_2(101)晶面(a);NO_3^-吸附于Ti原子上(b)

子之间形成双键(N=N)[图12.7(e)]。从(d)到(e)的这个过程的$\Delta_r G_m^\ominus$值(表12.4)为负值,说明质子链接到O_1、O_2促使N—O_1和N—O_2断裂的过程具有较高的反应活性。最后,N_2分子离开TiO_2(101)晶面,O原子则留在晶格氧的位置上[图12.7(f)],但O_{ad}具有较高的反应活性,会继续进行质子耦合光生电子的反应,生成2个H_2O分子,从而完成光催化反硝化反应。表12.4中整体光催化反应的$\Delta_r G_m^\ominus$值为负值,说明硝酸盐还原在1 atm和298.15 K的条件下是热力学可行的。对于温度的影响,计算分析。虽然随着温度的升高,有些基元反应的$\Delta_r G_m^\ominus$值增加,但对于总反应,温度的升高会使生物电辅助的光催化反硝

化过程更加有利(图12.10)。

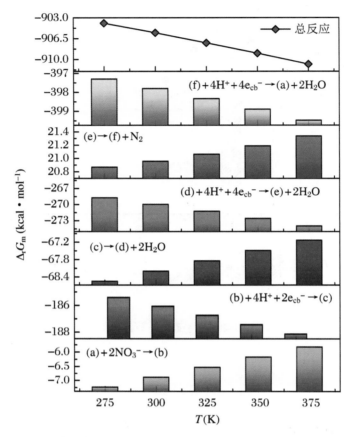

图 12.10　生物电辅助的光催化反硝化过程 Gibbs 自由能随温度的变化

系统的总能量随着结构的演变而发生变化,反应Ⅰ是 NO_3^- 在晶面上吸附的过程,所以没有进行动力学性质的计算。而反应Ⅴ是速率控制步骤,具有较高的能垒,为 3.115 eV,说明需要克服较高的能量断裂 N—O 键,从而在两个 N 原子之间形成第三个键。基元反应中能垒为负值,说明这个反应是典型的 barrierless 过程[31],能进行反应是因为表面形成了捕获分子的势阱[32]。因此,步骤Ⅳ和Ⅵ负的 E_a 值(图 12.8)表明质子耦合电子转移过程不需要克服能垒,能够快速地完成还原反应。与传统的反硝化机制相比,生物电辅助的光催化反硝化过程具有较低的能垒,表明完整的硝酸盐光催化还原为 N_2 的反应中不会积累中间产物。

2. 传统的光催化反硝化途径

传统的反硝化过程会产生中间产物亚硝酸(NO_2^-)、一氧化氮(NO)和一氧

化二氮(N_2O),如下列反应方程所示[33]:

$$2NO_3^- + 4H^+ + 4e^- \longrightarrow 2NO_2^- + 2H_2O \quad (12.8)$$

$$2NO_2^- + 4H^+ + 2e^- \longrightarrow 2NO + 2H_2O \quad (12.9)$$

$$2NO + 2H^+ + 2e^- \longrightarrow N_2O + H_2O \quad (12.10)$$

$$N_2O + 2H^+ + 2e^- \longrightarrow N_2 + H_2O \quad (12.11)$$

这里主要讨论在锐钛矿 TiO_2(101)晶面发生传统反硝化过程的热力学和动力学性质。表 12.5 的热力学数据表明,传统的反硝化过程是热力学可行的。首先,游离的 NO_3^- 和 H_3O^+ 在 TiO_2(101)晶面发生反应生成 NO_2^- 和 H_2O 分子[图 12.11(a)],虽然该过程是可以自发进行的($\Delta_r G_m^\ominus = -196.312 \text{ kcal} \cdot \text{mol}^{-1}$),但是反应的能垒非常高,$E_a$ 为 5.961 eV。此外,第二步生成 NO 过程的能垒也达到了 6.461 eV。如此高的能垒表明传统的反硝化过程在动力学上是不利于进行的,即反应非常缓慢,在实验中可能难以观察。然而,从目前的文献报道中发现,若在 TiO_2 中掺杂杂原子或负载过渡金属团簇,可以减小半导体的带隙,从而拓宽了光催化剂的应用领域[33]。

表 12.5 锐钛矿 TiO_2(101)晶面光催化传统反硝化过程的热力学(298.15 K,1 atm)和能垒

步骤	$\Delta_r G_m^\ominus$ (kcal·mol^{-1})	$\Delta_r H_m^\ominus$ (kcal·mol^{-1})	$\Delta_r S_m^\ominus$ (cal·mol^{-1} K)	E_a (eV)
$2NO_3^- + 4H^+ + 4e^- \rightarrow 2NO_2^- + 2H_2O$	−196.312	−192.153	13.948	5.961
$2NO_2^- + 4H^+ + 2e^- \rightarrow 2NO + 2H_2O$	−245.946	−239.451	21.785	6.461
$2NO + 2H^+ + 2e^- \rightarrow N_2O + H_2O$	−172.042	−162.614	31.619	2.625
$N_2O + 2H^+ + 2e^- \rightarrow N_2 + H_2O$	−156.258	−155.464	2.663	3.755
合计	−770.558	−749.682	70.015	—

除了对催化剂结构进行优化以外,另一个提高光催化性能的方法是在反应体系中添加牺牲剂,即空穴清除剂。但是空穴清除剂的添加,比如经常使用的甲酸盐($HCOO^-$)[34],可能会与 NO_3^- 发生竞争吸附,因而出现 NO_3^- 还原所需电子不足的现象,导致 NO_3^- 的不完全还原,从而产生 NO_2^-、NO 和 N_2O 等有毒副产物[35]。

为了进一步解析 NO_3^- 和 $HCOO^-$ 在 TiO_2(101)晶面竞争吸附过程,计算了两种离子在晶面上的吸附能(ΔE_{ad}):

$$\Delta E_{ad} = E_{TiO_2} + E_{anion} - E_{anion\text{-}TiO_2} \quad (12.12)$$

其中,E_{TiO_2} 和 E_{anion} 分别是 TiO_2(101)晶面和两种离子各自的总能量,$E_{anion\text{-}TiO_2}$ 是离子吸附于晶面后的体系总能量。

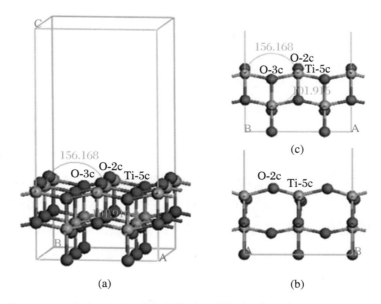

图 12.11 TiO₂(101)晶面光催化传统反硝化过程每一步的几何结构：
NO₃⁻，H₃O⁺ 在 TiO₂(101) 晶面上(a)；NO₃⁻ 还原为 NO₂⁻ (b)；
NO₂⁻ 和 H₃O⁺ 作为第二步的反应物(c)；H₃O⁺ 中的质子与
NO₂⁻ 的 O 原子结合形成 H₂O 和 NO(d)；H₃O⁺ 中的质子与
NO 的 O 原子结合形成 H₂O 和 N₂O(e)～(f)；H₃O⁺ 质子与
N₂O 的 O 原子反应(g)；N₂ 的生成(h)

图 12.12 给出了优化后的吸附构型和 ΔE_{ad} 值，发现空穴清除剂 HCOO⁻ 在 TiO₂(101) 晶面的吸附能几乎是 NO₃⁻ 的 5 倍。结果表明，过多的空穴清除剂会阻碍 NO₃⁻ 在催化剂表面的吸附，从而影响传统的反硝化反应。因此，通过生物电子来清除空穴，可以有效地避免清除剂占据催化剂表面的活性位点，也不会造成 NO₃⁻ 的不完全还原，从而进行具有选择性的高效光催化反硝化过程。

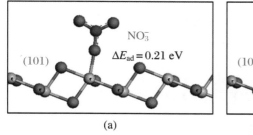 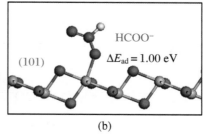

图 12.12 NO₃⁻ 和 HCOO⁻ 吸附在 TiO₂(101) 晶面的 Ti 原子：NO₃⁻ 在 TiO₂(101) 晶面的吸附能为 0.21 eV($\Delta E_{ad} = E_{TiO2} + E_{NO3} - E_{NO3\text{-}TiO2}$)(a)；HCOO⁻ 的 ΔE_{ad} 值几乎是 NO₃⁻ 的 5 倍(b)

参考文献

[1] Hooftman N, Messagie M, Van Mierlo J, et al. A review of the European passenger car regulations-real driving emissions vs. local air quality [J]. Renewable & Sustainable Energy Reviews, 2018, 86: 1-21.

[2] Geng L, Alexander B, Cole-Dai J, et al. Nitrogen isotopes in ice core nitrate linked to anthropogenic atmospheric acidity change [J]. Proceedings of the National Academy of Sciences of the United States of America, 2014, 111(16): 5808-5812.

[3] Hu Y, Ribbe M W. Nitrogenases-A tale of carbon atom(s) [J]. Angewandte Chemie-International Edition, 2016, 55(29): 8216-8226.

[4] Uekoetter F, Smil V. Enriching the earth: Fritz Haber, Carl Bosch and the transformation of world food production [M]. Cambridge: MIT Press, 2001.

[5] Yuan S J, Chen J J, Lin Z Q, et al. Nitrate formation from atmospheric nitrogen and oxygen photocatalysed by nano-sized titanium dioxide [J]. Nature Communications, 2013, 4: 2249.

[6] Guo Q, Zhou C, Ma Z, et al. Fundamentals of TiO_2 photocatalysis: Concepts, mechanisms, and challenges [J]. Advanced Materials, 2019, 31(50): 1901997.

[7] Tayel A, Ramadan A R, El Seoud O A. Titanium dioxide/graphene and titanium dioxide/graphene oxide nanocomposites: Synthesis, characterization and photocatalytic applications for water decontamination [J]. Catalysts, 2018, 8(11): 491.

[8] Lundberg J O, Carlstrom M, Weitzberg E. Metabolic effects of dietary nitrate in health and disease [J]. Cell Metabolism, 2018, 28(1): 9-22.

[9] Jones R R, Weyer P J, Dellavalle C T, et al. Nitrate from drinking water and diet and bladder cancer among postmenopausal women in Iowa [J]. Environmental Health Perspectives, 2016, 124(11): 1751-1758.

[10] Logan B E, Rossi R, Ragab A A, et al. Electroactive microorganisms in bioelectrochemical systems [J]. Nature Reviews Microbiology, 2019, 17(5): 307-319.

[11] Beegle J R, Borole A P. Energy production from waste: Evaluation of anae-

robic digestion and bioelectrochemical systems based on energy efficiency and economic factors [J]. Renewable & Sustainable Energy Reviews, 2018, 96: 343-351.

[12] Yang H G, Sun C H, Qiao S Z, et al. Anatase TiO$_2$ single crystals with a large percentage of reactive facets [J]. Nature, 2008, 453(7195): 638-634.

[13] Deiana C, Fois E, Coluccia S, et al. Surface structure of TiO$_2$ P25 nanoparticles: infrared study of hydroxy groups on coordinative defect sites [J]. Journal of Physical Chemistry C, 2010, 114(49): 21531-21538.

[14] Hussain A. A computational study of catalysis by gold in applications of CO oxidation [D]. Technische Universiteit Eindhoven, 2010.

[15] He Y, Tilocca A, Dulub O, et al. Local ordering and electronic signatures of submonolayer water on anatase TiO$_2$(101) [J]. Nature Materials, 2009, 8(7): 585-589.

[16] Vanderbilt. Soft self-consistent pseudopotentials in a generalized eigenvalue formalism [J]. Physical Review B, Condensed Matter, 1990, 41(11): 7892-7895.

[17] Segall M D, Lindan P J D, Probert M J, et al. First-principles simulation: ideas, illustrations and the CASTEP code [J]. Journal of Physics Condensed Matter, 2002, 14(11): 2717-2744.

[18] Perdew, Chevary, Vosko, et al. Erratum: Atoms, molecules, solids, and surfaces: Applications of the generalized gradient approximation for exchange and correlation [J]. Physical Review B, Condensed Matter, 1993, 48(7): 4978-4978.

[19] Monkhorst H J, Pack J D. Special points for Brillouin-zone integrations [J]. Physical Review B, 1976, 13(12): 5188-5192.

[20] Halgren T A, Lipscomb W N. The synchronous-transit method for determining reaction pathways and locating molecular transition states [J]. Chemical Physics Letters, 1977, 49(2): 225-232.

[21] Delley B. Fast Calculation of electrostatics in crystals and large molecules [J]. The Journal of Physical Chemistry, 1996, 100(15): 6107-6110.

[22] Perdew J P, Chevary J A, Vosko S H, et al. Atoms, molecules, solids, and surfaces: Applications of the generalized gradient approximation for exchange and correlation [J]. Physical Review B Condens Matter, 1992, 46(11): 6671-6687.

[23] Li Y F, Liu Z P, Liu L, et al. Mechanism and activity of photocatalytic oxygen evolution on titania anatase in aqueous surroundings [J]. Journal of the American Chemical Society, 2010, 132(37): 13008-13015.

[24] Valdés Á, Qu Z W, Kroes G J, et al. Oxidation and photo-oxidation of water on TiO_2 surface [J]. The Journal of Physical Chemistry C, 2008, 112(26): 9872-9879.

[25] Rossmeisl J, Qu Z W, Zhu H, et al. Electrolysis of water on oxide surfaces [J]. Journal of Electroanalytical Chemistry, 2007, 607(1): 83-89.

[26] Klamt A, Schüürmann G. COSMO: a new approach to dielectric screening in solvents with explicit expressions for the screening energy and its gradient [J]. Journal of the Chemical Society, Perkin Transactions 2, 1993, (5): 799-805.

[27] Kelly C P, Cramer C J, Truhlar D G. Aqueous solvation free energies of ions and ion-water clusters based on an accurate value for the absolute aqueous solvation free energy of the proton [J]. The Journal of Physical Chemistry B, 2006, 110(32): 16066-16081.

[28] Gong X Q, Selloni A, Dulub O, et al. Small Au and Pt clusters at the anatase TiO_2 (101) surface: Behavior at terraces, steps, and surface oxygen vacancies [J]. Journal of the American Chemical Society, 2008, 130(1): 370-381.

[29] Aschauer U, Chen J, Selloni A. Peroxide and superoxide states of adsorbed O_2 on anatase TiO_2 (101) with subsurface defects [J]. Phys Chem Chem Phys, 2010, 12(40): 12956-12960.

[30] Gong X Q, Selloni A, Batzill M, et al. Steps on anatase TiO_2 (101) [J]. Nature Materials, 2006, 5(8): 665-670.

[31] Alvarez-Idaboy J R, Mora-Diez N, Vivier-Bunge A. A Quantum chemical and classical transition state theory explanation of negative activation energies in OH addition to substituted Ethenes [J]. Journal of the American Chemical Society, 2000, 122(15): 3715-3720.

[32] Mozurkewich M, Benson S W. Negative activation energies and curved Arrhenius plots. 1. theory of reactions over potential wells [J]. The Journal of Physical Chemistry, 1984, 88(25): 6429-6435.

[33] Moura I, Moura J J G. Structural aspects of denitrifying enzymes [J]. Current Opinion in Chemical Biology, 2001, 5(2): 168-175.

[34] Zhang F, Pi Y, Cui J, et al. Unexpected selective photocatalytic reduction of nitrite to nitrogen on silver-doped titanium dioxide [J]. The Journal of Physical Chemistry C, 2007, 111(9): 3756-3761.

[35] Zhang F, Jin R, Chen J, et al. High photocatalytic activity and selectivity for nitrogen in nitrate reduction on Ag/TiO$_2$ catalyst with fine silver clusters [J]. Journal of Catalysis, 2005, 232(2): 424-431.

第 13 章

Pt/TiO₂光催化降解硝基苯的界面电子转移

13.1
贵金属与半导体复合催化剂的优势

硝基苯（NB）在染料、药物、炸药、农药等行业中被广泛应用。此类硝基化合物由于具有高毒性、致癌性以及残留时间卡，已经对水环境及人体健康产生了严重的危害[1,2]。因此有必要寻找有效、廉价且环境友好的方法来分解水环境中的硝基苯。已有许多方法被用于 NB 的转化和降解，包括生物还原[3-4]、Fenton 反应[5]、臭氧氧化[6]、稳态辐解[7]、漫游离解（roaming dissociation）[8]、生物电化学[9]以及光催化方法[10-11]等。在这些方法当中，利用纳米材料实现水体环境中有机物的光催化降解，可直接利用太阳能而不会引入新的污染物[12]，是一种很有前景的方法。在光催化过程中，锐钛矿相 TiO_2 由于具有良好的光敏性、适宜的能隙、无毒、难溶于水及表面稳定等性质[13-14]，所以是目前光催化降解中研究非常受关注的纳米材料。在光催化还原降解过程中，NB 被还原为毒性较低的中间产物苯胺（AN）[15]，而氧化降解过程中是通过·OH 的作用[16]。然而，锐钛矿 TiO_2 的光催化降解效率受限，往往是因为较宽的能隙引起的光生电子不足以及电子空穴对快速的重新复合。

为了促进光生电子-空穴对的产生和分离以提高其催化活性，通常采取对 TiO_2 表面进行适当的修饰。比如，有研究报道，在 TiO_2 表面修饰氨基酸能增强 NB 的吸附和光还原。在 TiO_2 表面修饰有机分子能加强 NB 与 TiO_2 之间的耦合，并有利于光生电子的转移[17,19]。此外，纳米贵金属团簇也是普遍使用的共催化剂，用于加速电子-空穴对的分离[20-22]。例如，有研究表明，在 TiO_2 上负载少量的 Ag 团簇能显著提高 TiO_2 对 NB 的光催化还原活性以及生成 AN 产物的选择性。这是由于 Au 和 TiO_2(001)晶面之间形成了强烈的界面键合力，导致金的价电子被高度离域化[23]，从而加速了硝基芳香烃的加氢反应[24-25]。类似的化学吸附与光催化活性的增强也可以通过在 TiO_2 晶面上沉积少量的 Pt 纳米团簇（NPs）作为共催化剂来实现[26-27]。

虽然已有的研究已证明金属 NPs-TiO_2 复合物具有对于 NB 的高效的光催化降解能力，并且提出了降解途径[28-29]，但是对降解过程中的分子机制仍然不清楚，尤其是催化剂表面 NB 和降解中间产物的吸附构型、催化剂表面吸附 NB 的活性位点，它们都会对 NB 与催化剂的相互作用产生重要的影响。此外，决定金属 NPs-TiO_2 光催化性能的因素还包括电子结构中的费米能级以及溶剂效应[30,31]，对此也需要进一步的探索。

因此，本章目的是为了探索有机分子与金属 NPs-TiO$_2$ 复合体系的相互作用以及对 NB 的光催化降解机制。Pt 团簇负载的纳米晶体 TiO$_2$（Pt/TiO$_2$）被选为代表性的催化剂。为了确定吸附的 NB 和中间产物分子的活性位点和构型，使用密度泛函理论（DFT）量化计算方法计算反应物（NB、H$_3$O$^+$ 和·OH）与催化剂表面的相互作用。基于 DFT 的分子模拟方法能够解析分子水平的相互作用，已用于确定硝基芳香烃化合物的构型变化，解析 NB 分子在 Si(001) 表面的排列[32]，阐明质子和水分子在 NB 分解中起到的作用[33]。另外，对 NB 在 Pt/TiO$_2$ 的 (001) 和 (101) 晶面降解的热力学和动力学方面的分析，有助于揭示降解过程中中间产物结构的演变、反应路径以及速率控制步骤。最后，通过光催化降解实验来验证计算结果的可靠性。

13.2
Pt/TiO$_2$ 催化剂制备与机制解析方法

13.2.1
Pt/TiO$_2$ 的制备与表征

锐钛矿 TiO$_2$ 十面体单晶按照文献报道的方法[34]通过水热方法合成。四氯化钛（2.67 mmol·L^{-1}）和氢氟酸（质量分数 10%）分别是前驱体和结晶控制剂。反应在以聚四氟乙烯内衬的高压釜中在 180 ℃下进行 22 h。反应后，通过过滤收集锐钛矿 TiO$_2$ 单晶，先用 NaOH（0.1 mol·L^{-1}）冲洗，再用去离子水冲洗 3 次，然后在真空中干燥过夜（90 ℃）。

Pt 纳米颗粒（质量分数 1%）通过化学还原法沉积在 TiO$_2$ 催化剂晶面上。沉积过程如下：首先把 0.2 g TiO$_2$ 粉末、100 mL 乙醇、H$_2$PtCl$_6$·6H$_2$O（2 mg Pt）和 50 mg 抗坏血酸依次加入到三口烧瓶中；接着将悬浮液加热回流，并不断搅拌。经过 1 h 的沉积，离心分离，60 ℃干燥 1 h，在空气中 450 ℃退火 1 h 后得到灰色的粉末。其中氟化铵（NH_4F）、乙醇、抗坏血酸、乙二醇（EG）、硝基苯（NB）和 $H_2PtCl_6 \cdot 6H_2O$ 都是分析纯，均无需进一步纯化。

Pt/TiO_2 纳米晶体的透射电镜（TEM）图像、高分辨（HR）的 TEM 图像和选区电子衍射（SAED）图案通过透射电子显微镜（JEM-2010，JEOL Co.，Japan）得到，加速电压为 200 kV。粉末 X 射线衍射（XRD）通过 MXPAHF 衍射仪（MacScience Co.，Japan）在 Cu Kα 辐射源（$\lambda = 1.54056$ Å）获得。扫描速率取决于样品的晶相，设定为 8°/min。加速电压和所施加的电流分别为 30 kV 和 300 mA。样品的 UV-Vis 漫反射光谱（DRS）通过 UV-Vis 分光光度计（SOL-ID3700，Shimadzu Co.，Japan）获得。样品的光致发光光谱由光致发光光谱仪（5301，Shimadzu Co.，Japan）测得。

13.2.2

NB 在 Pt/TiO_2 表面的光催化降解

利用 Xe 灯（CHF-XM-350W，Beijing Trusttech. Ltd.）为光源进行 NB 的光催化降解实验。通过紫外或可见滤光片来选择紫外光（$\lambda < 420$ nm）或可见光（$\lambda \geqslant 420$ nm）。可见光范围的光强是 160 mW·cm^{-2}，紫外光范围的光强是 1 min 内测得 157 mJ·cm^{-2}，即 2.6 mW·cm^{-2}。反应总体积为 25 mL，催化剂浓度为 1 mg·mL^{-1}，水相中 NB 的初始浓度为 10 mg·L^{-1}。光催化反应通过光照启动，在光催化降解的过程中，每隔 10 min 取一次样，用于分析 NB 的浓度以及降解中间产物。NB 的浓度通过高效液相色谱（HPLC-1100，Agilent Inc.，USA）配合 Hypersil-ODS 反相柱和 VWD 检测器来测定。流动相是 0.1%乙酸和甲醇（体积比为 40∶60）的水溶液，以 0.8 mL·min^{-1} 的流速输送。NB 的降解产物通过液质联用仪（LC/MS，Agilent 6460）分析。

13.2.3

催化剂表面

理论计算和实验检测都表明平衡态的锐钛矿 TiO_2 的（001）晶面具有最高的反应活性，可以进行吸附物的氧化反应，而（101）晶面是最稳定的，可以进行吸附物的还原反应[13,35-36]。锐钛矿 TiO_2 晶体的表面层的真空区域的厚度,即 slab 之间的距离是 15 Å。所有的(001)晶面的 slabs 包含 8 个 Ti 原子和 16 个 O 原子。晶面(001)上在两个 Ti 原子之间存在 2-配位的桥氧键(Ti—O—Ti 角度是156°),另外还有 5-配位的 Ti 原子(Ti-5c)[37]。锐钛矿 TiO_2 的(101)晶面呈现锯齿状,表面上存在 5-、6-配位的 Ti 原子(Ti-5c、Ti-6c)以及 2-、3-配位的 O 原子(O-2c、O-3c)。每一个 slab 表面包含 18 个 Ti 原子和 36 个 O 原子。反应物、中间产物以及最终产物仅吸附于 slab 的一面。为了 Pt 团簇对于催化性能的影响,在计算中用含有 4 个 Pt 原子的小团簇(Pt_4)来进行分析。Pt_4 团簇是最简单的 Pt 原子组成的三维系统,是被普遍使用的高分散的催化剂模型。在 Pt_4 团簇中,每一个邻位的 Pt 位点的配位数都是 3[38-39]。

13.2.4
第一性原理计算方法

第一性原理的 DFT 计算基于广义梯度近似(GGA),结合平面波基组和超软赝势[40],在 CASTEP 程序[41]中进行。交换相关能量和势能是通过 Perdew、Burke 和 Ernzerhof(PBE)泛函自洽描述[42]。计算的收敛标准设定为 Express,这种精度可以广泛地并精确地用于研究半导体、绝缘体和非磁性金属,既可以使计算结果达到较高的精度,也可以节约计算时间。能量收敛标准设定为 1×10^{-3} eV·$cell^{-1}$。这个设定比 Fine 高一个数量级,所以得到的结果用于探索性研究是可靠的。布里渊区积分通过 Monkhorst-Pack 算法产生的 k-points 进行,而 k-points 的取值则取决于晶胞的尺寸和形状[43]。采用线性同步度越(LST)/二次同步度越(QST)方法[44]进行过渡态搜索和能垒的计算。

为了促进 NB 的光催化降解效率,需要找出对于氧化和还原反应最优的表面结构。为了达到这个目的,利用 $DMol^3$ 程序[45-46],对不同催化剂表面上进行 NB 降解的热力学过程进行计算。选择的泛函是 GGA/PBE[47],双精度数值基组结合 d-轨道极化函数(DND)用于处理价电子,全电子核处理方法用于描述核电子。

结合能(E_b)[48]的计算用于解析 NB 分子与 Pt_4 团簇、TiO_2 晶面之间的相互作用以及 NB 在催化剂表面的吸附位点。Pt 团簇与 TiO_2 之间的相互作用通过相互作用能(ΔE_{int})进行表征,即吸附系统的总能量[$E_t(Pt/T_{hkl})$]与 TiO_2 晶面以及 Pt 团簇的能量之和的差值,即

$$\Delta E_{\text{int}} = E_{\text{t}(Pt/T_{hkl})} - E_{\text{t}(T_{hkl})} - E_{\text{t}(Pt_4)} \tag{13.1}$$

13.3
Pt/TiO$_2$催化剂对硝基苯的转化

13.3.1
锐钛矿型 TiO$_2$ 表面的 Pt$_4$ 团簇

锐钛矿型 TiO$_2$(001)和(101)晶面结合 Pt$_4$ 团簇的 slab 模型(图 13.1)用 DFT 方法来优化。考虑了几种 Pt$_4$ 团簇在表面上不同的初始位置,以确定最稳定的结构。

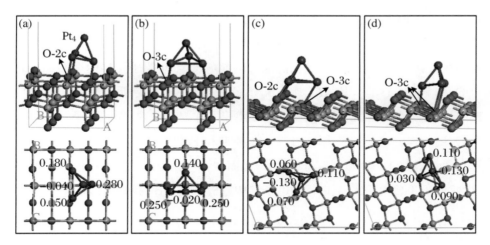

图 13.1 Pt$_4$ 在 TiO$_2$(001)和(101)晶面上负载的最低能量结构的立体图与俯视图:Pt/T$_{001}$-1(a);Pt/T$_{001}$-2(b);Pt/T$_{101}$-1(c);Pt/T$_{101}$-2(d)

为阐明吸附态团簇的电荷状态以及原子间的电荷分布,Mulliken 电荷[49]提供了系统中电荷再分配的定性衡量。Pt$_4$ 团簇各个 Pt 原子 Mulliken 电荷之和表明结合在 TiO$_2$ 晶面上的 Pt$_4$ 的电子密度减小(表 13.1),说明 Pt 被吸附后其电子趋向于TiO_2表面。在被吸附的Pt_4团簇中，3个直接与O-2c和O-3c原子连接的Pt原子一直荷正电，然而第四个Pt原子荷负电。Pt与邻近的氧原子之间的Mulliken population表明Pt和氧原子形成的键具有共价键的性质。显然在所有的构型中，Pt原子一旦被吸附就会因为电子的偏移而荷正电。从Pt团簇转移到(001)晶面的Mulliken电荷的数目多于转移到(101)晶面的数目。这说明同(101)晶面相比，(001)晶面聚集了更多荷正电的空穴，从而使得(001)晶面比(101)晶面更具有光氧化活性[50]。

表13.1 Pt_4团簇与锐钛矿TiO_2晶面的相互作用能(ΔE_{int})以及Pt_4团簇上的电荷分布

系统	ΔE_{int}(eV)	电荷Pt_4(e)
Pt/T_{001}-1	−6.55	0.57
Pt/T_{001}-2	−5.70	0.62
Pt/T_{101}-1	−4.37	0.11
Pt/T_{101}-2	−2.16	0.10

为了从不同角度来描述TiO_2表面与Pt团簇之间的相互作用，引入相互作用能(ΔE_{int})的概念，按照式(13.1)进行计算，$E_{t(T_{hkl})}$是纯TiO_2表面(hkl)的总能量，$E_{t(Pt_4)}$是Pt_4团簇的总能量，$E_{t(Pt/T_{hkl})}$是Pt_4-TiO_2体系的总能量。表13.1列出了Pt/TiO_2体系的ΔE_{int}随金属团簇位置变化的不同值。对于Pt/T_{001}体系，Pt_4团簇在锐钛矿型TiO_2(001)晶面的优选位置是3个Pt原子连接两个O-2c和一个O-3c[图13.1(a)]，其ΔE_{int} = −6.55 eV。如图13.1(c)所示的是较为稳定的Pt/T_{101}(ΔE_{int} = −4.37 eV)结构，其Pt团簇同样与(101)晶面的两个O-2c和一个O-3c相连。这些结果表明Pt/T_{001}-1模型和Pt/T_{101}-1模型更适合描述Pt团簇在锐钛矿型TiO_2(001)晶面和(101)晶面的特性。因此，之后的计算都是基于这两种Pt/TiO_2催化模型进行的电子结构分析以及NB的降解分析。

13.3.2
Pt的吸附对Pt和TiO_2电子结构的影响

Pt吸附原子和相关表面原子的局部态密度图(local density of states, LDOS)可用来进一步解释成键特性。在能量范围内，LDOS主要来源于氧的

p-轨道以及 Pt 和 Ti 的 d-轨道(图 13.2)。

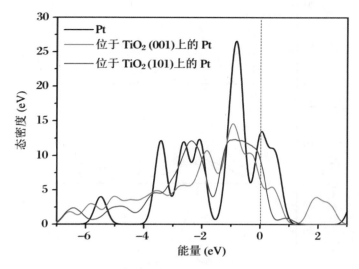

图 13.2　单独的 Pt 团簇和 Pt/TiO$_2$ 中 Pt 的局域态密度(LDOS)

图 13.2 给出了 Pt 团簇负载于 TiO$_2$(001)和(101)晶面的 LDOS，可以清晰地发现该电子比纯 Pt$_4$ 团簇的电子更加离域。图中能量坐标轴的零点对应费米能级，Pt 的电子与 TiO$_2$(001)和(101)晶面的 Ti 和 O 的电子混合充分，这毫无疑问地影响了 Pt 和 TiO$_2$ 的电子结构。因此，TiO$_2$ 表面的 Pt 原子与有机分子的反应活性更高。图 13.3 给出了纯 TiO$_2$ 和 Pt/TiO$_2$ 结构的 LDOS。在 Pt/T$_{001}$-1 体系和 Pt/T$_{101}$-1 体系中，表面带隙的金属诱导态主要是由 Pt 的 d 轨道电子贡献的，尽管也有一部分来自 O 的 p 轨道电子。对于 Pt 团簇被沉积于锐钛矿型 TiO$_2$(001)晶面，Pt 原子的电子状态进入费米能级附近，使表面的能隙从 1.5 eV 减小到不足 1 eV。

因此，通过负载少量的 Pt 团簇提高锐钛矿型 TiO$_2$ 在可见光下产生电子-空穴对的能力。在 Pt/T$_{001}$ 体系中 Pt 的 d 轨道和 O 的 p 轨道混合充分，暗示着 Pt 原子与 O 样子之间有较强的相互作用。在 Pt/T$_{101}$ 体系下，整个 LDOS 移向低能级，费米能级的电子云密度提高，带隙消失，这些都表明该体系金属化。可以推测，Pt/T$_{001}$ 和 Pt/T$_{101}$ 体系对 NB 的氧化和还原有更好的催化活性。

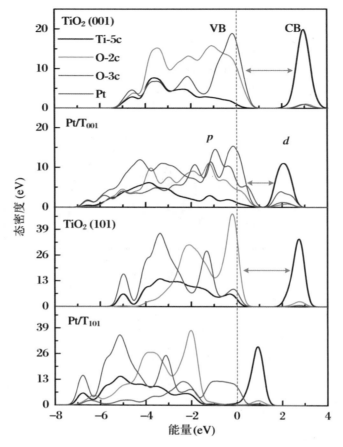

图 13.3 TiO$_2$(001)晶面表面原子(a);Pt/T$_{001}$-1 系统(b);
TiO$_2$(101)晶面(c)和 Pt/T$_{101}$-1 系统(d)的 LDOS

13.3.3

Pt/TiO$_2$ 光催化剂的特性

根据 DFT 关于有效催化性能的计算,合成了 Pt/TiO$_2$ 催化剂来验证计算结果。如图 13.4(a)所示,所制备 TiO$_2$ 的 XRD 图谱可以与标准卡片对比,是锐钛矿型 TiO$_2$。在 25.32°、37.93°、48.02°、53.98°和 55.04°处的衍射峰分别对应着锐钛矿型 TiO$_2$ 的(101)、(004)、(200)、(105)和(211)晶面。Pt/TiO$_2$ 复合材料的衍射峰表明化学沉积的 Pt 是立方相,对应着立方体的 Pt(JCPDS 4-802)。

为进一步确认 Pt 团簇在 TiO$_2$ 表面的位置,用透射式电子显微镜进行了 TiO$_2$ 结构特征的表征。TEM 显示合成的锐钛矿 TiO$_2$ 暴露了(001)和(101)晶面

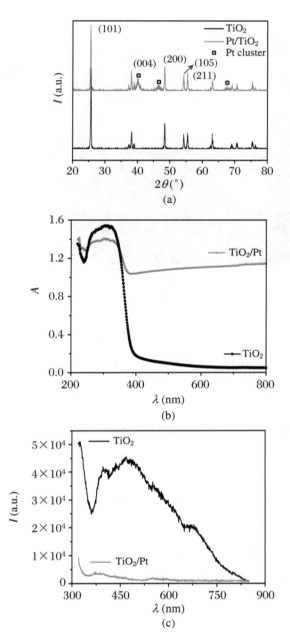

图 13.4 TiO$_2$ 与 Pt/TiO$_2$ 的 XRD(a); UV/vis DRS(b) 和 PL 光谱(c)

的十面体[图 13.5(a)]。而 HRTEM 可用于直接观察 Pt 团簇在 TiO$_2$ 晶面上的位置。图 13.5(b)显示了纳米晶体的晶格间距是 0.19 nm,恰好与十面体 TiO$_2$ 的(020)和(200)晶面的层间距相符。SAED 图案[图 13.5(b)中的插图]进一步证实的确形成了锐钛矿的纳米晶体,证实了晶面(200)和晶面(020)的夹角是 90°。而且,晶格间距与 XRD 的结果非常符合。此外,化学沉积形成的 Pt 团簇的尺寸非常均一,且在 TiO_2 不同晶面上均有分布,包括(001)和(101)晶面。

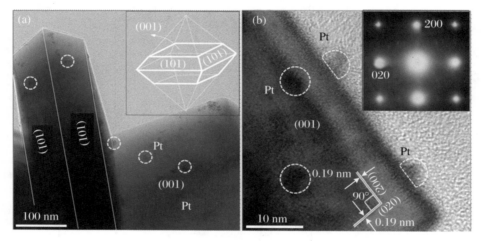

图 13.5　Pt/TiO_2(001)和(101)晶面的 TEM (a)和 HRTEM (b)。(a)中的插图:晶体晶面的示意图;(b)中的插图:锐钛矿 TiO_2 SAED 图

图 13.4(b)是光催化剂的紫外-可见漫反射光谱(DRS)。同纯 TiO_2 晶体相比,Pt/TiO_2 复合材料在可见区域有明显的吸收峰。Pt 纳米颗粒对 TiO_2 晶体的表面等离子体共振可增强对可见光的吸收[51]。这与 LDOS 的结果相一致,即带隙因 Pt 的 d 轨道的贡献而减小。光致发光(photoluminescence,PL)是由光生空穴和电子的复合引起的。PL 的低强度说明空穴和电子复合的几率减少,从而说明光生电子的存在时间被延长。本研究中,检测了若干样品在 300~850 nm 之间的荧光光谱。如图 13.4(c)所示,Pt 负载于 TiO_2 晶体表面导致了很大程度的荧光淬灭。纯 TiO_2 的 PL 远强于 Pt/TiO_2 的 PL,表明 Pt 的负载抑制了电子-空穴的复合。可以从 LDOS 结果得出 Pt 团簇的负载加快了光生电子从 TiO_2 导带的分离,而 PL 的实验结果也正好验证了这一点。

13.3.4

NB 在 Pt/TiO_2 表面的光催化降解

为了进一步验证以上的结果,分别以 TiO_2 单晶和 Pt/TiO_2 作为催化剂进行 NB 光催化降解。图 13.6 展示了降解中 NB 浓度随时间的变化,符合一级动力学。

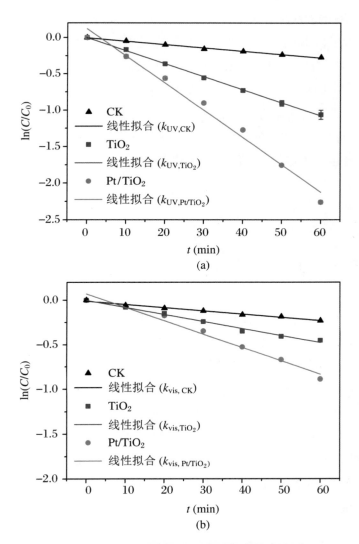

光源	系统	$k(\min^{-1})$	R^2
	CK	−0.005	0.994
UV紫外光	TiO_2	−0.018	0.999
	Pt/TiO_2	−0.037	0.983
	CK	−0.004	0.987
可见光	TiO_2	−0.008	0.988
	Pt/TiO_2	−0.015	0.976

图 13.6 水溶液中利用 P25，TiO_2 单晶和 Pt/TiO_2 光催化降解 NB，分别在 UV 紫外光(a)和可见光(b)条件下，C_0 和 C 分别是 NB 在 $t=0$ 和 $t=t$ 的浓度，k 表示反应速率常数

因此，NB 的表观降解速率常数(k)能够从 $\ln(C/C_0)$-t 线性拟合的斜率得到，其中 C 表示 NB 的浓度。通过比较发现，使 NB 降解最快的是在紫外光照射下，以 Pt/TiO_2 为催化剂时。无论是在紫外光还是可见光的条件下，降解速率 k_{Pt/TiO_2} 都是 k_{TiO_2} 的两倍，对于 1 h 紫外光照射下，Pt/TiO_2 系统的总体 NB 去除效率比 TiO_2 系统高了 30%。所有这些结果表明 Pt 团簇明显促进了 NB 的光催化降解。值得注意的是，在没有光催化剂存在时，NB 的浓度也会有轻微地减少，这可能是由于直接的光降解或是挥发造成的。

13.3.5
NB 在 Pt/TiO_2 表面的吸附

为进一步评估不同晶面的光催化活性，进行了一系列的 DFT 计算来研究 NB 在 Pt/T_{001} 和 Pt/T_{101} 晶面的吸附和降解。首先需要确定 NB 在催化剂表面的吸附位置。利用 DFT 计算来确定 NB 吸附于 Pt_4 团簇表面时的结构，以便深入研究 NB 与 Pt_4 团簇的相互作用。图 13.7 显示了复合物模型的结构并对其进行了优化。第一个模型认为 NB 的两个氧原子接到 Pt_4 团簇的一个 Pt 原子上 [图 13.7(a)，NB-Pt]。第二个模型表示两个氧原子各自连接一个 Pt 原子 [图 13.7(b)，NB-2Pt]。NB 和 Pt 团簇优化后的结构参数见表 13.2。同自由态 NB 分子相比，吸附态 NB 的 N—O 键从 1.24 Å 增加到 1.27 Å，而 C—N 键和 O—N—O 键的键角减小了。

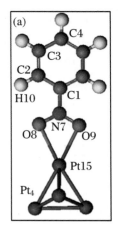
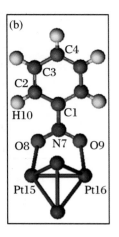

图 13.7　Pt_4 上 NB 可能的吸附构型：NB-Pt(a)；NB-2Pt(b)

表 13.2　NB 与 NB-Pt 复合物的几何结构参数(键长的单位是 Å，键角和二面角的单位是°)

	NB-Pt	NB-2Pt	NB
C1—C2	1.405	1.404	1.398
C2—C3	1.394	1.395	1.395
C3—C4	1.402	1.402	1.401
C1—N7	1.430	1.435	1.496
N7—O8	1.270	1.271	1.241
N7—O9	1.267	1.271	1.241
Pt15—O8	2.497	2.464	—
Pt15—O9	2.551	—	—
Pt16—O9	—	2.456	—
O8—N7—O9	119.985	121.822	123.532
C1—N7—O8—Pt15	178.833	116.280	—
C1—N7—O9—Pt15	−178.860	—	—
C1—N7—O9—Pt16	—	−115.698	—

稳定的吸附结构可通过 NB 和 Pt_4 的键能(E_b)来确定。定义 NB 分子从 Pt 团簇上无限分离的 $E_b = 0$。E_b 越负，结构越稳定。优化后的 NB-Pt 结构如图 13.7(a)所示，其键能为 −0.615 eV(表 13.3)。从能量方面考虑，相比于 NB-2Pt 结构，这种结构不利于反应，因此不再进一步讨论。

表 13.3　NB 在 Pt_4 团簇上两种结合构型的结合能(E_b)

系统	E_t(Ha)*	E_b(eV)
NB-Pt	−69776.151582	−0.615
NB-2Pt	−69776.146498	−0.477
NB	−436.442029	—
Pt_4	−69339.686939	—

* E_t 是系统的总能量。

Pt_4 团簇是电子供体，NB 是电子受体。用 Mulliken 布居分析估算出 Pt_4 团簇的电荷为 +0.114，NB 的电荷是 −0.26(图 13.8)。这与 Ag_4-NB 体系的结果相一致[33]。元素电荷从 Pt_4 团簇转移到 NB，这有利于光生电子从 TiO_2 导带转出。

根据前线轨道理论，反应最可能发生在 Pt(电子供体)的最高占据分子轨道

（HOMO）与 NB（电子受体）的最低未占据分子轨道（LUMO）之间（图 13.9）。因此，电荷从纳米簇 Pt 原子的 HOMO 转移到 NB 的 LUMO。这说明 Pt₄ 团簇与硝基上的两个氧原子相连，从而吸附 NB。这也印证了 Pt 团簇上吸附基团的电子密度分布分析以及稳定结构结合能分析。

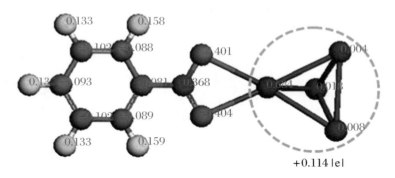

图 13.8　Mulliken 布居分析得到的 NB-Pt 吸附构型的电荷分布

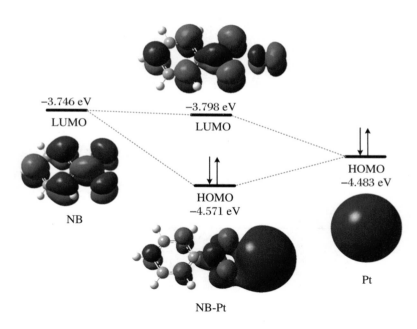

图 13.9　NB-Pt 电荷转移复合的前线轨道机制图（等值面 = 0.02）

对于 NB 在催化剂表面吸附位点的分析是通过 NB 在不同吸附位置的吸附能（ΔE_{ad}）进行判断的。这里计算了 NB 在 Pt/T₁₀₁-1 的（101）晶面上 Pt 以及 Ti-5c 位点的吸附能，以此来阐明 NB 最佳的吸附位置。图 13.10 展示了可能的吸附构型的能量最小化几何结构，ΔE_{ad} 越负表明越有利于 NB 的吸附。NB 在 Pt 位点吸附时的 ΔE_{ad} 为 -0.913 eV，比在 Ti-5c 位点吸附时（-0.321 eV）更负，所以 NB 主要吸附在 Pt/TiO₂ 催化剂表面的 Pt 位点上。

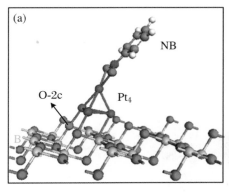 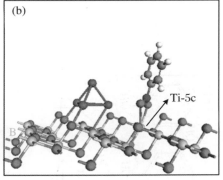

图 13.10　NB 在催化剂表面不同吸附位点的能量最小化几何结构，Pt/T_{101}-1 的 (101)晶面 Pt 位点(a)；Ti-5c 位点(b)

13.3.6
光催化机制以及在不同晶面上的结构变化

在光反应中，锐钛矿型 TiO_2 的(001)晶面主要发生氧化反应，而(101)晶面发生还原反应，这是因为在不同晶面上光致载流子的优先转移方向不同[52-54]。而·OH 是光催化过程中重要的氧化剂之一，NB 的氧化降解途径与之相关[34]。基于以上的分析，可以推出 NB 在 Pt/T_{001} 和 Pt/T_{101} 的降解机制（图13.11）。NB 在 Pt/T_{001} 表面的氧化降解过程如式(13.2)～式(13.5)所示，在 Pt/T_{101} 表面的还原路径分三步，中间产物有亚硝基苯（nitrosobenzene，NSB）和苯基羟胺（hydroxyl-aminobenzene，HAB）[2,28]，如式(13.6)～式(13.8)所示。

$$Pt/TiO_2 + h\nu \longrightarrow Pt/TiO_2(e_{CB}^- + h_{VB}^+) \tag{13.2}$$

$$Pt/T_{001}(h_{VB}^+) + H_2O \longrightarrow Pt/T_{001}(\cdot OH) + H^+ \tag{13.3}$$

$$Pt/T_{001}\text{-}NB + \cdot OH \longrightarrow Pt/T_{001}\text{-}NP + H\cdot \tag{13.4}$$

$$Pt/T_{001}\text{-}NP + \cdot OH \longrightarrow Pt/T_{001}\text{-}NO_2\cdot + HQ \tag{13.5}$$

$$Pt/T_{101}\text{-}NB + 2H_3O^+ + 2(e_{CB}^-) \longrightarrow Pt/T_{101}\text{-}NSB + 3H_2O \tag{13.6}$$

$$Pt/T_{101}\text{-}NSB + 2H_3O^+ + 2(e_{CB}^-) \longrightarrow Pt/T_{101}\text{-}HAB + 2H_2O \tag{13.7}$$

$$Pt/T_{101}\text{-}HAB + 2H_3O^+ + 2(e_{CB}^-) \longrightarrow Pt/T_{101} + AN + 3H_2O \tag{13.8}$$

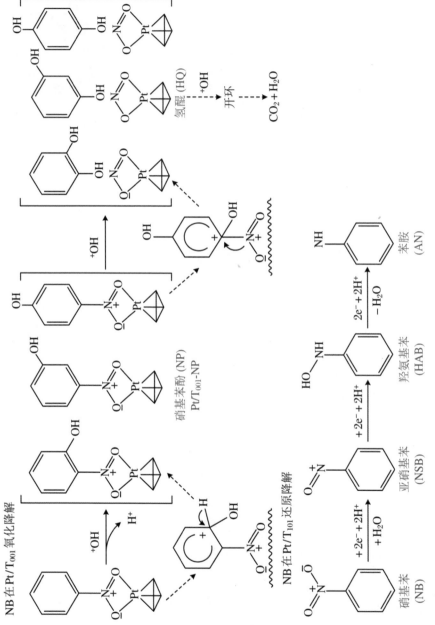

图 13.11　NB 在 Pt/T_{001} 和 Pt/T_{101} 晶面上光催化降解的机理

图 13.12 描述了中间产物有机分子在 Pt/TiO$_2$ 催化剂表面吸附后的能量最小化的几何结构。NB 光催化氧化和还原降解的反应自由能和能垒(E_a)在表 13.4 中进行了总结。每个基元反应的 Gibbs 自由能(ΔG)用下式计算:

$$\Delta G = \Delta E + \Delta ZPE - T\Delta S \tag{13.9}$$

其中,ΔE 是系统的总能量,ΔZPE 是零点能,T 是温度,ΔS 是熵变。由于 ΔZPE 和 $T\Delta S$ 的值远小于 ΔE,它们对于 ΔG 的贡献可以忽略[38],因此方程(6.9)可以写成 $\Delta G \approx \Delta E$。

NB 分子通过 NO$_2$ 基团上的 O 原子连接在 Pt/TiO$_2$ 的 Pt 上,这一点可以通过 NB 在 Pt$_4$ 团簇上的吸附结构证明。NB 的降解可以分为氧化和还原两种机制。

表 13.4 NB 在 Pt/TiO$_2$ 的(001)和(101)晶面光催化降解过程基元反应的自由能变化和能垒

步骤	ΔG(eV)	E_a(eV)
OX-1: Pt/T$_{001}$-NB + ·OH ⟶ Pt/T$_{001}$-(o-NP) + H·	−1.85	0.39
OX-1′: Pt/T$_{001}$-NB + ·OH ⟶ Pt/T$_{001}$-(m-NP) + H·	−2.72	1.85
OX-1″: Pt/T$_{001}$-NB + ·OH ⟶ Pt/T$_{001}$-(p-NP) + H·	−1.89	0.05
OX-2: Pt/T$_{001}$-(o-NP) + ·OH ⟶ Pt/T$_{001}$-NO$_2$· + o-HQ	−1.99	0.16
OX-2′: Pt/T$_{001}$-(m-NP) + ·OH ⟶ Pt/T$_{001}$-NO$_2$· + m-HQ	−2.27	1.20
OX-2″: Pt/T$_{001}$-(p-NP) + ·OH ⟶ Pt/T$_{001}$-NO$_2$· + p-HQ	−3.32	0.43
RED-1: Pt/T$_{101}$-NB + 2H$_3$O$^+$ + 2(e$_{CB}^-$) ⟶ Pt/T$_{101}$-NSB + 3H$_2$O	−0.60	2.69
RED-2: Pt/T$_{101}$-NSB + 2H$_3$O$^+$ + 2(e$_{CB}^-$) ⟶ Pt/T$_{101}$-HAB + 2H$_2$O	−0.10	无能垒
RED-3: Pt/T$_{101}$-HAB + 2H$_3$O$^+$ + 2(e$_{CB}^-$) ⟶ Pt/T$_{101}$ + AN + 3H$_2$O	−1.02	0.67

对于 NB 的氧化降解途径,·OH 主要攻击吸附态 NB 苯环上 NO$_2$ 基团的邻、间、对位,各自的 ΔG 分别为 −1.85 eV、−2.72 eV 和 −1.89 eV,从而形成吸附态的硝基苯酚[Pt/T$_{001}$-NP,图 13.12(b),(c)和(d)]。体系中·OH 来源于吸附水分子的间接氧化或者价带空穴(h$_{VB}^+$)]上的羟基。然后另一个·OH 攻击连接 NO$_2$ 基团的 C 位点,使 C—N 键断裂,从而形成邻、间、对苯二酚[其中 meta-HQ 如图 13.12(e)所示]。

然而在还原降解路径中,在 Pt/T$_{101}$ 表面的 NB 与两个 H$_3$O$^+$ 和光生电子(e$_{CB}^-$)反应,生成 Pt/T$_{101}$-NSB[图 13.12(g)]和 3 个水分子,其 ΔG 和 E_a 的值分别为 −0.56 eV 和 2.69 eV(表 13.4),都比氧化降解路径的第一步[式(6.4)]的值要大。这表明相比于还原途径的 NB 降解,氧化路径(OX-1,OX-1′和 OX-1″)的第一步(表 13.4)在热力学上更容易发生,特别是 OX-1′,而且也比还原途径

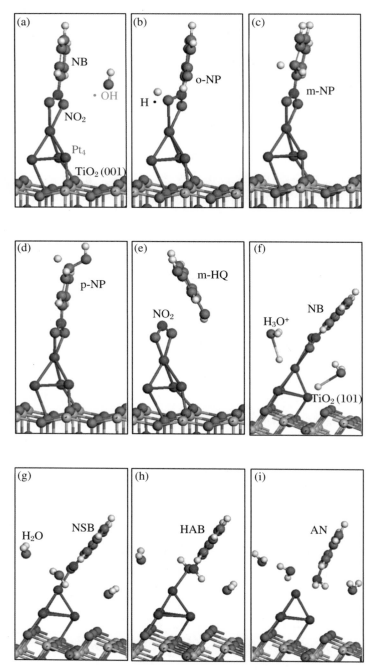

图 13.12 NB 光催化降解过程中 NB 及中间产物在 Pt/TiO$_2$ 的吸附构型：Pt/T$_{001}$ 晶面的 Pt/T$_{001}$-NB(a)；Pt/T$_{001}$-(o-NP)(b)；Pt/T$_{001}$-(m-NP)(c)；Pt/T$_{001}$-(p-NP)(d)；Pt/T$_{001}$-NO$_2$·(e)；Pt/T$_{101}$ 晶面的 Pt/T$_{101}$-NB(f)；Pt/T$_{101}$-NSB(g)；Pt/T$_{101}$-HAB(h) 和 Pt/T$_{101}$＋AN(i)

具有更高的反应速率。这进一步说明了光生空穴 h_{VB}^+ 更有利于 NB 光催化降解,即 NB 在 Pt/TiO₂ 上更趋向于在(001)晶面按照氧化途径进行降解。

NB 降解的中间产物进一步通过 LC/MS 分析确认,如图 13.13 所示。经过 1 h 的光照,NB 的浓度降低了 90%,由图 13.13(b)可以看出 10 min 左右存在两种中间产物,并标注了 1 和 2,最早出现的峰极性非常大,可能是残留在测试系统中的杂质。接着用 MS 进一步表征了两种中间产物,由分子量可以确定都是硝基苯酚,羟基在不同的取代位点,可能是间位和对位衍生物,邻位硝基苯酚由于位阻效应不易生成,所以系统中只检测到了硝基苯酚中间产物,进一步表明了 NB 在 Pt/TiO₂ 光催化降解是通过·OH 氧化进行的。该结果与理论计算通过

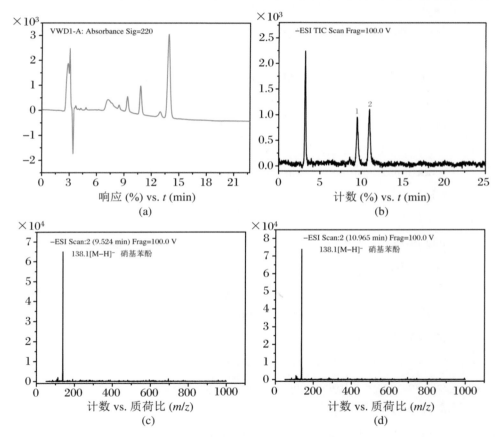

图 13.13　NB 在 Pt/TiO₂ 光催化降解的中间产物的 LC/MS 色谱图:UV spectra(a);LC(b);在 9.524 min 中间产物 1 的 MS(c)和在 10.965 min 处的中间产物 2 的 MS(d),[M-H]⁻ =138,中间产物可能是邻硝基苯酚,间硝基苯酚和对硝基苯酚,在3.208 min 时出现的是底物,LC/MS 条件;column:Zic-HILIC(150 mm × 2.0 mm i.d.),流速:0.2 mL/min,流动相:甲醇/0.05%甲酸=50/50,温度:30 ℃,注入体积:1.0 L,与探测条件:MS-ESI-m/z (10~300)

热力学和动力学计算预测得到的结果是一致的,即 NB 的光催化降解是在 Pt 负载的锐钛矿 TiO$_2$(001)晶面上通过氧化方式进行的。

参考文献

[1] Pang S, Zhang Y, Su Q, et al. Superhydrophobic nickel/carbon core-shell nanocomposites for the hydrogen transfer reactions of nitrobenzene and N-heterocycles [J]. Green Chemistry, 2020, 22 (6): 1996-2010.

[2] Li J, Song S, Long Y, et al. Investigating the hybrid-structure-effect of CeO$_2$-encapsulated Au nanostructures on the transfer coupling of nitrobenzene [J]. Advanced Materials, 2018, 30 (7): 1704416.

[3] Luan F, Liu Y, Griffin A M, et al. Iron(Ⅲ)-bearing clay minerals enhance bioreduction of nitrobenzene by *Shewanella putrefaciens* CN32 [J]. Environmental Science & Technology, 2015, 49 (3): 1418-1426.

[4] Wang H, Zhao H P, Zhu L. Role of pyrogenic carbon in parallel microbial reduction of nitrobenzene in the liquid and sorbed phases [J]. Environmental Science & Technology, 2020, 54(14): 8760-8769.

[5] Bai J, Liu Y, Yin X, et al. Efficient removal of nitrobenzene by Fenton-like process with Co-Fe layered double hydroxide [J]. Applied Surface Science, 2017, 416: 45-50.

[6] Latifoglu A, Gurol M D. The effect of humic acids on nitrobenzene oxidation by ozonation and O$_3$/UV processes [J]. Water Research, 2003, 37 (8): 1879-1889.

[7] Zhang S J, Jiang H, Li M J, et al. Kinetics and mechanisms of radiolytic degradation of nitrobenzene in aqueous solutions [J]. Environmental Science & Technology, 2007, 41 (6): 1977-1982.

[8] Hause M L, Herath N, Zhu R, et al. Roaming-mediated isomerization in the photodissociation of nitrobenzene [J]. Nature Chemistry, 2011, 3 (12): 932-937.

[9] Zhang E, Wang F, Zhai W, et al. Efficient removal of nitrobenzene and concomitant electricity production by single-chamber microbial fuel cells with activated carbon air-cathode [J]. Bioresource Technology, 2017, 229: 111-118.

[10] Wang W K, Chen J J, Li W W, et al. Synthesis of Pt-loaded self-interspersed anatase TiO_2 with a large fraction of (001) facets for efficient photocatalytic nitrobenzene degradation [J]. ACS Applied Materials & Interfaces, 2015, 7 (36): 20349-20359.

[11] Yu W, Guo X, Song C, et al. Visible-light-initiated one-pot clean synthesis of nitrone from nitrobenzene and benzyl alcohol over CdS photocatalyst [J]. Journal of Catalysis, 2019, 370: 97-106.

[12] Zhong S, Xi Y, Chen Q, et al. Bridge engineering in photocatalysis and photoelectrocatalysis [J]. Nanoscale, 2020, 12 (10): 5764-5791.

[13] Yang H G, Sun C H, Qiao S Z, et al. Anatase TiO_2 single crystals with a large percentage of reactive facets [J]. Nature, 2008, 453 (7195): 638-641.

[14] Zhao Y, Ma W, Li Y, et al. The surface-structure sensitivity of dioxygen activation in the anatase-photocatalyzed oxidation reaction [J]. Angewandte Chemie-International Edition, 2012, 51 (13): 3188-3192.

[15] Donlon B A, Razo-Flores E, Lettinga G, et al. Continuous detoxification, transformation, and degradation of nitrophenols in upflow anaerobic sludge blanket (UASB) reactors [J]. Biotechnology and Bioengineering, 1996, 51 (4): 439-449.

[16] Wahab H S, Koutselos A D. Computational modeling of the adsorption and (OH)—O-center dot initiated photochemical and photocatalytic primary oxidation of nitrobenzene [J]. Journal of Molecular Modeling, 2009, 15 (10): 1237-1244.

[17] Huang H, Zhou J, Liu H, et al. Selective photoreduction of nitrobenzene to aniline on TiO_2 nanoparticles modified with amino acid [J]. Journal of Hazardous Materials, 2010, 178 (1-3): 994-998.

[18] Cropek D, Kemme P A, Makarova O V, et al. Selective photocatalytic decomposition of nitrobenzene using surface modified TiO_2 nanoparticles [J]. Journal of Physical Chemistry C, 2008, 112 (22): 8311-8318.

[19] Makarova O V, Rajh T, Thurnauer M C, et al. Surface modification of TiO_2 nanoparticles for photochemical reduction of nitrobenzene [J]. Environmental Science & Technology, 2000, 34 (22): 4797-4803.

[20] Hossein-Babaei F, Lajvardi M M, Alaei-Sheini N. The energy barrier at noble metal/TiO_2 junctions [J]. Applied Physics Letters, 2015, 106 (8): 083503.

[21] Huang Z-F, Song J, Li K, et al. Hollow cobalt-based bimetallic sulfide polyhedra for efficient all-pH-value electrochemical and photocatalytic hydrogen evolution [J]. Journal of the American Chemical Society, 2016, 138 (4): 1359-1365.

[22] Liu R, Sen A. Controlled synthesis of heterogeneous metal-titania nanostructures and their applications [J]. Journal of the American Chemical Society, 2012, 134 (42): 17505-17512.

[23] Sun C, Smith S C. Strong interaction between gold and anatase TiO_2 (001) predicted by first principle studies [J]. Journal of Physical Chemistry C, 2012, 116 (5): 3524-3531.

[24] Cardenas-Lizana F, Gomez-Quero S, Idriss H, et al. Gold particle size effects in the gas-phase hydrogenation of m-dinitrobenzene over Au/TiO_2 [J]. Journal of Catalysis, 2009, 268 (2): 223-234.

[25] Kiyonaga T, Fujii M, Akita T, et al. Size-dependence of Fermi energy of gold nanoparticles loaded on titanium(Ⅳ) dioxide at photostationary state [J]. Physical Chemistry Chemical Physics, 2008, 10 (43): 6553-6561.

[26] Huang H, Feng J, Zhang S, et al. Molecular-level understanding of the deactivation pathways during methanol photo-reforming on Pt-decorated TiO_2 [J]. Applied Catalysis B-Environmental, 2020, 272:118980.

[27] Wang Z, Ma P, Zheng K, et al. Size effect, mutual inhibition and oxidation mechanism of the catalytic removal of a toluene and acetone mixture over TiO_2 nanosheet-supported Pt nanocatalysts [J]. Applied Catalysis B-Environmental, 2020, 274:118963.

[28] Wang A J, Cheng H Y, Liang B, et al. Efficient reduction of nitrobenzene to aniline with a biocatalyzed cathode [J]. Environmental Science & Technology, 2011, 45 (23): 10186-10193.

[29] Li Y P, Cao H B, Liu C M, et al. Electrochemical reduction of nitrobenzene at carbon nanotube electrode [J]. Journal of Hazardous Materials, 2007, 148 (1-2): 158-63.

[30] Taing J, Cheng M H, Hemminger J C. Photodeposition of Ag or Pt onto TiO_2 nanoparticles decorated on step edges of HOPG [J]. ACS Nano, 2011, 5 (8): 6325-6333.

[31] Tada H, Kiyonaga T, Naya S i. Rational design and applications of highly efficient reaction systems photocatalyzed by noble metal nanoparticle-loaded

titanium (IV) dioxide [J]. Chemical Society Reviews, 2009, 38 (7): 1849-1858.

[32] Peng G, Seo S, Ruther R E, et al. Molecular-scale structure of a nitrobenzene monolayer on Si(001) [J]. Journal of Physical Chemistry C, 2011, 115 (7): 3011-3017.

[33] Tada H, Ishida T, Takao A, et al. Kinetic and DFT studies on the Ag/TiO_2-photocatalyzed selective reduction of nitrobenzene to aniline [J]. Chemphyschem: A European Journal of Chemical Physics and Physical Chemistry, 2005, 6 (8): 1537-1543.

[34] Yang L, Luo S, Li Y, et al. High efficient photocatalytic degradation of p-nitrophenol on a unique Cu_2O/TiO_2 p-n heterojunction network catalyst [J]. Environmental Science & Technology, 2010, 44 (19): 7641-7646.

[35] Gong X Q, Selloni A. Reactivity of anatase TiO_2 nanoparticles: the role of the minority (001) surface [J]. The Journal of Physical Chemistry. B, 2005, 109 (42): 19560-19562.

[36] Ohno T, Sarukawa K, Matsumura M. Crystal faces of rutile and anatase TiO_2 particles and their roles in photocatalytic reactions [J]. New Journal of Chemistry, 2002, 26 (9): 1167-1170.

[37] Hussain A. A computational study of catalysis by gold in applications of CO oxidation[D]. Eindhoven: Technische Universiteit Eindhoven, 2010.

[38] Roudgar A, Eikerling M, van Santen R. Ab initio study of oxygen reduction mechanism at Pt_4 cluster [J]. Physical Chemistry Chemical Physics, 2010, 12 (3): 614-620.

[39] Parreira R L T, Caramori G F, Galembeck S E, et al. The Nature of the Interactions between Pt-4 Cluster and the Adsorbates H-center dot, (OH)-O-center dot, and H_2O [J]. Journal Of Physical Chemistry A, 2008, 112 (46): 11731-11743.

[40] Vanderbilt. Soft self-consistent pseudopotentials in a generalized eigenvalue formalism [J]. Physical Review. B, Condensed Matter, 1990, 41 (11): 7892-7895.

[41] Segall M, Lindan P J, Probert M a, et al. First-principles simulation: ideas, illustrations and the CASTEP code [J]. Journal of Physics: Condensed Matter, 2002, 14 (11): 2717.

[42] Perdew J P, Burke K, Ernzerhof M. Generalized gradient approximation

made simple [J]. Physical Review Letters, 1996, 77 (18): 3865.

[43] Monkhorst H J, Pack J D. Special points for Brillouin-zone integrations [J]. Physical Review B, 1976, 13 (12): 5188.

[44] Halgren T A, Lipscomb W N. The synchronous-transit method for determining reaction pathways and locating molecular transition states [J]. Chemical Physics Letters, 1977, 49 (2): 225-232.

[45] Delley B. Fast Calculation of Electrostatics in Crystals and Large Molecules [J]. The Journal of Physical Chemistry, 1996, 100 (15): 6107-6110.

[46] Delley B. From molecules to solids with the DMol$_3$ approach [J]. The Journal of Chemical Physics, 2000, 113 (18): 7756-7764.

[47] Perdew J, Burke K, Ernzerhof M. Generalized Gradient Approximation Made Simple [J]. Physical Review Letters, 1996, 77: 3865-3868.

[48] Vayssilov G N, Migani A, Neyman K. Density functional modeling of the interactions of platinum clusters with CeO_2 nanoparticles of different size [J]. Journal of Physical Chemistry C, 2011, 115 (32): 16081-16086.

[49] Han Y, Liu C J, Ge Q. Interaction of Pt clusters with the anatase TiO_2 (101) surface: a first principles study [J]. The Journal of Physical Chemistry. B, 2006, 110 (14): 7463-7472.

[50] Roy N, Sohn Y, Pradhan D. Synergy of low-energy {101} and high-energy {001} TiO_2 crystal facets for enhanced photocatalysis [J]. ACS Nano, 2013, 7 (3): 2532-2540.

[51] Tanaka A, Sakaguchi S, Hashimoto K, et al. Preparation of Au/TiO_2 with metal cocatalysts exhibiting strong surface plasmon resonance effective for photoinduced hydrogen formation under irradiation of visible light [J]. ACS Catalysis, 2013, 3(1): 79-85.

[52] Tachikawa T, Wang N, Yamashita S, et al. Design of a highly sensitive fluorescent probe for interfacial electron transfer on a TiO_2 surface [J]. Angewandte Chemie-International Edition, 2010, 49 (46): 8593-8597.

[53] Tachikawa T, Yamashita S, Majima T. Evidence for crystal-face-dependent TiO_2 photocatalysis from single-molecule imaging and kinetic analysis [J]. Journal of the American Chemical Society, 2011, 133 (18): 7197-7204.

[54] Tachikawa T, Majima T. Photocatalytic oxidation surfaces on anatase TiO_2 crystals revealed by single-particle chemiluminescence imaging [J]. Chemical Communications, 2012, 48 (27): 3300-3302.